# Studies in Political Economy

**Series editor**
Norman Schofield, Washington University in Saint Louis,
Saint Louis, MO, USA

More information about this series at http://www.springer.com/series/13366

Maria Gallego • Norman Schofield
Editors

# The Political Economy of Social Choices

Springer

*Editors*
Maria Gallego
Department of Economics
Wilfrid Laurier University
Waterloo
Ontario, Canada

Norman Schofield
Department of Political Science
Washington University in St. Louis
Saint Louis
Missouri, USA

ISSN 2364-5903 ISSN 2364-5911 (electronic)
Studies in Political Economy
ISBN 978-3-319-40116-4 ISBN 978-3-319-40118-8 (eBook)
DOI 10.1007/978-3-319-40118-8

Library of Congress Control Number: 2016950530



Printed on acid-free paper

This Springer imprint is published by Springer Nature
The registered company is Springer International Publishing AG Switzerland

# Introduction

This volume draws from papers presented at *The Political Economy of Social Choices* workshop organized by Maria Gallego and Norman Schofield held in Oaxaca, Mexico, in July 2015 and funded by the Banff International Research Station for Mathematical Innovation and Discovery (BIRS), Banff, Calgary, Canada, and by Casa Matematica Oaxaca (CMO), Oaxaca, Mexico.

The workshop brought together political economists and social choice theorists and empiricists to Oaxaca to share their current research. The group included prominent senior scholars as well as junior scholars doing their Ph.D. at Washington University in St. Louis.

In the Introduction to this volume, we first give a brief overview of the field before providing a brief summary of the papers included in this volume.

## 1 Overview of the Field

The topics covered in the workshop and in this volume are at the intersection of two broad fields: *political economy* and *social choice*. These two fields combine economic and political science to examine how groups of people and societies make collective decisions on how to allocate scarce resources among competing social needs.

Political economy examines how economic theory and methods influence political ideology and the decisions made by governments and how politics shapes and determines the economic environment in which firms and individuals, groups of people, or societies operate. The questions addressed in political economy combine economics, law, political science, and sociology while taking into account how political institutions, the political environment, and the economic system (capitalist, socialist, or mixed) interact to determine the choices of governments or groups of agents and how these institutions evolve under different political, social, and economic systems. These decisions depend on the political institutions under which agents operate.

Social choice complements political economy as its framework specifies how the opinions, preferences, interests, or welfare of individuals within a group or in a society is aggregated to reach collective decisions or some level of social welfare in a sense specified by the questions or issues being studied recognizing that the interests of all members of the group may not be perfectly aligned and that some may have opposing preferences. Social choice specifies the properties that models must have in order for the model to generate an internally consistent aggregation of the well-being of a group of individuals, e.g., the elites under autocracy or citizens under democracy. It also identifies the properties that these preference aggregation rules must have to obtain certain desired outcomes.

Political economy and social choice study a wide range of questions in different areas using many different mathematical, game theory, and statistical methodologies and actual data about individuals to study many social issues. In this framework, political agents—be it voters, politicians, parties, and/or interest groups—maximize their payoff or utility functions taking into account that their decisions are made in an interdependent world.

The research presented in this volume focuses on developing or testing models in which economic policy and political institutions are the outcome of interactions between different agents with perhaps opposing preferences operating under different economic and political institutions. The analytical frameworks of economics and political science are jointly used as researchers in these areas believe that if economic recommendations are made to governments or political agents without taking into account the political institutions under which they operate, governments or politicians may not find it in their interest to implement these recommendations or the recommendations may not be politically implementable as powerful groups within society may block their implementation.[1] Moreover, if political leaders make political decisions without taking into account how these interact with the economy and the incentives these decisions give to different groups of agents—be it individual voters, groups of agents, or firms—then these political decisions may lead to catastrophic economic circumstances in the future.[2] To address and incorporate these two sides of the problem into the models, political economists integrate the economic and political characteristics of agents, decisions into their models.

Given that agents may have opposing preferences, political economy models use social choice aggregation rules to determine decisions at the aggregate societal or group level. Moreover, using mathematical, statistical, and game theory tools, political economy models the strategic interaction of political agents—voters

[1] As happened in Greece recently where the recommendations or "demands" made by the European Commission, the European Central Bank, and the IMF (the Troika) on the economic and policy reforms Greece must implement in order to receive various bailout packages have led to the resignation of Prime Minister Antonis Samaras and to early elections after massive demonstrations in the streets of Athens.

[2] For example, the oversized Greek public sector with a large number of employees who may retire at age 55 has been identified as one of the culprits of the recent Greek financial crisis. This crisis has threatened the stability of the Euro and put at risk the European Monetary Union.

and political leaders or parties and interest groups—under various political and economic institutions and shows how agents may use these institutions to their advantage and perhaps to the detriment of other members of society. Using these models, it is possible to examine what happens as the institutions under which these decisions are made vary and study how agents make decisions taking the decisions of other agents into account. It is also possible to model new institutions or variation of existing ones using mechanism design to explore the properties that social aggregation rules must have in order to minimize the negative incentives these institutions generate.

Furthermore, in order to understand how the large number of moving pieces in these models—the decisions of many agents with perhaps opposing preferences—interact in these highly complex multi-agent multidimensional policy models, theoretical and empirical models make extensive use of mathematical and game theory tools.

Unforeseen random events may affect the decisions of some or all agents and thus the aggregate social choice (e.g., a recession in China affects US consumers and thus the outcome of the presidential election in the USA), or agents may not have complete information on how their decisions affect other agents (e.g., candidates do not know with certainty how voters will vote). Under incomplete and/or imperfect information, researchers model events as being stochastic assuming that players have an implicit understanding of the distribution of stochastic events affecting players' decisions. This approach has been used, for example, to model the decisions of undecided voters where it is assumed that each voter's utility function is affected by a random shock that is known only to the voter with parties or candidates knowing only the distribution of the shocks affecting voters' preferences. This is an approach that has been used in empirical analysis as researchers have less than perfect information when estimating agents' decisions.

The objective of deterministic or stochastic models is to find the equilibria that will be implemented, i.e., the political economy solutions of these complex social choice problems as they pertain to the issue studied. Using these models and the equilibria they generate, it is then possible to examine what happens as the assumptions or the parameters of these models change. These comparative statics not only increase our understanding of how these models work but also generate testable predictions.

In order to see if these political economy and social choice theories and their predictions reflect the social phenomena being modeled, these theories must be tested. One way of testing these theories is to examine if the predictions generated by these theories are observed in the real world, that is, to use actual data to test these predictions based on the observed behavior of voters, parties, interest groups, or firms.

Alternatively, some test these predictions using laboratory experiments. In these experiments, researchers vary the assumptions in their models to examine the behavioral responses of individuals and investigate if the actual behavioral responses of individuals correspond to those predicted by the theory. Others experiment with the response of individuals as they vary the institutional setup under which decisions are made.

Rather than testing theories using empirical data and sophisticated econometric techniques, some prefer to test their theories by using simulations making use of the parameter values estimated in other research. Variations in these parameter values lead to a deeper understanding of how theories work and of how their predictions relate what others have observed in real life or estimated in their empirical models. These simulations are also used to examine what happens as the parameters of model or institutions change. These counterfactuals allow researchers to evaluate, for example, whether changes to institutions produce an outcome deemed desirable when a parameter changes.

Different empirical techniques and models have been developed to gain greater understanding of the hypotheses and predictions generated by theoretical models. If these predictions do not accord with what is observed, then the theory is missing some important aspect of reality which usually leads to the theoretical model being modified. There are different ways of modifying these models with the most common being that of relaxing some of the simplifying assumptions embedded in the base model.

Empirical models where a theoretical framework is applied to different countries with different political institutions are included in this volume. Other works compare predictions of theoretical models under different political regimes to examine if political leaders value policies differently under different political institutions and political regimes.

This volume contains papers embedded within the political economy and social choice traditions. There are theoretical and empirical papers, with some papers using actual data and empirical tests and others using laboratory experiments or simulations. While some study specific issues, others examine broader social issues.

We now provide a broad overview of the papers included in this volume. Then group the papers along different themes so as to give a general sense of the topics and issues covered in this volume.

## 2 Issues Covered in This Volume

The topics covered in this volume address social issues from either a theoretical or an applied framework or use theories to guide applied work. Theoretical papers develop models from a social choice and/or a political economy perspective; empirical ones take institutions as given and so are mostly in political economy. This section highlights that the papers included in this volume cover a great variety of topics and issues.

In "Autocratic Health Versus Democratic Health: Different Outcome Variables for Health as a Factor Versus Health as a Right," Rosenberg and Shvetsova document that autocracies and democracies implement different healthcare policies under these two political regimes.

In "Comparison of Voting Procedures Using Models of Electoral Competition with Endogenous Candidacy," Bol, Dellis, and Oak survey the literature to compare the theoretical properties of different voting procedures when candidates choose whether or not to run in the election.

Gomberg, Gutierrez, and Thepris, in their paper "Negative Advertising During Mexico's 2012 Presidential Campaign," use a unique data set to illustrate the effect of negative campaign advertising on the presidential election.

In "Legislative Leaders as Condorcet Winners? The Case of the US Congress," Erikson and Ghitza examine the probability with which a Condorcet winner is elected in open pairwise vote (tournament) under the US congressional institutions.

Ferris, Winer, and Grofman study electoral competitiveness when members of parliament are elected in single-member districts under plurality rule in multiparty settings in "The Duverger-Demsetz Perspective on Electoral Competitiveness and Fragmentation: With Application to the Canadian Parliamentary System, 1867–2011."

In "Modelling the Effect of Campaign Advertising on US Presidential Elections," Gallego and Schofield examine candidates' policy platform and advertising (ad) campaign choices.

Morton, Tyran, and Wengström investigate why women tend to be more leftist than men in their political choices in "Personality Traits and the Gender Gap in Ideology."

In "Statistical Utilitarianism," Pivato shows that social welfare can be estimated with a certain degree of accuracy in societies with a large number of individuals.

Barutt and Schofield study non-candidate and traditional campaign expenditures in the 2014 US congressional elections in "Measuring Campaign Spending Effects in Post-Citizens United Congressional Elections."

Kim and Schofield study the role of activists in "Spatial Model of US Presidential Election in 2012" examining the effect that changes in campaign law had on the election.

In "Modeling Elections and Referenda in Ireland," Schofield and Simoneau examine how the Irish reacted to the streamlining of European institutions proposed in the Lisbon Treaty in the 2007 Irish election and in the 2008 and 2009 referenda.

We now give a more detailed summary of these papers while also linking them according to common themes.

## 3 Thematic Topics

The papers included in this volume can be grouped into three major themes. Section 3.1 gives an overview of the topics dealing with the well-being of individuals examining measures of social welfare, political differences across genders, and public policy differences across political regimes. The performance of different electoral systems is examined in Sect. 3.2. The role of campaign advertising and expenditures on candidates' policy platforms and voters' choice of candidate is examined in Sect. 3.3.

## 3.1 Social Choices: Welfare, Gender Differences, and Healthcare

This section summarizes papers dealing with preference aggregation in society under incomplete information, differences in political preferences across genders, and on the differences in healthcare preferences across regimes.

### 3.1.1 "Statistical Utilitarianism" by Pivato

Within the "utilitarian" approach to measuring social welfare, Pivato shows that if social welfare is calculated as the average of individuals' cardinal utility functions in a society with a large number of individuals whose utility functions satisfy certain statistical properties, then it is possible to get an accurate estimate of a utilitarian social welfare function. He shows that averaging utility data from a large population of voters gives a good approximation of utilitarianism with high probability. This is true even when the utility functions of individuals are miscalibrated or noisy.

Pivato's results hold even when he assumes that utilities are subject to multiplicative or additive noise or when there are measurement errors (with zero mean and bounded variance) associated with the utilities of different voters. These results hold even when voters' utilities are highly correlated which leads to correlation among the preference intensities and measurement errors of the utilities of these voters (while imposing constraints of the covariance of the random errors). Under these circumstances, he shows that the probability of a socially suboptimal decision not only decays to zero in large population but does so quickly when the number of voters becomes large.

Pivato's positive results give credence to the utility functions used in the various theoretical and empirical studies presented in this volume and elsewhere in attempts to measure the well-being of individuals or societies under various political regimes and voting systems. These utilities are then used to formulate how agents make decisions and are aggregated in electoral processes.

### 3.1.2 "Personality Traits and the Gender Gap in Ideology" by Morton, Tyran, and Wengström

Morton, Tyran, and Wengström model personality traits as mediating variables (indirect effects) between the effects of gender and political ideological preferences in a sample of Danish citizens. Their objective is to estimate the indirect effect of gender on ideological preferences through personality trait differences between males and females.

Their results indicate that—for their sample and what they argued probably for other developed countries—the ideological gender gap can be largely explained by differences in personality traits between women and men. In particular, they find that women are more open to experience, more agreeable, and less emotionally stable than men. They note that these women with these trait differences tend to be more leftist, largely through a direct effect on ideology but also indirectly through the negative effects these traits have on income. Thus, their results also suggest that women are more leftist than men because of these trait differences, that is, because they have different personalities which shape their ideology. Moreover, they find that after controlling for personality traits, women tend to be more leftist because they earn less. In addition, their results measure the effects that personality trait differences have on ideological differences between the sexes that tend to be larger (over three quarters of the observed gender gap in general ideological preferences) than those independent of personality trait differences (such as income or education differences) and so outweigh the non-personality trait effects.

### 3.1.3 "Autocratic Health Versus Democratic Health: Different Outcome Variables for Health as a Factor Versus Health as a Right" by Rosenberg and Shvetsova

Rosenberg and Shvetsova examine the healthcare policies of autocracies and democracies. The basic premise of this empirical paper is that healthcare policies in autocracies and democracies differ because the government's healthcare objectives differ under these two political regimes. They argue that autocrats value healthcare because it complements their economic policies as they help maintain the health of those working for the economic elites on whose support dictator depends to stay in office. In democracies, politicians face regularly scheduled elections and so are accountable to broader coalitions and thus cannot target the health needs of specific groups of people.

Using disaggregated data on mortality from specific diseases, Rosenberg and Shvetsova show that, other things being equal, while autocracies deal more efficiently with diseases that "damage" the elites' workforce, at the expense of other health problems, democracies do not have such bias, with their healthcare policy priorities being less clear. That is, they find that improved mortality from workforce-affecting diseases in autocracies that have extensive labor markets relatively to nonlabor-intensive autocracies, but not in democracies, under similar conditions. In addition, their results show that the public investment made by the autocrat increases mortality from old-age diseases in these countries and is highly significant jointly with autocratic labor force participation.[3] Thus, they find substantial evidence that it is the economic elites' preferences that are reflected in health outcomes in autocracies with labor-dependent economies. They also show that in democracies healthcare policies depend on the preferences of their coalitions supporting the governments in office.

The papers in this section dealt with measuring the well-being of individuals in a society, how gender traits influence differences in political ideology across genders, and with differences in healthcare policies across autocracies and democracies. We now turn to papers dealing with the effect that differences in voting systems have on candidates and on electoral outcomes.

## 3.2 *Performance of Electoral Systems*

In this section, we describe the contributions in this volume pertaining to general voting rules. The first is a survey paper comparing candidates entrance under different voting rules, the second studies the probability that the Condorcet winner is chosen under US congressional institutions, and the third analyzes electoral competitiveness under plurality rule.

### 3.2.1 "Comparison of Voting Procedures Using Models of Electoral Competition with Endogenous Candidacy" by Bol, Dellis, and Oak

Bol, Dellis, and Oak survey the literature to examine the predictions made by unidimensional policy models on the number of candidates running for election and on the degree of policy polarization among candidates. Their main focus is on understanding how different voting procedures affect the number of candidates running in the election and the policies they adopt in their effort to win votes. To do so, they classify models according to the assumptions made on candidates' policy or win motivation objectives and on the timing of their entry into the election. They concentrate on models in which there is a national election in a single district and examine variations on the type of ballots voters cast in the election distinguishing whether voters rank or not candidates in their ballot. Their survey focuses on three voting procedures: plurality, runoff, and approval voting.

By comparing the properties of alternative voting procedures between these families of models, Bol, Dellis, and Oak highlight the advantages that endogenous candidacy models have—at the theoretical and empirical levels—over the standard

[3]This results accords with the theory developed by Gallego and Pitchik (2004) and the evidence provided by Gallego (1998) that the fate of autocrats (their survival in office) depends on the well-being of the elites which she measures through the sum of public and private investment as they are the only ones capable of investing in developing countries.

Hotelling-Downsian model. Theoretically, they argue that these models can provide more satisfactory micro-foundations for the emergence and/or stability of a specific configuration of parties or candidates under different voting procedures. Empirically, they highlight that these models offer a better account of actual electoral results—namely, that countries using plurality rule (e.g., the UK and the USA) tend to hover around Duverger's prediction of a two-party system, whereas countries using plurality runoff rules (e.g., France) or proportional representation tend to have multiparty systems. They also examine the degree of policy polarization between the parties—differences in one-dimensional policy platforms on a left-right scale—generated by different voting procedures.

### 3.2.2 "Legislative Leaders as Condorcet Winners? The Case of the US Congress" by Erikson and Ghitza

Using historical data starting from 1789 for the US House of Representatives and the US Senate of the US Congress, Erikson and Ghitza simulate the selection of a Condorcet winner. For each chamber and Congress, they identify the preferences of the legislators using the DW-Nominate scores and examine whether a Condorcet winner exists in the two-dimensional space created by these scores. In addition, using post-World War II data, they examine the existence of a Condorcet winner within each political party for each chamber and Congress. Their objective is to find the frequency with which a Condorcet winner exists and the closeness to the center of policy space of tournament winners in open pairwise elections.

Using congressional roll call voting modeled in two dimensions with many members (currently, 100 in the Senate and 435 in the House), their results show that for the US House as a whole, Condorcet winners usually do *not* exist; for the Senate as a whole and for each party in the House and the Senate, a Condorcet winner exists a little over *half* the time. They conclude that if congressional party caucuses were to choose a winner solely based on who is closest to their views, there would be a clear winner at least half the time. They also show that in the recent past, half of the actual party leaders in the Senate were predicted by their model to be the Condorcet winners in at least one Congress prior to their ascent to the leadership position.

### 3.2.3 "The Duverger-Demsetz Perspective on Electoral Competitiveness and Fragmentation: With Application to the Canadian Parliamentary System, 1867–2011" by Ferris, Winer, and Grofman

The innovation of Ferris, Winer, and Grofman's paper is to take Duverger's law (1954) and combine it with Demsetz's (1968) theory of natural monopoly to define the competitiveness of electoral system as depending on the contestability of the election. They argue that competitiveness declines in plurality systems as party fragmentation exceeds the long-run level predicted by Duverger's law. To do so, they develop a new index of electoral contestability and discuss its properties.

To show how their index fares with other measures of competitiveness, they examine the relationship between their index and the concentration of vote shares during the history of the Canadian parliamentary system from 1867. After compiling riding-level electoral data, Ferris, Winer, and Grofman build different competitiveness indices for each legislative assembly and compare their competitiveness measure with that provided by Laakso and Taagepera's (1979) index on the effective number of parties in a legislature. They also compare it with "first versus second place vote margins" at both the constituency and national party level. They show the evolution of party competitiveness in Canada over the last century and a half. In particular, they examine periods in which a large number of parties competed in the elections in different regions of the country. As a by-product, their study also shows the evolution of Canada's federal parties and how competitiveness has influenced their evolution.

The papers in this section examined the performance of different electoral systems by comparing different voting procedures in national elections and by looking at the probability that the Condorcet winner is chosen in the US federal institutional setting and at the degree of electoral competitiveness and fragmentation in multiparty parliamentary countries using plurality rule. The next section summarizes papers dealing with the political campaigns undertaken by candidates in different countries.

## *3.3 The Role of Campaign Advertising in Elections*

The papers included in this section examine the effect that campaign advertising and expenditures have on electoral outcomes in elections first at the theoretical level and then at the empirical level.

### 3.3.1 "Modelling the Effect of Campaign Advertising on US Presidential Elections" by Gallego and Schofield

Gallego and Schofield extend Schofield's (2007) model to examine the effect that candidates' abilities to directly communicate with voters through campaign ads—delivered directly to their smart phones or social media accounts—have on voters' choices and on candidates' policy positions. In this theoretical paper, voters are endowed with policy preferences as well as preferences over candidates' advertising campaign relative to their *campaign tolerance level*, i.e., their preferences over how many times they wish to be contacted by candidates. In addition, voters' choices are also affected by the *composite valence*—voters' non-policy evaluation of candidates is measured as the sum of the sociodemographic, traits, and competency valences—common to all voters. The *sociodemographic valence* identifies the voting propensities among groups of voters with common sociodemographic characteristics, the *traits valence* measures the effect that candidates' traits (age, gender, race, etc.) have on voters' choices, and the *competency valence* measures voters' beliefs on candidates' competency or ability to govern.

Given evidence that candidates adopt different policy platforms, they study the conditions under which candidates would adopt the same electoral campaign, i.e., the same policy and ad campaign. They show that if candidates adopt the same campaign, they adopt the electoral mean[4] as their campaign strategy. They then derive the sufficient and necessary conditions for candidates to converge to the electoral mean. They find very intuitive results: The sufficient (necessary) condition for convergence to the electoral mean is that the expected vote share of *all candidates* should be larger than the sufficient (necessary) pivotal vote share. Moreover, they show that if the expected vote share of the candidate with the *lowest composite valence* is less than the necessary pivotal vote share, this candidate does *not* adopt the electoral mean as its campaign strategy as by adopting a different strategy the candidate increases its vote share. In this case, other candidates may also adopt a strategy that is not at the electoral mean. The electoral mean is then not a local Nash equilibrium of the election. They show that when voters give greater weight to candidates' policies or advertising campaigns or when the distribution of voters' ideal policy and/or campaign tolerance levels becomes more dispersed, it is less likely that candidates adopt the electoral mean as their campaign strategy.

The following papers empirically investigate the effect that campaign advertising has on the elections in various countries. The first three papers examine the effect of changes in the electoral campaign laws (in Mexico and the USA) on the election; the last one looks at the effect that advertising had on the Irish general election and on the two referenda to ratify the Lisbon Treaty in Ireland.

### 3.3.2 "Negative Advertising During Mexico's 2012 Presidential Campaign" by Gomberg, Gutierrez, and Thepris

In "Negative Advertising During Mexico's 2012 Presidential Campaign," Gomberg, Gutierrez, and Thepris study the effect that negative campaign advertising had on the presidential election. To assess this effect, they exploit changes to the Mexican electoral law adopted prior to the 2012 election centralizing the allocation of political advertising through the Federal Electoral Institute (FEI) with the law setting limits on candidates' advertising expenses according to the party's electoral results in the previous election. In addition, the law also made the FEI responsible for administering the campaign ads' air times for each day of the campaign. For this study, Gomberg, Gutierrez, and Thepris use the data collected by the FEI on

[4]The electoral mean is defined by the mean of voters' ideal policies and campaign tolerance levels.

campaign advertising ads of all four candidates/parties including the ad content during each hour of every day of the campaign.

Interested in measuring the effect that negative ads have on the various candidates in a multi-candidate race with front-runners, Gomberg, Gutierrez, and Thepris identify negative ads as those where the ad explicitly mentions the other competing candidates. They develop a series of graphical illustrations of the evolution of each candidate's negative advertising strategies over the course of the campaign. Then they relate the evolution of candidates' negative campaigns to their expected vote shares using pre-election polls. Their results indicate that the timing of candidates' negative advertising strategies accords with a model in which ads affect voting intentions with negative (positive) ads having a negative (positive) effect on the vote share of the party mentioned in the ad and a positive (negative) effect on all other candidates. Their results also suggest that the candidate that consistently ranked lowest in voting intentions throughout the entire campaign never engaged in negative advertising. They argue that the reason for this is that this candidate's objective was to get past the cutoff rule that determines government financial support after the election.

### 3.3.3 "Measuring Campaign Spending Effects in Post-Citizens United Congressional Elections" by Barutt and Schofield

Given the Supreme Court's ruling in *Citizen United* that struck down limitations on campaign contributions, Barutt and Schofield compare the effect that independent, noncandidate-related, campaign expenditures and traditional campaign expenditures had on the 2014 US congressional elections.

Their results show differences between independent and traditional campaign expenditures. Independents target only the most competitive elections. Whereas incumbents rely mostly on campaign expenditures, even in the most competitive races, challengers benefit mostly from independent expenditures. Moreover, their results also show that while incumbents have greater campaign expenditures on aggregate, independents favor challengers in contested races. They also find that challengers' campaign expenditures are significantly more productive than those of incumbents and argue that this is mainly due to the endogeneity bias caused by spending levels affecting vote shares and expectations about vote shares affecting spending levels. They also find that independent expenditures exhibit symmetrical marginal productivities exerting similar effects on incumbents and challengers.

### 3.3.4 "Spatial Model of US Presidential Election in 2012" by Kim and Schofield

Kim and Schofield study the role of activists due to this being the first election after the Supreme Court's *Citizens United* ruling eliminating the limit on campaign contributions. This decision allowed for an unprecedented increase in political spending and the creation of "Super PACs" that can accept unlimited amounts of political funds from contributors.

Kim and Schofield examine how *Citizens United* affected the policy position of candidates and voters' choices in the 2012 election. They find that the exogenous increase in campaign contributions led to voters' ideological difference with candidates becoming more important in their choice of candidate but that candidates' valence, the non-policy evaluation of candidates, did not play a significant role. They argue that these findings suggest that the exogenous increase in campaign contribution has emphasized the role of ideological distance in voting behavior while reducing the effect of valence on voters' choice of candidate.

#### 3.3.5 "Modeling Elections and Referenda in Ireland" by Schofield and Simoneau

Schofield and Simoneau examine the 2007 Irish election and the Irish Lisbon Treaty referenda of 2008 and 2009 that were to ratify the proposed changes to the European Union institutions. They examine voters' choices and parties' policy positions after a period in which the Celtic Tiger had been growing at an accelerated rate beginning in 1990 but was seriously affected by the 2008 global financial crisis. Analyzing these two referenda is important because they occurred just before and just after the global financial crisis.

By examining differences in voters' response to the referenda before and after the crisis, they are able to examine the effect that the financial crisis had on how citizens voted in the referenda. This is particularly important as the Irish rejected the treaty in the first referendum but passed it in the second.

Their results show that leaders' valences—voters' non-policy evaluation of the ability of leaders to govern—played a significant role in the outcome of the election and in the referenda. They find that there is a significant decrease in the valence of the Yes campaign between the two referenda and that this effect was more pronounced once the electorate's view on the economic effect of the treaty was controlled for. They attribute the success of the Yes campaign in 2009 to the fear the electorate had of not being able to weather the financial crisis without assistance from the European Union.

The papers included in this section highlight that advertising affects the policies candidates' adopt during elections. The theoretical model highlights the conditions under which candidates converge or not to the electoral mean. The empirical papers estimate the effect that changes in campaign laws had on the elections in Mexico and the 2012 US presidential and the 2014 congressional elections or how proposed changes to the European institutions affected the 2007 Irish elections and the outcome of the two referenda.

## 4 Conclusion

The papers included in this volume study a wide range of social issues and could be grouped into three general themes: social choices, electoral performance and the role of campaign advertising on elections.

There were three papers examining differences in social choices. Pivato investigates the conditions under which aggregating the well-being of individuals can lead to a meaningful social welfare function. Morton, Tyran, and Wengström study the effect of trait differences across genders on the gender ideological bias. Rosenberg and Shvetsova estimate the difference of social choices across autocracies and democracies as it pertains to their healthcare policy choices.

Three papers analyzed electoral performance under different voting systems. Bol, Dellis, and Oak surveyed the literature on candidate entrance under plurality, runoff, and approval voting. Erikson and Ghitza study the probability that the US congressional institutions select the Condorcet winner. Ferris, Winer, and Grofman develop a new measure of competitiveness for plurality rule using the contestability of the election in multiparty countries such as Canada.

Finally, several papers examine the effect of campaign advertising on candidates' policy platforms at the theoretical and empirical level. Gallego and Schofield's theoretical paper examines how candidates choose not only their policy platform but also their campaign advertising. Four empirical models estimate candidates' policy platforms in different elections and countries. Gomberg, Gutierrez, and Thepris illustrate the effect of negative campaign advertising on the 2012 Mexican presidential election. Barutt and Schofield examine the effect of campaign expenditure on the 2014 US congressional election. Kim and Schofield estimate candidates' positions in the 2012 US presidential election. Schofield and Simoneau study the Irish people's response to the changes proposed in the Lisbon Treaty to European governing institutions.

Waterloo, ON, Canada Maria Gallego
St. Louis, MO, USA Norman Schofield

## References

Demsetz, H. (1968). Why regulate utilities? *Journal of Law and Economics, 11*(1), 55–65.

Duverger, M. (1954) *Political parties: Their organization and activities in the modern state*. New York: Wiley.

Gallego, M. (1998). Economic performance and leadership accountability: An econometric analysis. *Economics and Politics, 10*(3), 249–296.

Gallego, M., & Pitchik, C. (2004). An economic theory of leadership turnover. *Journal of Public Economics, 88*, 2361–2382.

Laakso, M., & Taagepera, R. (1979). Effective number of parties: A measure with application to West Europe. *Comparative Political Studies*, 12, 3–27.

Schofield, N. (2007). The mean voter theorem: Necessary and sufficient conditions for convergent equilibrium. *The Review of Economic Studies, 74*(3), 965–980.

# Contents

**Autocratic Health Versus Democratic Health: Different Outcome Variables for Health as a Factor Versus Health as a Right** ...... 1
Dina Rosenberg and Olga Shvetsova

**Comparison of Voting Procedures Using Models of Electoral Competition with Endogenous Candidacy** ...... 21
Damien Bol, Arnaud Dellis, and Mandar Oak

**Negative Advertising During Mexico's 2012 Presidential Campaign** ...... 55
Andrei Gomberg, Emilio Gutiérrez, and Zeev Thepris

**Legislative Leaders as Condorcet Winners? The Case of the U.S. Congress** ...... 73
Robert S. Erikson and Yair Ghitza

**The Duverger-Demsetz Perspective on Electoral Competitiveness and Fragmentation: With Application to the Canadian Parliamentary System, 1867 –2011** ...... 93
J. Stephen Ferris, Stanley L. Winer, and Bernard Grofman

**Modelling the Effect of Campaign Advertising on US Presidential Elections** ...... 123
Maria Gallego and Norman Schofield

**Personality Traits and the Gender Gap in Ideology** ...... 153
Rebecca Morton, Jean-Robert Tyran, and Erik Wengström

**Statistical Utilitarianism** ...... 187
Marcus Pivato

**Measuring Campaign Spending Effects in Post-Citizens United Congressional Elections** ...... 205
Brandon Barutt and Norman Schofield

**Spatial Model of U.S. Presidential Election in 2012** ........................ 233
Jeong Hyun Kim and Norman Schofield

**Modeling Elections and Referenda in Ireland** .............................. 243
Norman Schofield and William Simoneau

# Contributors

**Brandon Barutt** is a Ph.D. candidate in the Department of Political Science at Washington University in St. Louis.

Mailing address: Department of Political Science, Washington University in St. Louis, St. Louis, MO, USA

**Damien Bol** is a lecturer (assistant professor) in Political Behavior in the Department of Political Economy, King's College London. His research lies at the intersection of political behavior, comparative politics, and political institutions. He uses observational and experimental methods to study elections, voting behavior, and party strategies.

Mailing address: Department of Political Economy, King's College London, London, UK

**Arnaud Dellis** is a professor in the Department of Economics at the University of Quebec in Montreal (UQAM). He is a member of the Centre for the Study of Democratic Citizenship (CSDC) and of the Centre Interuniversitaire sur le Risque, les Politiques Economiques et l'Emploi (CIRPEE). He is specialized in political economics and public economics. His current research interests include the comparison of voting procedures in spatial models of electoral competition with endogenous candidacy and the interaction between informational lobbying and agenda formation.

Mailing address: Department of Economics, Université du Québec à Montréal, Montréal, QC, Canada

**Robert S. Erickson** is a professor of political science at Columbia University. He is a past editor of two journals: *American Journal of Political Science* and *Political Analysis*. He has written or coauthored numerous articles and books, including *Statehouse Democracy*, *The Macro Polity*, *The Timeline of Presidential Elections*, and *American Public Opinion*. A ninth edition of *American Public Opinion* was published in 2014.

Mailing address: Columbia University, New York, NY, USA

**J. Stephen Ferris** received his undergraduate degree from the University of Toronto and his Ph.D. from UCLA; has taught at Simon Fraser, York, and Carleton Universities; and is now a distinguished research professor of economics at Carleton. His published work typically intersects the macro side of public finance and politics, usually involving transaction cost and coordination problems, and can be found in such journals as the *Quarterly Journal*, *Canadian Journal of Economics*, *Canadian Journal of Political Science*, *Public Choice*, *European Journal of Political Economy, and Journal of Economic Behavior and Organization.*

Mailing address: Department of Economics, Carleton University, Ottawa, ON, Canada

**Maria Gallego** is an associate professor of economics at Wilfrid Laurier University in Waterloo, Canada. Gallego has studied leadership accountability in political economy situations and how the strategic interactions between political and economic agents determine policy outcomes. She has developed theoretical and empirical studies of leadership transitions in democracies and dictatorships. She worked on extending Nash's bargaining model by allowing agents to have single-peaked rather than increasing preferences over the policy space, and then applied this theory to model intergovernmental negotiations. Gallego and Schofield have modeled parties' multidimensional policy positions in elections in various countries with different political institutions including democracies and anocracies. More recently, she and Schofield developed models to study the party's policy decisions when different sets of parties compete in different regions of the country and have developed models of US presidential elections where candidates make policy and advertising campaign decisions. She is the coauthor with Schofield of *Leadership or Chaos* (Springer, 2014). Gallego has been a visiting scholar at the Toulouse School of Economics and the Center in Political Economy at Washington University in St. Louis. She has an M.A. and Ph.D. in economics from the University of Toronto.

Mailing address: Department of Economics, Lazaridis School of Business and Economics, Wilfrid Laurier University, Waterloo, ON, Canada

**Yair Ghitza** is chief scientist at Catalist LLC, a political consulting and analysis firm in Washington, DC. He leads Catalist's statistical analysis and research efforts, producing statistical "microtargeting models" and research projects related to elections at all major levels of government. He holds a Ph.D. from Columbia University, where he conducted extensive public opinion research, specializing in the development of statistical and data visualization tools for understanding American politics.

Mailing address: Catalist, Washington, DC, USA

**Andrei Gomberg** (born in Moscow in 1974) is an associate professor of economics at ITAM, Mexico City. He obtained his Ph.D. from New York University in 2000. His theoretical and empirical research in political economy has been published, among others, in the *Journal of Economic Theory, Economic Theory*, the *International Journal of Game Theory*, and *Social Choice and Welfare.*

Mailing address: ITAM, CIE, Mexico City, Mexico

**Bernard Grofman** is a professor of political science and Jack W Peltason endowed chair of democracy studies at the University of California, Irvine, and former director of the UCI Center for the Study of Democracy. His research deals with topics such as voting rights, electoral rules, theories of representation, behavioral social choice, and political satire. He is a coauthor of five books (four from Cambridge University Press and one from Yale University Press) and coeditor of 23 other books, with over 300 research articles and book chapters, including ten in the *American Political Science Review*. A member of the American Academy of Arts and Sciences since 2001, he has been a scholar-in-residence at universities and research centers in the USA, Canada, France, Germany, Hungary, Italy, Japan, the Netherlands, Spain, and the UK, and he has an honorary Ph.D. from the University of Copenhagen. His work has been cited in more than a dozen US Supreme Court cases over the course of the past four decades.

Mailing address: School of Social Sciences, University of California at Irvine, Irvine, CA, USA

**Emilio Gutiérrez** is an assistant professor of economics at Instituto Tecnológico Autónomo de México (ITAM), in Mexico City. He received a Ph.D. in economics from Brown University. He works primarily on areas related to political economy, and environmental, development, and health economics. He has published in several journals, including the papers and proceedings issue of the *American Economic Review*, the *Journal of Population Economics*, and the *World Bank Economic Review*.

Mailing address: ITAM, CIE, Mexico City, Mexico

**Jeong Hyun Kim** is a Ph.D. candidate in the Department of Political Science at Washington University in St. Louis. Her research focuses on electoral politics and public opinion in advanced democracies.

Mailing address: Washington University in St. Louis, St. Louis, MO, USA

**Rebecca Morton** is a professor of politics at New York University (both in New York City and Abu Dhabi) and director of the Social Science Experimental Laboratory at NYU Abu Dhabi.

Her research focuses on voting processes as well as experimental methods. She is the author or coauthor of four books and numerous journal articles, which have appeared in noted outlets such as the *American Economic Review*, *American Journal of Political Science*, *American Political Science Review*, *Journal of Law and Economics*, *Journal of Politics*, and *Review of Economic Studies*.

Mailing address: Department of Politics, NYU NYC and NYU Abu Dhabi, New York, NY, USA

**Mandar Oak** is an associate professor in the School of Economics at the University of Adelaide, Australia. His areas of research specialization are political economy, development economics, and applied game theory. He has published several journal articles in these areas. He has a master's degree from the Delhi School of Economics and a Ph.D. from Cornell University, Ithaca, NY.

Mailing address: School of Economics, University of Adelaide, Adelaide, SA, Australia

**Marcus Pivato** received a B.Sc. in mathematics from the University of Alberta in 1994 and a Ph.D. in mathematics from the University of Toronto in 2001. From 2002 to 2015, he was a professor in the Department of Mathematics at Trent University in Canada. Since 2015, he has been a professor in the Department of Economics at the Université de Cergy-Pontoise in France. He also holds the Labex MME-DII Chaire d'Excellence. His main research interest is normative economics, especially social choice, social welfare, and decision theory.

Mailing address: THEMA/UFR d'Economie et Gestion, Université de Cergy-Pontoise, Cergy-Pontoise Cedex, France

**Dina Y. Rosenberg** is an assistant professor of political science at the National Research University-Higher School of Economics in Moscow. She holds a Ph.D. (2013) in political science from Binghamton University and studies the interaction of innovations and political institutions, comparative political economy, and the politics of health.

Mailing address: National Research University-Higher School of Economics, Moscow, Russia

Binghamton University, Binghamton, NY, USA

**Norman Schofield** is the Taussig professor in political economy at Washington University in St. Louis. He has published numerous articles and books in this field and holds Ph.D.'s in economics and political science from Essex University, a Litt. D. from Liverpool University, and a D.Sc. from the University of Caen.

Mailing address: Department of Political Science, Washington University in St. Louis, St. Louis, MO, USA

Dr. William Taussig Professor of Political Economy, Washington University in St. Louis, St. Louis, MO, USA

**Olga Shvetsova** works in the field of constitutional political economy and studies the determinants of agents' strategies in the political process. She coauthored *Designing Federalism*, with Mikhail Filippov and Peter Ordeshook (2004, Cambridge University Press), and *Party System Change in Legislatures Worldwide: Moving Outside the Electoral Arena*, with Carol Mershon (2013, Cambridge University Press), and published on institutional design, party systems and their determinants, distributive content of political and economic institutions, and comparative political economy of health. She is a professor of political science and economics.

Mailing address: Binghamton University, Binghamton, NY, USA

**William Simoneau** is a Ph.D. candidate in the Department of Political Science at Washington University in St. Louis.

Mailing address: Washington University in St. Louis, St. Louis, MO, USA

**Zeev Thepris** holds a bachelor's degree in economics from Instituto Tecnológico Autónomo de México (ITAM). He currently works as an analyst in the Investment Banking Department at Credit Suisse.

Mailing address: Credit Suisse, Zürich, Switzerland

**Jean-Robert Tyran** is professor of public economics at the University of Vienna and director of the Vienna Center for Experimental Economics. He is an associate editor of the *Journal of Behavioral and Experimental Economics* and a member of editorial boards (*Experimental Economics, European Journal of Political Economy, Journal of Experimental Political Science*) and a member of the Board of the Society for the Advancement of Behavioral Economics. He is a research fellow at various institutions (CEPR, London; EPRU, U Copenhagen; CAMA, Australian National University) and has held numerous visiting positions (Harvard Kennedy School, London School of Economics, among others).

Before moving to Vienna in September 2010, he was at U Copenhagen since 2004, and at U St. Gallen since 1997. He has earned his Ph.D. in economics at the U of Zurich finishing in 1997.

In his research, Jean-Robert Tyran mainly investigates how institutions like markets and democracy are shaped by bounded rationality and social preferences. A current project is to establish iLEE, an Internet-based platform for running large-scale economic experiments. Jean-Robert Tyran has published in field journals such as the *Journal of Public Economics* as well as in general interest journals such as *American Economic Review*, *Econometrica*, and the *Review of Economic Studies and Science.*

Mailing address: Department of Economics, Faculty of Business, Economics and Statistics, University of Vienna, Vienna, Austria

Department of Economics, University of Copenhagen, Copenhagen, Denmark

**Erik Wengström** is an associate professor of economics at Lund University and a research fellow at the University of Copenhagen and the Vienna Center for Experimental Economics. He received his Ph.D. at Lund University in 2007 and thereafter worked at the University of Copenhagen before returning to Lund in 2011. His research is primarily behavioral and experimental in nature and covers a broad range of topics including decision-making under risk, cooperation, and communication in games. His research has been published in leading economics journals such as the *American Economic Review*, *Management Science*, and the *Journal of Public Economics*.

Mailing address: Department of Economics, Lund University, Lund, Sweden

Department of Economics, University of Copenhagen, Copenhagen, Denmark

**Stanley L. Winer** is the Canada Research Chair professor in public policy in the School of Public Policy and Administration and the Department of Economics at Carleton University, Ottawa. He is a CESifo research associate at the University of Munich. He was the Fulbright-Duke University visiting chair in 2003 and has also been a visiting professor or visiting research professor at Carnegie Mellon, Western, Renmin, Australian National, Montreal, U.C. Irvine, Eastern Piedmont, University of Economics Prague, Rennes I, and Hitotsubashi University. He was executive vice-president of the International Institute of Public Finance from 2002 to 2005.

He has published widely, with a focus on empirical models of the structure and evolution of fiscal systems in mature democracies as well as in the world as a whole. This work includes *Coercion and Social Welfare in Public Finance*, coedited with Jorge Martinez-Vazquez for Cambridge University Press (2014), and *Democratic Choice and Taxation: A Theoretical and Empirical Analysis*, coauthored with Walter Hettich, also for Cambridge (1999). His work includes the study of the relationship between interregional migration and public policy, most recently *Interregional Migration and Public Policy in Canada* (with Kathleen Day), published by McGill-Queen's University Press, which received the Purvis Memorial Prize from the Canadian Economics Association in 2013. Recent research, with colleagues in Canada, the USA and India, addresses the meaning, measurement, and consequences for policy of electoral competitiveness in these countries. Professor Winer holds M.A. and Ph.D. degrees in economics from the Johns Hopkins University.

Mailing address: School of Public Policy, Carleton University, Ottawa, ON, Canada

# Autocratic Health Versus Democratic Health: Different Outcome Variables for Health as a Factor Versus Health as a Right

**Dina Rosenberg and Olga Shvetsova**

## 1 Introduction

In this essay, we argue that there is the theoretically meaningful and empirically well-defined contrast between the ways democracies and autocracies set their healthcare priorities. Within the framework of the theory of political coalitions and with the data on disease mortality, we show that the direction of healthcare policies in autocracies is distinct from that in democracies. We argue that autocracies pursue healthcare policy as part of their economic policy. Specifically, they maximize the contribution via healthcare to developing the labor force as a production factor. The same is not true in democracies: there, the nexus of objectives in healthcare does not include the consideration of health as an economic factor. With the data on mortality from diseases which are specific to the in- and out-of-workforce demographic groups, we show that autocracies perform well in dealing with the diseases that "damage" the workforce, at the expense of other areas of health improvement. Democracies, in contrast, do not have such biases.

We build our theoretical argument on the premise that political agents in decisive coalitions under the two regime types differ in their goals with regard to public health. Their divergent goals come into play, as politicians design institutional

---

We would like to thank participants of the *Political Economy of Social Choices Conference* (Oaxaca, Mexico. July 26–31, 2015) and of the *Workshop in Comparative Politics* at Binghamton University for their helpful suggestions and critique.

D. Rosenberg
National Research University - Higher School of Economics, Moscow, Russia

O. Shvetsova (✉)
Binghamton University, Binghamton, NY, USA
e-mail: shvetsova@gmail.com



M. Gallego, N. Schofield (eds.), *The Political Economy of Social Choices*, Studies in Political Economy, DOI 10.1007/978-3-319-40118-8_1

mechanisms to regulate the healthcare sector. Those mechanisms, otherwise known as health policy, in turn provide various forms of *economic* incentives (including via targeted resource allocation) to patients, healthcare providers, financial institutions, and various economic actors in the complex infrastructure that surrounds the industry of health. Thus in our theoretical framework we view the political process as primary and argue, that p*olitical* institutions (autocracy versus democracy) set countries' healthcare systems on the path to eventually divergent policy outcomes. Political institutions do so by determining the composition of the decisive coalitions and, therefore, the ultimate goals of the policy maker. In its most general form, our theory can be extended beyond healthcare, and could apply to policy issues such as education, infrastructure development, arts and culture, and defense.

## 2 Theory: Autocratic vs. Democratic Health

### *2.1 The State of the Field*

Global health is improving; according to the World Health Organization (2013), the recent decade saw reduction in child and maternal mortality, improved nutrition, and decreased morbidity due to HIV, tuberculosis, and malaria (World Health Statistics Report 2013). In OECD countries, life expectancy at birth exceeded 80 years in 2011 (*OECD Health Statistics 2013*). These aggregates, however, conceal the fact that the benefits of recent improvements are distributed unevenly: both across the world and among groups within individual societies.

Aside from the large historical gaps between the developed and developing nations and between different demographics within many countries, there exist, we argue, politically determined *divergent trends* in health development between autocracies and democracies. Our theory leads us to expect such divergence even when accounting for other relevant variables, such as the variation in wealth, culture, demographics, and etc., as well as for the usual indicators of good governance.

Our argument sits well with the extant literature, even though, unlike many scholars, we arrive to our conclusions based on the theory of political coalitions. Much of the political science literature attributes the cross-national gap in health outcomes to the differences in political regimes. Many studies claim that democracies outperform nondemocracies in human well-being indicators other than economic development, such as health and education outcomes.[1] Mostly, the of democracies' success is being attributed to politicians being institutionally constrained and thus highly accountable to voters who demand more public goods and of higher quality, including healthcare services. Specifically, democracy or the democratic stock (i.e., the democracy's age), have been shown empirically to exert a

[1]See, e.g., Baum and Lake (2001), Lake and Baum (2003), Besley and Kudamatsu (2006), Franco et al. (2004), Vollmer and Ziegler (2009) and Zweifel and Navia (2000, 2003).

positive and statistically significant impact upon health outcomes measured as infant mortality rate (IMR), child mortality rate, life expectancy and/or health-adjusted life years (HALE).[2]

Recent scholarship, however, increasingly focuses on the redistributive nature of the way the healthcare good is defined (Ossa and Towse 2004). That is, we do not really know any longer what it means to provide "more healthcare." There is a concern over the question "for Who?" rather than "How much?" The political choice question then becomes two-dimensional, where not only the amount of public investment in health, but also the priorities assigned to those funds must be taken into consideration. Political coalitions, we argue, have preferences on both these dimensions.

Wigley and Akkoyunlu-Wigley (2011) suggest, that distributive theory is not the only viable reason for democracies to have better health outcomes than autocracies. The authors argue that democracies allow more people to participate in the decision-making process, better protect individual rights, promote social capital, and facilitate collective action. The first mechanism suggests that taking part in decision-making process translates into an individual's greater sense of autonomy and control over her own life, which is identified as an important pro-health factor by biomedical researchers. The second mechanism is that self-autonomy is better protected in democracies because individual civil and political rights are secured, and politicians are held accountable for violating those (Cingranelli et al. 2014; Cingranelli and Filippov 2010). Democracies promote social capital, as they increase marginal utility from entering different types of pressure groups and associations (Farrer 2014), which is argued to alleviate human misery. Furthermore, it is easier to overcome collective action problems in democracies, because the latter provide the necessary stimuli (e.g., protection of workers' rights). Finally, better access to information allows spreading health-related knowledge among different segments of population.

Yet in empirical studies, the positive impact of democracy on health outcomes is far from being established. Many scholars fail to find any connection or claim it to be negligible (Gauri and Khaleghian 2002; McGuire 2006; Muntaner et al. 2011; Pande 2003; Ross 2006; Shandra et al. 2004). We conjecture, that this is because the literature relies on general health outcomes indicators intended to evaluate the *overall* level of protection of citizens' health. Such indicators smooth over and thus conceal possible differences in health outcomes with regard to the numerous aspects of health of relevance to different groups in the population and, as we argue here, different decisive coalitions. We argue that democracies and autocracies prioritize different aspects of health, and that we should expect systematic differences to emerge, as long as the policy outcomes are measured in a finer way so as to trace the difference in implemented priorities. To this end, we measure policy outcomes as

[2]More on this in Gerring et al. (2012), Ghobarah et al. (2004), Klomp and de Haan (2008), Lin et al. (2012), Mackenbach et al. (2013), McGuire (2013), Muntaner et al. (2011) and Wigley and Akkoyunlu-Wigley (2011).

disease-specific mortality, because different priority groups are affected by different sets of diseases.

The theoretical basis to expect differences in disease-specific mortality is that the decisive political coalitions in charge of health policy in autocracies and democracies are consistently different. In autocracies, decision-making leans heavily towards the national economic elites, and the selectorate uses public healthcare budgets with an eye to maximizing the efficiency of health as a production factor. Meanwhile, broader decisive coalitions in democracies select public policy towards maximizing their members' private consumption of health as a good. Below we discuss the consequences of the preferences of policy-setting agents in autocracies and democracies for the ensuing health outcomes.

## 2.2 Health as a Production Factor

To conceptualize the health of a nation as a production factor is exciting for an economist. Bloom and Canning (2000) argue that not only the size of available workforce but its relative productivity is increased with improvements in health. Furthermore, they argue, better health leads to improvement in education and so not only to additional increases in labor productivity, but also to the reduction of the cost of healthcare. All in all, in their story, health is an excellent investment from the point of view of macroeconomic efficiency (Bloom and Canning 2000).

But does there exist a general category of political agents who specifically seek to maximize macroeconomic efficiency via healthcare? Probably, not as such. There is, instead, a category of agents who stand to benefit from an increase of the productivity of labor as a factor, and in autocracies, the decisive political coalitions are tilted heavily in favor of such economic agents: these are the national economic elites. Healthier workforce benefits national economic elites because it positively influences economic development (Acemoglu and Johnson 2007; Bhargava et al. 2001; Bloom et al. 2004; Lee and Tuljapurkar 1997; Narayan et al. 2010; Zhang and Zhang 2005). Healthcare increases the marginal productivity of workforce and prevents early withdrawal for retirement or due to disability (Cutler et al. 2006: 114). In the cost-benefits calculations of national economic elites, the benefits of increased marginal productivity multiply by the weight of the country's labor factor in the national economy.

Autocratic leaders' survival depends on the support of the so called "selectorate"—the winning coalition, or those key supporters necessary to keep a leader in power (Bueno de Mesquita et al. 2005; see also Gallego and Pitchik 2004; Luttwak 1979; Olson 1993, 2000; Tullock 1987). These people usually assume important positions in government and business and more often both. In order to please members of this coalition, and since they are not many, unlike in democracies, autocrats use state budgets in ways that benefit this inner circle (Gallego 1998; Gallego and Pitchik 2004; Wintrobe 1998). For instance, Gallego (1998) finds compelling evidence that, in developing countries, dictators provide the version of public goods that benefits the elites—"targeted", or even "club" goods. More so, autocrats are held accountable by their backers for the delivery of such targeted goods (Gallego and Pitchik 2004). Such relationships often result in the collusion of political and economic power (Claessens et al. 2000b; Fisman 2001; Sonin 2003; Vatikiotis 1998). Scholars gave this autocratic phenomenon labels such as crony capitalism (Kang 2002; Krueger 2002) and conjugal dictatorship (Thompson 1998). To illustrate, Claessens et al. (2000b) report that most of the political power and corporate assets in Indonesia as well as in Philippines belong to the top ten families. Apart from the direct fusion of political and economic elites, political connections are a must for the firms in the autocratic world (Fisman 2001). To paraphrase the famous quote, what is good for the national economic elites is good for an autocratic incumbent. In sum, decisive political coalitions in autocracies are often national economic elites. Of course, democracies are also vulnerable to capture by vested interests (Olson 1993), but due to fewer restrictions on the minimal size of the decisive coalition, autocracies are much worse.

## *2.3 Implications of Prioritizing Labor Factor Productivity for Healthcare Policy*

The aim to maximize the factor productivity of labor in the use of public healthcare funds, which we theorize for autocracies, implies the choice of specific resource allocation priorities. We would expect such healthcare system to focus on patients who either belong to the labor force or would likely enter the labor force in the future. This means prioritizing primary and secondary care,[3] which include preventative measures such as immunizations and regular check-ups, low-cost medications and easy access to medical services sufficient for curing common illnesses, heavy investment in prenatal and post-natal care, and vigorous approach to containing communicable diseases. Consider such anecdotal evidence as establishment of the local doctors system in Cuba or government-led mobilization of Chinese villagers to stop contagious diseases (Cutler et al. 2006), the Semashko (Soviet) healthcare system's defeat of tuberculosis (Golinowska and Sowa 2007), and the drop in infant mortality in Chile under Pinochet (McGuire 2001). Furthermore, Gauri and Khaleghian (2002) report that autocracies have higher immunization rates than democracies.

It is also relatively clear what is *not*, in theory at least, a public healthcare priority for an autocratic incumbent. A political coalition of national economic elites would not prioritize the specialized care needs of the demographic groups that are past the

[3]We follow the conventional terminology, identifying as primary care the prevention of illness, as secondary—curative care, as tertiary—maintenance of improved but not restored health, and finally custodial—care that improves patient's comfort but not health status (see, e.g., Andersen and Newman 1973).

age of active workforce participation. Thus we should expect lower priority given to life-saving technologies and treatments for diseases that occur late in life. Tertiary care which does not fully restore patients' productivity should also be a low funding priority. And, for the same reason (that economic elites drive policy decisions), we should expect public spending on health to be less generous than would be preferred by a broader selectorate or by the median voter.[4] "Kornai and McHale (1999) estimate that for industrialized countries a 1 percentage point increase in the share of the population between 65 and 74 raises national health-spending growth by 0.7 percent" (Kornai and Eggleston 2001: 65).

## 2.4 *Health as a Distributive Good*

In the case of a democratic incumbent, on the contrary, the concern with healthcare as an economic factor is unlikely to drive policy priorities. This is because of the majoritarian nature of political coalitions, whereby at least some of participants in a winning coalition will demand the healthcare policy to be needs-based rather than goal-oriented. Musgrove (1996) posits that there exist universally shared preferences to save fellow human beings from dying of a disease that is customarily treatable, as well as to make healthcare available to the most sick in the society. Kornai and Eggleston (2001) echo this claim, while Shvetsova and Sieberg (2013) argue that this claim has experimental support. Supposing that the median voter has such preferences, there is no reason to expect that policy effort would be limited to protection from only some specific subset of all treatable diseases.[5]

Thus the best we can say about the preferences of the democratic winner is that she prefers that everyone would have access to life-saving medicine in life-threatening situations, and also that she might under-invest in some non-life-threatening conditions which are unlikely to strike the members of its own democratic coalition. In both of these regards, healthcare is just another distributive decision for a democracy. Insofar as there is consensus on the need to save lives and sustain the very sick, as per Musgrove's (1996) and Kornai and Eggleston's (2001) premises, the redistribution is (1) from wealthy to poor *and* (2) from healthy to sick. Sen (1981, 1999) notes that famines have never occurred in a "functioning democracy" (Sen 1999: 16). His explanation of this regularity is political, that poor citizens affected by a famine would be able to punish incumbents as a part

[4]A populist autocrat is somewhat accountable to a broader national constituency. Still, the same priorities in healthcare that we listed as serving national economic elites also appeal to the populist sentiment, e.g., to a winning coalition of a large supermajority. Mobilizing supermajority against the autocratic health strategy would require a shared alternative objective, which is unlikely to exist. It takes smaller coalitions, with intense issue preferences, which can be brought together by competitive electoral campaigns, to challenge the healthy-workforce health policy.

[5]The reverse, however, is not true, and it is possible in a democracy (autocracy) to under-prioritize some diseases, e.g., neglected tropical diseases are a case in point (Hotez et al. 2007).

of a decisive coalition.[6] The second consideration, notably, features a new form of redistribution—the redistribution in the multidimensional space of health. It combines the obvious dimension of healthy versus sick, whereby "the most costly 10 percent of patients accounts for as much as 75 percent of total spending" (Kornai and Eggleston 2001: 55–56), and also multiple dimensions of being either sick or not with various *specific* diseases/conditions (Shvetsova and Sieberg 2013).

As a consequence of the complexity of the space of health, predicting what the democratic winning coalition would demand in terms of healthcare priorities is less precise than for autocracies. Because of the fluidity of the democratic "winner," resource allocation within the public health sector remains subject to never-ending political bargaining in a democracy.[7] Coalitions on the dimension of health can form in favor of consuming different types of healthcare good and promoting policies that are favorable for the accelerated development and delivery of that good, and at the expense of other, also healthcare-related, products and innovations. We will illustrate this connection below with the American mid-twentieth century case.

Institutions in place, both political and economic, can generate path-dependent preferences on healthcare and foster formation of unique, not immediately intuitive bargaining coalitions. In sum, there may be latent or even active political coalitions in wealthy democracies that are based on health rather than wealth that lose out, and therefore are underserved by the status-quo healthcare policy. American pre-ACA (Affordable Care Act) healthcare system is a nice illustration of how preferences of political coalitions can impart both the political and the economic bases for the industry to focus on certain types of health problems. The status-quo public sector in the US healthcare used to combine those who were either the poorest or the sickest (needing the most expensive care) or both. This is because it encompassed the Medicare patients who were expensive because of their older age, and Medicaid patients, who, by definition, were either very poor or at the extreme of ill health and disability. Notice also, that both medical disability and life in retirement correlate negatively with one's wealth, making a large portion of this combined constituency into non-taxpayers. This constituency would likely have very specific preferences for the budget allocation between different types of health goods. It would be biased in favor of life-saving and tertiary research at the expense of research in costly high-end preventative technologies. Not paying much in taxes, this constituency would be less sensitive to the cost of the type of treatments that it demands. And if this constituency has significant electoral weight, public health policy would likely take its preferences into consideration. Public sector demand for advanced treatments

[6] Ross (2006) shows that, although democracies redistribute more than autocracies, this does not necessarily benefit the poorest.

[7] The decades-long struggle over the healthcare reform in the US, the reversals of fortunes of various reform proposals, and the slim margin of eventual passage of the Affordable Health Care Act in November 2009 with the 220-215 vote in the House of Representatives—all illustrate this complexity of democratic coalitions on health.

would generate a corresponding supply from the medical industry and cause greater improvements in health outcomes from that group's "preferred" diseases.

Arguments in the extant literature that rich countries are privileged in terms of healthcare do not logically preclude the fact that some groups-by-diseases are privileged as well. Raw evidence indeed suggests that this is the case. For example, Trouiller et al. (2002) argue that, for the new drugs approved in the last quarter of the twentieth century, the "quantitative distribution in different therapeutic areas shows a bias toward high-income countries," (p. 2188), leaving "the needs of millions in the developing world . . . ignored" (p. 2193). Their argument is striking and persuasive, yet their evidence in fact shows the existence of health-conditions privilege first, and wealthy-countries privilege only by implication. Their data show the bias towards older-age diseases and life-saving maintenance drugs and away from research into preventative and curative medicines.

Predicting the disease-type focus of public policy in a democracy is therefore difficult. A different makeup of political coalitions would lead to a different sort of priorities in funded demand. Of course, the size of such winning coalitions will depend on the political institutions of the democracy in question. Additionally, the ability of health-based coalitions to enact serious redistribution from other groups will depend on whether healthcare policy is entirely politically regulated and subject to majoritarian whims, or some fundamentals are constitutionally prescribed and therefore the distributive bargaining over health is constrained. A separate enquiry is warranted into the question of the losing democratic coalitions on the dimension of health. Here we but scratch the surface of the theoretical problem of differences among democracies in the priorities of their winning coalitions in public health. We leave a more detailed exploration of this complex and fascinating problem to future research. The one insight that is important for us here is the lack of majoritarian political support in democracies for treating healthcare as a tool to increase the labor productivity.

## 3 Research Design and Data

Our theoretical expectations regarding autocratic goals in public health spending are clear: to maximize labor factor productivity. It is, however, impossible to measure directly the incumbents' effort on behalf of groups in the population, and so the most appropriate operationalization is unavailable. Equally unobservable is the effort on behalf of specific diseases and services that contribute the most to the productivity of the current and future workforce. Short of issuing edicts to serve only a certain type of patients and only from a specific list of diseases, the intent of the policy-maker cannot be inferred from the organization of the healthcare system. While we posit that the way healthcare is organized and delivered does reflect the incumbent's political objectives, the industry is complex and the technology of health is decentralized, with multiple agents and services junctions along extensive timelines accounting for each patient's health outcomes. Furthermore, there are a plethora of substitutionalities and complementarities, whereby different healthcare systems can achieve equivalent health results via alternative means. Therefore, to code some healthcare mechanisms (e.g., single-payer) as prioritizing the workforce and others as not could only produce a very crude and questionable indicator. In order to overcome these difficulties, we adopt the design of multiple outcome indicators, with different theoretical expectations about regimes' performance with regard to different types of health outcomes. Our general testable proposition then becomes as follows: *autocracies with high reliance on labor as productive factor would perform relatively well on health outcomes that directly influence the health of the active workforce, at the expense of addressing diseases unrelated to labor productivity.*

Thus in our research design, we operationally equate policy goals of authoritarian and democratic winning coalitions with success in policy outcomes of heightened relevance for those coalitions. We operationalize the goals of the autocratic incumbent as protecting the citizens form diseases that diminish the workforce (as shown in the left column of Table 1), and not from the diseases in the control group of *dependent* variables, which affect primarily older individuals and/or imply long-term disability (the right column of Table 2). We measure these dependent variables as diseases-specific mortality. Since some diseases detract from the labor factor productivity more than others, we can evaluate our theory by comparing democracies' and autocracies' performance with regard to outcome variables of mortality from different diseases.

Our dependent variables of disease-specific mortality include both autocrat-preferred and non-autocrat-preferred indicators. The expectation is that autocrats would strive to minimize the former, but would be less motivated regarding the latter. Specifically, treating health as an economic factor implies the use of policy mechanisms that prioritize treatment and prevention of the diseases of the "young"—of those entering or participating in the workforce (see Table 1). The diseases of the "old" should receive relatively low priority, as should debilitating illnesses that preclude workforce participation even if controlled. Therefore, in autocracies, we expect to see better outcomes with regard to the first than the second. Meanwhile, in democracies, we cannot identify such bias (or any cross-nationally consistent bias, for that matter), and so we expect to see better outcomes in the second category of diseases as compared to autocracies.

With country-level mortality from each of these types of diseases as in Table 1 as our dependent variables, we now turn to testing the following hypotheses:

*H1: Mortality from diseases affecting active workforce will be lower in autocracies than in democracies, other things, including health expenditures, equal.*

**Table 1** Research design: diseases affecting active workforce versus older demographic cohorts

| Active workforce | Older cohorts |
|---|---|
| • Infectious diseases• Tuberculosis | • All cancers• Diabetes• Cardio-vascular disease |

**Table 2** Percentage of government spending directed to public health, 1995–2011

| | Autocracies | | Democracies | | All | | | | |
|---|---|---|---|---|---|---|---|---|---|
| | 1 | 2 | 3 | 4 | 5 | 6 | 7 | 8 | 9 |
| GDP per capita (Constrained to exceed $300 in 2005 USD.) | −0.0001*** (0.000) | −0.0001*** (0.000) | 0.0001*** (0.000) | 0.0001*** (0.000) | 0.0001*** (0.000) | 0.0001*** (0.000) | 0.0001*** (0.000) | 0.00004*** (0.000) | 0.00004*** (0.000) |
| Govt. final consumption (% of GDP) | 0.103*** (0.025) | 0.093*** (0.027) | 0.075*** (0.019) | 0.078*** (0.019) | 0.079*** (0.015) | 0.082*** (0.016) | 0. 089*** (0.015) | 0.091*** (0.016) | 0.064*** (0.016) |
| Labor force participation (%) | 0.048*** (0.011) | 0.045*** (0.011) | 0.040*** (0.014) | 0.040*** (0.016) | – | 0.066** (0.008) | 0.047*** (0.009) | 0.049*** (0.010) | 0.055*** (0.009) |
| Polity index | – | 0.090*** (0.025) | – | 0.147* (0.067) | – | – | – | – | 0.200*** (0.013) |
| Polity squared | – | – | – | – | – | – | – | 0.007* (0.004) | 0.013*** (0.003) |
| Democracy | – | – | – | – | 2.546*** (0.182) | – | 2.588*** (0.184) | 2.509*** (0.215) | – |
| Constant | 4.790*** (0.819) | 5.166*** (0.836) | 7.341*** (0.993) | 6.290*** (1.173) | 7.627*** (0.240) | 4.691*** (0.619) | 4.478*** (0.646) | 3.971*** (0.691) | 4.585*** (0.651) |
| R-squared | 0.07 | 0.09 | 0.18 | 0.17 | 0.20 | 0.14 | 0.22 | 0.23 | 0.25 |
| N obs. | 848 | 792 | 1219 | 1106 | 2123 | 2625 | 2067 | 1898 | 2407 |

$^*p < 0.5$, $^{**}p < 0.1$, $^{***}p < 0.01$

*H1a: In autocracies, mortality from diseases affecting active workforce will be the lower, and from other diseases—higher, the larger the size of the labor market, other things, including health expenditures, equal.*
*H1b: In autocracies, mortality from diseases affecting active workforce will be the lower, and from other diseases—higher, the larger the per capita economic investments, other things, including health expenditures, equal.*
*H2: Mortality from diseases that weakly affect active workforce (older cohort diseases) will be higher in autocracies than in democracies, other things equal.*

The main theoretically predicted effect is that of the political regime and so of the goals in healthcare policy set by the decisive political coalitions behind that regime.

Our data span 14 years, from 1995 to 2008. Since our theory dictates that different political coalitions should prioritize and, as a result, improve upon different sets of health outcomes while acting under budget constraints, in order to test our hypotheses we employ multivariate multiple regression. This method implies that several dependent variables are jointly regressed on the predictors, in contrast to the ordinary least squared (OLS) regression when different dependent variables are regressed on predictors separately. The coefficients obtained through the multivariable regression are the same as in the OLS regression analysis, but the former allows testing our coefficients *across* the different dependent variables, in our case, different diseases. With time-series data, we lag almost all our independent variables by 1 year to account for the long duration of treatment of most diseases that we include, infections being an exception. The dependent variable for infections in the multiple variable regression, thus, is also lagged by 1 year, making the hypothesized determinants of infection mortality concurrent with that outcome.

We use the data in the World Health Organization Mortality Database (http://apps.who.int/healthinfo/statistics/mortality/whodpms/param.php) for the diseases (or rather groups of diseases) as in Table 1: infectious diseases, tuberculosis, all cancers, diabetes, and cardio-vascular disease. Table 1 categorizes which of these we consider as affecting active workforce.

Infections and tuberculosis are the two types of diseases that we include as affecting the health of the workforce. Mortality rates for them are calculated as mortality per 100,000 of population per year. For the diseases associated with older cohorts, outside the workforce: cancers, diabetes, and cardio-vascular, we use the indicators of total mortality from a disease in reported year per 100,000 of the population aged 65 or higher. An even better indicator would be age-standardized mortality rate from specific diseases, yet data issues, such as missing yearly observations and the reliance on *estimates* of disease-specific outcomes in demographic cohorts lead us to prefer our current approach.

Our most important explanatory variables are *political institutions,* specifically, autocracy, as it interacts with indicators of labor-intensity of the national economy. The latter we measure in two ways, as *total investment*, and *rate of workforce participation* in appropriate age cohorts, 15–64. We measure political institutions as autocracy versus democracy in the *Democracy* variable which takes the value 1 for democracies and 0 for autocracies.[8] We measure workforce participation as the percentage of the population in the 15–64 years of age demographic cohort which is in the active workforce. Our *total investment* variable comes from the International Monetary Fund data and is available for 173 countries on an annual basis, 1980–2013 (IMF data).

Since poorer populations are going to be in worse health (Preston 1975), and since spending more money on providing medical care should improve health outcomes in its own right (Filmer and Pritchett 1999), we control in our estimations for the countries' per capita GDP in constant US dollars (estimated in PPP, purchasing power parity) as well as for the public sector's health expenditures In our analysis both variables come from the Quality of Government Standard Dataset.

Because extant studies suggest that quality of governance positively influences policy effectiveness of government spending,[9] we include the control for *Quality of Governance.* We use the International Country Risk Guide (ICRG) indicator constructed as the mean value of three indicators: corruption, law and order, and bureaucracy quality. The variable is available for 146 countries and takes values from 0 to 1.

Since we use investment as one of our main explanatory variables, we include such covariates as *general government final consumption expenditure* (% of GDP), which Gallego (1998) uses as an additive component with *private investment*, and also *economic globalization* (again, paralleling Gallego (1998) export indicator).[10]

*Globalization* is shown to positively influence health outcomes through increased economic growth (Dreher 2006); foreign aid (Gomanee et al. 2005); health sector-specific foreign aid (Mishra and Newhouse 2009); technology transfers and knowledge spillovers (Deaton 2004). Our hypothesis with regard to globalization in light of the theorized priorities in autocracies pointing away from investing in high-end life-saving and life-maintaining treatments, is that exposure to globalization permits patients to seek either private, or international, or charitable options for treatment of diseases that are underserved by authoritarian public health.

*H3: Autocracies with greater exposure to globalization will have lower mortality from non workforce-prevalent diseases that require advanced treatment.*

Statistical models below selectively include a number of interaction terms, which we use to express conditionality of specific effects, e.g., "but only if in an autocracy." We also selectively employ a GDP/per capita restriction on the sample rather than include fixed effect variables to that end. That approach is driven by the expectation

---

[8]Source: Democracy and Dictatorship Revisited dataset by Cheibub et al. (2010).

[9]See: Baldacci et al. (2008), Cingranelli et al. (2014), Holmberg and Rothstein (2011a), Lewis and Lloyd-Sherlock (2009), McGuire (2006), Rajkumar and Swaroop (2008) and Wagstaff and Claeson (2004).

[10]General government final consumption expenditure serves as a proxy for public goods provision by the state. The latter is a component of Dreher's globalization index (2006) and is measured as actual goods and investments flows as well as the existing restrictions (e.g., tariffs).

that the data generating mechanism is different at different income strata. We do so to exclude the high variance at extremely low income levels due to the inevitable issues with data acquisition in those countries.

## 4 Results

Statistical analysis lends support to our theory. We indeed find the structural difference of improved mortality from workforce-affecting diseases in autocracies with extensive labor markets relatively to non-labor intensive autocracies, but not in democracies, under similar conditions. We report the findings in Tables 2 and 3 below.

We begin in Table 2 by comparing the spending on public health in autocracies and democracies. This comparison is of interest to us as it reveals the differences in the nature of the budget constraint in health spending that different countries have to work under. Table 2 shows that the share of the budget spent on healthcare increases with income in democracies, but *decreases* in autocracies (income is measured as GDP per capita). While wealthier autocracies might be spending more on health in absolute terms, the increase of government spending in other categories outpaces the spending on health. This finding is consistent with the theoretical account of authoritarian public health objectives as limited to achieving factor productivity goals. Meanwhile, democracies *consume* public health at an increasing rate with wealth, and generally more than comparable autocracies. Regardless of the type of the political regime, the Polity IV index correlates positively with the weight of public healthcare spending, i.e., a comparable democracy consumes more public healthcare than autocracy.

We claim that, because protecting labor factor productivity is the healthcare priority of the winning coalition in autocracies, the level of working age diseases protection should not directly depend on the size of healthcare budget. Instead, other health services should occupy the residual place. Therefore, beyond a certain level of public expenditure, additional allocation to healthcare does not serve the economic objective of the incumbent any longer. While Table 2 results are consistent with that expectation, in Table 3 we specifically focus on diseases which either should, or should not be the autocratic healthcare priority according to our theory. The patterns of protection from these diseases in autocracies are sharply different, compared to democracies, indicating divergent trends in health sector orientation.

The model in Table 3 addresses mortality from five types of diseases. Mortality from infectious diseases and tuberculosis—our selected working-age diseases—is statistically significantly reduced by the interaction term of labor force participation with autocratic government. Substantively, this means that consistent with H1a, possessing a large labor market *and* being an autocracy significantly improves survival in autocracies when affected with infectious diseases and tuberculosis. Strikingly, but consistent with our theory, this effect reverses for cardiovascular diseases and cancers, where the impact on labor factor productivity should be

**Table 3** Multiple variables disease mortality model with regime type, labor force participation, and total investment, per 100,000, GDP per capita > $1000

| | Infect (t-1) | TB | Diabetes | CVD | Neoplasms |
|---|---|---|---|---|---|
| *L. f. part. in 15–64 cohort (%) × (1-Democracy), t-1* | ***−1.11****** ***(0.08)*** | ***−1.46****** ***(0.22)*** | ***−19.26****** ***(4.41)*** | *6.38* *(5.31)* | ***10.70**** ***(4.27)*** |
| *Investment × (1-Democracy), t-1* | $-2.36 \times 10^{-10}$ $(6.27 \times 10^{-10})$ | $1.60 \times 10^{-9}$ $(1.68 \times 10^{-9})$ | $2.81 \times 10^{-8}$ $(2.66 \times 10^{-8})$ | $\mathbf{2.18 \times 10^{-7}}$*** $\mathbf{(4.04 \times 10^{-8})}$ | $\mathbf{8.17 \times 10^{-8}}$* $\mathbf{(3.38 \times 10^{-8})}$ |
| *Labor force participation in 15–64 cohort (%), t-1* | *0.05* *(0.04)* | *0.09* *(0.10)* | *−2.25* *(1.52)* | *−3.12* *(2.32)* | *−3.70** *(1.94)* |
| *Investment, t-1* | $1.27 \times 10^{-10}$ $(2.07 \times 10^{-10})$ | $5.42 \times 10^{-10}$ $(5.55 \times 10^{-10})$ | $1.67 \times 10^{-8}$* $(8.76 \times 10^{-9})$ | $-4.10 \times 10^{-8}$** $(1.33 \times 10^{-8})$ | $1.31 \times 10^{-8}$ $(1.12 \times 10^{-8})$ |
| *Democracy* | *−74.74**** *(6.10)* | *−98.03**** *(16.37)* | *−386.64* *(258.19)* | *768.88** *(393.10)* | *553.11** *(328.77)* |
| Government Healthcare spending (% of GDP), t-1 | −0.02 (0.02) | −0.07 (0.047) | −1.53* (0.76) | 0.12 (1.12) | 4.10*** (0.938) |
| Gov. Health. Sp. (% of GDP) × (1-Democracy), t-1 | −0.08** (0.03) | −0.36*** (0.08) | 5.21*** (1.20) | 9.39*** (1.82) | −0.76 (1.52) |
| Gov. final consumption (% of GDP), t-1 | 0.01 (0.06) | 0.120 (0.152) | 1.23 (2.42) | −2.14 (3.69) | 11.84*** (3.09) |
| Gov. final consumption (% of GDP) × (1-Democracy), t-1 | 0.29** (0.10) | 1.28*** (0.28) | 9.33* (4.41) | 0.89 (6.71) | −21.84*** (5.61) |
| Economic globalization, t-1 | 0.03 (0.02) | 0.04 (0.06) | 3.29*** (0.90) | −3.04* (1.36) | 2.45* (1.14) |
| Econ. Globalization × (1-democracy), t-1 | 0.04 (0.04) | 0.18 (0.11) | 6.19*** (1.78) | 2.13 (2.71) | −0.74 (2.27) |
| GDP per capita, t-1 | $-2.02 \times 10^{-6}$ $(1.49 \times 10^{-5})$ | $-6.02 \times 10^{-5}$ $(3.99 \times 10^{-5})$ | $4.07 \times 10^{-4}$ $(6.29 \times 10^{-4})$ | $-7.07 \times 10^{-3}$*** $(9.58 \times 10^{-4})$ | $-2.20 \times 10^{-3}$** $(7.81 \times 10^{-4})$ |

| | | | | | |
|---|---|---|---|---|---|
| Quality of government, t-1 | −5.30** (1.88) | −14.35** (5.06) | −130.50 (79.72) | −60.56*** (121.38) | 26.35** (85.15) |
| Population aged 65 or older, % of total | −0.22*** (0.06) | 0.21 (0.16) | −29.62*** (2.59) | 52.95*** (3.94) | 7.95* (3.30) |
| Year | −0.08 (0.06) | −0.07 (0.17) | 1.54 (2.64) | −18.92*** (4.01) | −10.64** (3.36) |
| Constant | $2.35 \times 10^2$* ($1.26 \times 10^2$) | $2.35 \times 10^2$ ($3.37 \times 10^2$) | $-2.04 \times 10^3$ ($5.31 \times 10^3$) | $3.83 \times 10^4$*** ($8.09 \times 10^3$) | $2.15 \times 10^3$*** (676.78) |
| Observations | 794 | 794 | 794 | 794 | 794 |
| "R-squared" | 0.40 | 0.29 | 0.41 | 0.46 | 0.36 |

*p < 0.1, **p < 0.01, ***p < 0.001

minimal. Our statistical specification enables us to demonstrate that within autocracies labor force participation simultaneously decreases mortality from infections when labor is important and increases mortality from CVD and neoplasms. The *autocratic investment* variable increases mortality from old-age diseases in autocracies (CVD and neoplasms) and is highly significant jointly with *autocratic labor force participation*. We thus see substantial evidence that economic elites' preferences are reflected in health outcomes in autocracies with labor-dependent economies. In light of dual goals of budget minimization and protection of the labor factor productivity, autocratic effort on diseases that affect labor force steals away from the effort directed to life-maintenance technologies, consistent with Hypothesis 2. Quality of government, across the board, helps reduce mortality from diseases that require consistent and accessible but inexpensive primary and secondary care such as infections and tuberculosis, but, interestingly, not diabetes (Table 3).

The results with budget allocation per se (government healthcare spending as a percent of GDP) in autocracies parallel that picture. In autocracies, the increased share of state budget allocated to healthcare seems to lower labor-affecting tuberculosis and infections mortality (H1), but increase mortality from CVD, and have no impact on mortality from both diabetes and "all cancers" (H2). The latter might be because cancers are simply not diagnosed since diagnosis itself is rather costly. Since cancer treatment is more technologically demanding than other included diseases, estimations give us a glimpse at the differential willingness of the political incumbents in autocracy and democracy to invest in high-end and life-saving medical technology.

With regard to H3, which suggests the remedial role of exposure to globalization for health outcomes, our findings are essentially negative. Globalization may create alternative paths for accessing medical technology, but it seems to also instill unhealthy lifestyle and act as a trigger for the spread of communicable diseases (Saker et al. 2006), leading to higher disease-related mortality via increase in morbidity. This is a possibly fruitful direction for further exploration, which we are limited from pursuing here.

## 5 Conclusion

In his seminal paper, Ross (2006) shows that democracies do not necessarily improve the welfare of the poor. Our subject here, healthcare, is inelastic in demand, which should make the poor the primary victims of its underproduction: income-constrained, the poor could be cut off from consumption. Yet there is a debate in the literature on whether or not autocracies are so inferior in terms of health outcomes. Our theory in this essay calls for the breaking down of what we understand to be relevant health outcomes, as we argue that some of those outcomes may be autocrats' primary targets while others may not. The evidence indicates that following such an approach is helpful in separating the things in healthcare that autocracies do well from those that they simply do not address (notice, however, that we found no support for our H1 that autocracies deal with workforce-affecting diseases better than democracies in absolute terms). Perhaps even more importantly, our approach of selecting policy indicators consistent with the specific theorized goals of political incumbents allows us to test hypotheses with regard to the causal political mechanisms that create the divergence between autocratic and democratic health.

Empirically, Ross (2006) shows that poor autocracies tend to obscure their good performance in order to, say, receive international help, and, after this trend is taken care of, democracies do not outperform autocracies in reducing IMR or CMR, all else equal. As regards other channels of influence, "Democracy unquestionably produces noneconomic benefits for people in poverty, endowing them with political rights and liberties. But for those in the bottom quintiles, these political rights produced few if any improvements in their material well-being. This troubling finding contradicts the claims made by a generation of scholars" (Ross 2006: 872).

We here share Ross' assumption that the effect of political institutions on health outcomes is more subtle and probably conditional. We do not, however, emphasize different beneficiaries of health services, but rather choose to focus on differences in health outcomes e.g., certain diseases that interfere with the objectives of the political incumbent.

The recent heated debate in the United States over the choice of the healthcare system and the salient question of reforming public healthcare sector in many European and East Asian countries, the urgency of which is reinforced by the "pension bomb" problem, demonstrates how important healthcare policy choices are and how controversial. The primary reason behind conflict and controversy is that these choices affect vital outcomes (literally), and different groups in the society possess distinct, if not conflicting, interests over which health outcomes to prioritize. In this paper we take up the theory of political coalitions and show that, ceteris paribus, political coalitions in autocracies, which often coincide with the national economic elites, prioritize health outcomes that protect labor force. Autocrats do this at the expense of other healthcare outcomes such as diseases that occur later in life. In contrast, in democracies, electoral calculus pushes the incumbent to spread the public health effort over a more diverse range of needs, and not to aim for raising the productivity of the economy. Thus, national healthcare systems in democracies generally do not de facto neglect elderly and most "expensive" patients.

Importantly, the unwillingness and, as a result, the inability of autocratic national healthcare systems to efficiently cure diseases that affect people at older age might continue to broaden the existing gap between countries in their utilization and invention of modern medical technologies, given that these diseases happen to be the most expensive, research- intensive and technology demanding.

## References

Acemoglu, D., & Johnson, S. (2007). Disease and development: The effect of life expectancy on economic growth. *Journal of Political Economy, 115*(6), 925–985.

Andersen, R., & Newman, J. F. (1973). Societal and individual determinants of medical care utilization in the United States. *The Milbank Memorial Fund Quarterly. Health and Society, 51*(1), 95–124.

Baldacci, E., Clements, B., Gupta, S., & Cui, Q. A. (2008). Social spending, human capital, and growth in developing countries. *World Development, 36*(8), 1317–1341.

Baum, M. A., & Lake, D. A. (2001). The invisible hand of democracy: Political control and the provision of public services. *Comparative Political Studies, 34*(6), 587–621.

Besley, T. J., & Kudamatsu, M. (2006). Health and democracy. *American Economic Review, 96*(2), 313–318.

Bhargava, A., Jamison, D. T., Lau, L. J., & Murray, C. J. L. (2001). Modeling the effects of health on economic growth. *Journal of Health Economics, 20*(3), 423–440.

Bloom, D. E., & Canning, C. (2000). The health and wealth of nations. *Science, 287*(5456), 1207–1209.

Bloom, D. E., Canning, C., & Sevilla, J. (2004). The effect of health on economic growth: A production function approach. *World Development, 32*(1), 1–13.

Bueno de Mesquita, B., Morrow, J. D., Siverson, R. M., & Smith, A. (2005). *The logic of political survival*. Cambridge: MIT Press.

Cheibub, J. A., Gandhi, J., & Vreeland, J. R. (2010). Democracy and dictatorship revisited. *Public Choice, 143*(2-1), 67–101.

Cingranelli, D., Fajardo-Heyward, P., & Filippov, M. (2014). Principals, agents and human rights. *British Journal of Political Science, 44*(3), 605–630.

Cingranelli, D., & Filippov, M. (2010). Electoral rules and incentives to protect human rights. *The Journal of Politics, 72*(1), 243–257.

Claessens, S., Djankov, S., & Lang, L. (2000). The separation of ownership and control in East Asian corporations. *Journal of Financial Economics, 58*, 81–112.

Cutler, D., Angus, D., & Lleras-Muney, A. (2006). The determinants of mortality. *Journal of Economic Perspectives, 20*(3), 97–120.

Deaton, A. (2004). *Health in an age of globalization*. NBER working paper series. Working Paper № 10669. Available from http://www.nber.org/papers/w10669 [accessed 22 November 2013].

Dreher, A. (2006). Does globalization affect growth? Evidence from a new index of globalization. *Applied Economics, 38*(10), 1091–1110.

Farrer, B. (2014). A theory of organizational choice: Interest groups and parties as substitutable influence mechanism. *Party Politics, 20*(4), 632–645.

Filmer, D., & Pritchett, L. (1999). The impact of public spending on health: Does money matter? *Social Science and Medicine, 49*(1), 1309–1323.

Fisman, R. (2001). Estimating the value of political connections. *American Economic Review, 91*(4), 1095–1102.

Franco, A., Alvarez-Dardet, C., & Ruiz, M. T. (2004). Effect of democracy on health: Ecological study. *British Medical Journal, 329*(7480), 1421–1423.

Gallego, M. (1998). Economic performance and leadership accountability: An econometric analysis. *Economics and Politics, 10*(3), 249–296.

Gallego, M., & Pitchik, C. (2004). An economic theory of leadership turnover. *Journal of Public Economics, 88*, 2361–2382.

Gauri, V., & Khaleghian, P. (2002). Immunization in developing countries: Its organizational and political determinants. *World Development, 30*(12), 2109–2132.

Gerring, J., Thacker, S. C., & Alfaro, R. (2012). Democracy and human development. *Journal of Politics, 74*(1), 1–17.

Ghobarah, H. A., Huth, P., & Russett, B. (2004). Comparative public health: The political economy of human misery and well-being. *International Studies Quarterly, 48*(1), 73–94.

Golinowska, S., & Sowa, A. (2007). *Investing in health institutions in transition countries*. Studies and Analyses № 348. Center for Social and Economic Research.

Gomanee, K., Morrissey, O., Morley, P., & Verschoor, A. (2005). Aid, government expenditure, and aggregate welfare. *World Development, 33*(3), 355–370.

Holmberg, S., & Rothstein, B. (2011). Dying of corruption. *Health Economics, Policy, and Law, 6*(4), 529–547.

Hotez, P. J., Molyneuw, D. H., Fenwick, A., Kumaresan, J., Sachs, S. E., Sachs, J. D., et al. (2007). Control of neglected tropical diseases. *New England Journal of Medicine, 357*(10), 1018–1027.

Kang, D. C. (2002). *Crony capitalism: Corruption and development in South Korea and the Philippines*. Cambridge, England and New York: Cambridge University Press.

Klomp, J., & de Haan, J. (2008). Effects of governance on health: A cross-national analysis of 101 countries. *Kyklos, 61*(4), 599–614.

Kornai, J., & Eggleston, K. (2001). *Welfare, choice, and solidarity in transition: Reforming the health sector in Eastern Europe*. Cambridge: Cambridge University Press.

Krueger, A. (2002). Why crony capitalism is bad for economic growth. In S. Haber (Ed.), *Crony capitalism and economic growth in Latin America: Theory and evidence* (pp. 1–24). Stanford, CA: Hoover Institution Press.

Lake, D. A., & Baum, M. A. (2003). The political economy of growth: Democracy and human capital. *American Journal of Political Science, 47*(2), 333–347.

Lee, R., & Tuljapurkar, S. (1997). Death and taxes: Longer life, consumption, and social security. *Demography, 34*(1), 67–81.

Lewis, C. M., & Lloyd-Sherlock, P. (2009). Social policy and economic development in South America: An historical approach to social insurance. *Economy and Society, 38*(1), 109–131.

Lin, R. T., Chen, Y. M., Chien, L. C., & Chan, C. C. (2012). Political and social determinants of life expectancy in less developed countries: A longitudinal study. *BMC Public Health, 12*, 85.

Luttwak, E. (1979). *Coup d'état: A practical handbook*. Cambridge, MA: Harvard Univ. Press.

Mackenbach, J. P., Hu, Y., & Looman, C. W. N. (2013). Democratization and life expectancy in Europe, 1960–2008. *Social Science & Medicine, 93*, 166–175.

McGuire, J. W. (2001). Social policy and mortality decline in East Asia and Latin America. *World Development, 29*(10), 1673–1697.

McGuire, J. W. (2006). Basic health care provision and under-5 mortality: A cross-national study of developing countries. *World Development, 34*(3), 405–425.

McGuire, J. W. (2013). Political regime and social performance. *Contemporary Politics, 19*(1), 55–75.

Mishra, P., & Newhouse, D. (2009). Does health aid matter? *Journal of Health Economics, 28*(4), 855–872.

Muntaner, C., Borrell, C., Ng, E., Chung, H., Espelt, A., Rodriguez-Sanz, M., et al. (2011). Politics, welfare regimes, and population health: Controversies and evidence. *Sociology of Health and Illness, 33*, 946–964.

Musgrove, P. (1996). Public and private roles in health. Theory and financing patterns. *World Bank Discussion Paper No. 339*. Washington, DC.

Narayan, S., Narayan, P. K., & Mishra, S. (2010). Investigating the relationship between health and economic growth: Empirical evidence from a panel of 5 Asian countries. *Journal of Asian Economics, 21*(4), 404–411.

OECD. (2013). OECD Health Statistics 2013. Paris: OECD Publishing. doi:10.1787/health-data-en.

Olson, M. (1993). Dictatorship, democracy, and development. *American Political Science Review, 87*(3), 567–576.

Olson, M. (2000). *Power and prosperity; outgoing communist and capitalist dictatorships*. NY: Basic Books.

Ossa, D. F., & Towse, A. (2004). Genetic screening, health care and the insurance industry: Should genetic information be made available to insurers? *The European Journal of Health Economics, 5*(2), 116–121.

Pande, A. (2003). Taking time and space into consideration: Correcting Zweifel and Navia's (2000) model on the effect of regime on infant mortality rates. Unpublished manuscript.
Preston, S. H. (1975). The changing relation between mortality and level of economic development. *Population Studies (Camb), 29*(2), 231–248.
Rajkumar, A. S., & Swaroop, V. (2008). Public spending and outcomes: Does governance matter? *Journal of Development Economics, 86*(1), 96–111.
Ross, M. L. (2006). Is democracy good for the poor? *American Journal of Political Science, 50*(4), 860–874.
Saker, L., Lee, K., & Cannito, B. (2006). Infectious disease in the age of globalization. In Kawachi I., & Wamala S. (Eds.), *Globalization and health* (pp. 19–38). New York: Oxford University Press.
Sen, A. (1981). *Poverty and famines: An essay on entitlement and deprivation*. New York: Oxford University Press.
Sen, A. (1999). *Development as freedom*. New York: Alfred A. Knopf.
Shandra, J. M., Nobles, J., London, B., & Williamson, J. B. (2004). Dependency, democracy, and infant mortality: A quantitative, cross-national analysis of less developed countries. *Social Science and Medicine, 59*(2), 321–333.
Shvetsova, O., & Sieberg, K. K. (2013). Deciding how to choose the healthcare system. In N. Schofield (Ed.), *Advances in political economy* (pp. 145–166). Berlin, Heidelberg: Springer.
Sonin, K. (2003). Why the rich may favor poor protection of property rights. *Journal of Comparative Economics, 31*(4), 715–731.
Thompson, M. R. (1998). The Marcos regime in the Philippines. In H. E. Chehabi & J. J. Linz (Eds.), *Sultanistic regimes* (pp. 206–229). Baltimore: Johns Hopkins University Press.
Trouiller, P., Olliaro, P., Torreele, E., Orbinski, J., Laing, R., & Ford, N. (2002). Drug development for neglected diseases: A deficient market and a public-health policy failure. *The Lancet, 359*(9324), 2188–2194.
Tullock, G. (1987). *Autocracy*. Boston, MA: Kluwer Academic Publishing.
Vatikiotis, M. R. J. (1998). *Indonesian politics under Suharto: The rise and fall of the new order. Politics in Asia series* (3rd ed.). London: Routledge.
Vollmer, S., & Ziegler, M. (2009). *Political institutions and human development: Does democracy fulfill it's "constructive" and "instrumental" role?* Policy Research Working Paper № 4818. Washington, DC: World Bank. Available from http://elibrary.worldbank.org/doi/pdf/10.1596/1813-9450-4818 [accessed 5 September 2013].
Wagstaff, A., & Claeson, M. (2004). *Rising to the challenge. The Millennium goals for health*. Washington, DC: The World Bank.
Wigley, S., & Akkoyunlu-Wigley, A. (2011). The impact of regime type on health: Does redistribution explain everything? *World Politics, 63*(4), 647–677.
Wintrobe, R. (1998). *The political economy of dictatorship*. Cambridge, UK; New York, NY: Cambridge University Press.
World Bank. (2013). *World development indicators online*. Washington, DC: The World Bank Group. Available from: http://www.qog.pol.gu.se/data/ [accessed 5 April 2013].
World Health Statistics Report. (2013). *World Health Organization*.
Zhang, J., & Zhang, J. (2005). The effect of life expectancy on fertility, saving, schooling and economic growth: Theory and evidence. *The Scandinavian Journal of Economics, 107*(1), 45–66.
Zweifel, T., & Navia, P. (2000). Democracy, dictatorship, and infant mortality. *Journal of Democracy, 11*(2), 99–114.
Zweifel, T., & Navia, P. (2003). Democracy, dictatorship, and infant mortality revisited. *Journal of Democracy, 14*(3), 90–103.

# Comparison of Voting Procedures Using Models of Electoral Competition with Endogenous Candidacy

**Damien Bol, Arnaud Dellis, and Mandar Oak**

**Subject Classification:** C72; D72; H11

## 1 Introduction

This article surveys the latest advances in the literature on the comparative properties of different voting procedures in the context of political elections. In particular, we focus on the various attempts at understanding how different voting procedures affect the number of candidates and the positions they adopt. In public debate as well as academic scholarship, the comparative properties of different voting procedures has been a topic of enduring interest. This interest dates as far back as the late eighteenth century French philosopher-mathematicians Condorcet and Borda (and even earlier with the works of Lull and Cusanus), building up to the classic works of Arrow, May and Gibbard-Satthertwaite in the mid-twentieth century. The work in

D. Bol
Department of Political Economy, King's College London, Strand Building, London WC2R 2LS, UK
e-mail: damien.bol@kcl.ac.uk

A. Dellis (✉)
Department of Economics, Université du Québec à Montréal, 320 rue Sainte-Catherine Est, Local DS-5915, Montréal, QC, Canada H2X 1L7
e-mail: dellis.arnaud@uqam.ca

M. Oak
School of Economics, University of Adelaide, 10 Pulteney Street, Level 3, Adelaide, SA 5005, Australia
e-mail: mandar.oak@adelaide.edu.au



M. Gallego, N. Schofield (eds.), *The Political Economy of Social Choices*, Studies in Political Economy, DOI 10.1007/978-3-319-40118-8_2

this tradition is largely confined to a rather abstract treatment of voting procedures, over an exogenously given set of alternatives.

The works of Harold Hotelling and Anthony Downs embed voting into a larger game, one of electoral competition, thereby enabling the study of how a voting procedure affects the policy positions adopted by the political parties.[1] However, the canonical Hotelling-Downs model assumes two political parties/candidates. Hence, by construction, the set of choices available to the voters is restricted to having at most two elements. It was conjectured by Maurice Duverger that the number of political parties emerging is systematically related to the voting procedure used.[2] This conjecture took the form of Duverger's law, which states an empirical regularity that the plurality rule[3] favors a two party system, and Duverger's hypothesis, which suggests that the plurality runoff rule[4] and proportional representation favor a multiparty system. The literature we review here consists of analytical approaches to studying conjectures similar to those of Duverger's by modeling political competition under alternative voting procedures.

A voting procedure is comprised of three elements: the first element, *the ballot structure*, refers to the type of ballot a voter is permitted to cast (numbers of votes each voter must and is allowed to cast, whether each voter must rank-order candidates or not, and so on); the second element, *the allocation rule*, refers to the formula by which a set of election winners is determined from the ballots cast by the voters; finally, the third element, *the district magnitude*, refers to the number of candidates to be elected in an electoral district. In this paper we do not study variations in the district magnitude; we consider elections for a single office (a single district of magnitude one) and focus on variations in the first two elements of a voting procedure.[5]

While the focus of our survey is primarily on the theoretical literature, it is instructive to look at some stylized facts about elections around the world. Consistent with the theme of our survey, we will look at the stylized facts about the number of political parties and the extent of polarization of party policies in several democracies.[6] In particular, we look at data on five democracies of which

---

[1]See Hotelling (1929), Downs (1957).

[2]See Duverger (1954).

[3]Under the plurality rule, each voter votes for one candidate and the candidate who gets the most votes is elected.

[4]The plurality runoff rule takes several forms. In a standard form, each voter votes for one candidate. If a candidate gets a majority of votes, he is elected. Otherwise, all but the top two vote-getters are eliminated, and voters are invited to vote for one of the two candidates still standing. The one of these two candidates who then gets a majority of votes is elected.

[5]There is a small literature on models of endogenous candidacy in multi-member district elections which we briefly refer to in the concluding section.

[6]The numbers presented below are based on our own calculations on data from the comparative manifesto dataset (Volkens et al. 2015). The dataset can be accessed at https://manifestoproject.wzb.eu/

**Table 1** Effective number of parties

| Country | # of elections | *ENP* (s.d.) |
|---|---|---|
| Canada | 21 | 3.07 (0.44) |
| UK | 17 | 2.37 (0.33) |
| US | 17 | 1.99 (0.02) |
| Australia | 27 | 2.60 (0.27) |
| France | 14 | 4.31 (0.70) |

three (Canada, UK and US) use the plurality rule for parliamentary elections, while the other two (Australia and France) use runoff rules.[7]

Lets first look at the number of political parties. A standard measure of the number of parties that is used in the political science literature is the Effective Number of Parties. Table 1 below presents the average Effective Number of Parties (*ENP*) across all national parliamentary elections held between 1946 and 2013 in five democracies. The *ENP* is an adjusted measure of the number of parties that accounts for the relative strength of each party. It is a commonly used indicator of party system fragmentation (Laakso and Taagepera 1979). The *ENP* in a given election is obtained by taking the inverse of the sum of squared vote shares of the participating parties ($v_i$s); in effect it is the inverse of the *Herfindahl Index* of concentration. The precise formula is given by:

$$ENP = \frac{1}{\Sigma_{i=1}^{M} v_i^2}$$

where $M$ is the number of parties participating in the election. Thus, the *ENP* varies from 1 (when one party obtains all the votes) to $M$ (when votes are spread equally across parties).

The numbers obtained are arguably consistent with Duverger's law (except for Canada which is, and with India, a well-known exception to Duverger's law). There also appear to be mixed evidence regarding Duverger's hypothesis in that France has a greater *ENP* than any of the plurality rule democracies while Australia has an *ENP* not too different from a plurality rule democracy such as the UK.

Turning to the issue of policy polarization, in Table 2 we present for the same set of countries and time period as above, the average degree of polarization across elections. Polarization is a measure of how different the platforms of the competing parties are, along the left-right dimension.[8] A party system is said to be heavily

[7]Australia uses the alternative vote rule, an example of instant runoff rule, in elections held to elect members of the House of Representatives; France uses the plurality runoff rule in its Presidential elections and in elections held to elect members of the Assemblée Nationale.

[8]We use manually coded data from the comparative manifesto dataset (Volkens et al. 2015). In this dataset, each sentence is coded according to its theme, in particular, a left-wing theme (e.g., a sentence in favor of multiculturalism) and a right-wing theme (e.g., a sentence in favor of maintaining law and order). The left-right position of each party is calculated by taking the

**Table 2** Polarization

| Country | # of elections | Polarization (s.d.) |
|---|---|---|
| Canada | 21 | 0.10 (0.05) |
| UK | 17 | 0.15 (0.10) |
| US | 17 | 0.08 (0.06) |
| Australia | 27 | 0.16 (0.08) |
| France | 14 | 0.21 (0.08) |

polarized if all the parties are located at the extremes, while polarization is said to be absent if all the parties are located at the center (Sartori 1976). Formally, polarization is calculated as the weighted sum of squared distances between each party's position on a left-right scale and the center of gravity of the party system, which is itself the weighted average of all the parties' positions on the left-right scale (e.g., see Dalton 2008 and Lupu 2015). Thus, we have

$$Polarization = \sum_{i=1}^{M} v_i \cdot (p_i - \bar{p})^2$$

where $v_i$ is party $i$'s vote share and $p_i$ is the policy position of party $i$ on the left-right policy spectrum with the center of gravity, denoted by $\bar{p}$, which is obtained by the formula

$$\bar{p} = \sum_{i=1}^{M} v_i \cdot p_i$$

The above table shows that, on average, it is usual to see between 10 and 20 % polarization (100 % being complete polarization). This is a rather substantial degree of polarization, even in plurality rule elections[9]; moreover, elections using runoff rules appear to lead to a greater degree of polarization.

In our survey we will be reviewing the literature on endogenous candidacy models of electoral competition keeping the following questions in mind: (1) to what extent do the models of endogenous candidacy provide predictions consistent with the stylized findings presented above; (2) if and in what sense are these models an improvement over the exogenous candidacy models in explaining the stylized facts presented above. The theoretical literature on voting procedures extends beyond the comparison of the plurality and runoff rules. More generally one would like to understand the effect of allowing voters to cast multiple votes on the

difference between the number of left and right sentences, and dividing this difference by the sum of left and right sentences altogether. This measure of the left-right position varies between $-1$ (extreme left) to $+1$ (extreme right).

[9]For instance, even in the most moderate case of the US, the polarization index of 0.08 means that party manifestos have, on average, an imbalance of about 28 % of sentences towards left or right compared to the "center of gravity" of the party system.

electoral outcomes. Hence, our survey will also try to assess (3) whether endogenous candidacy models provide in general a better paradigm for comparative analysis of voting procedures (for single office) over the exogenous candidacy models.

The remainder of the paper is organized as follows: in Sect. 2 we present a unified model of electoral competition that encompasses, as special cases, models of electoral competition with exogenous candidacy as well as various types of endogenous candidacy models reviewed in the survey. Section 3 presents a model due to Cox (see Cox 1987, 1990) which serves as our benchmark for understanding the exogenous candidacy approach to studying alternative voting procedures. Section 4 begins with a discussion of the shortcoming of Cox's approach, and provides us the rationale for endogenous candidacy models. In Sect. 5 we discuss various models of electoral competition that endogenize candidacy and the conclusions they reach regarding the properties of alternative voting procedures. Section 6 discusses the findings from Sect. 5, with particular emphasis on how the comparative properties of voting procedures are influenced by the modeling choices made under different models discussed in the previous sections. Section 7 concludes with some general remarks about the endogenous candidacy approach, its shortcomings and directions for future research.

## 2 A Unifying Framework

In this section we present a canonical spatial model of electoral competition that provides a unifying framework for the different families of models reviewed in this survey.

We consider a community that must elect a policymaker to choose a policy. The policy space $X$ is a non-empty and closed interval on the real line $\mathbb{R}$, say, $X = [0, 1]$. The community consists of a continuum of voters, $\mathcal{N}$, the size of which is normalized to unity. Each voter has single-peaked preferences over the policy space $X$. We denote the (unique) ideal policy of voter $n \in \mathcal{N}$ by $x_n \in X$. Voters' ideal policies are distributed according to a continuous and strictly increasing distribution function $F$, with corresponding density function $f$ and support $[0, 1]$. We denote by $m$ the policy at which $F(m) = 1/2$, and refer to $m$ as the median voter's ideal policy. Voter $n$'s utility from policy $x$ is given by $u^n(x) = u(|x - x_n|)$, where $u : \mathbb{R}_+ \to \mathbb{R}$ is a strictly quasi-concave function and $|x - x_n|$ is the distance between policy $x$ and voter $n$'s ideal policy, $x_n$.

There is a finite set of $M \geq 2$ potential candidates, $\mathcal{M}$, who must decide whether to contest the election and, if so, at which position.[10] We denote potential candidate $i$'s decision by $e_i \in \mathcal{E}_i \subseteq X \cup \{\emptyset\}$, where $e_i = x_i \in X$ if potential candidate $i$ contests the election at position $x_i$, and $e_i = \emptyset$ if he does not run for office. There

[10]Depending on the family of models, the set of potential candidates is disjoint from the set of voters ($\mathcal{M} \cap \mathcal{N} = \emptyset$) or is a subset of the set of voters ($\mathcal{M} \subset \mathcal{N}$).

is a cost $\delta$ that a potential candidate must bear if he chooses to contest the election. Depending on the family of models, potential candidates are either Downsians or Partisans. Downsians are purely office-motivated, i.e., are interested only in their electoral performance (e.g., vote share or winning probability) and the rent they can extract when holding office. In other words, Downsians do not care about the policy outcome; rather, they treat policymaking as an instrument towards gaining office and extracting rents. By contrast, Partisans are policy-motivated, i.e., care intrinsically about the policy outcome.[11] Partisans treat policymaking not just as an instrument towards gaining office, but also as an instrument towards choosing policy.

The policymaking process is modeled as a three stage game. At the first stage, each potential candidate makes his candidacy decision. At the second stage, voters vote simultaneously over the set of candidates. In most of this review, we assume that voters are sincere, i.e., they report directly their preferences over the set of standing candidates, without taking account of the electoral prospects of the candidates. Ties are broken equiprobably. Finally, at the third stage of the game, the elected candidate chooses and implements policy. In case no candidate is running for election, a default policy $x_0$ is implemented which, to facilitate exposition, we assume provides every voter $n$ with a utility $u^n(x_0) = -\infty$.

The solution concept is pure-strategy (subgame perfect) Nash equilibrium.

## 3 The Cox Model

Cox (1987) adapts the standard Hotelling-Downsian framework to study equilibria under alternative voting procedures.[12] In this model, the set of candidates is exogenous and given by $\mathcal{M}$. Candidates compete by simultaneously choosing their position on $X$. The set of feasible actions for each candidate $i$ is thus given by $\mathcal{E}_i = X$, where $e_i \in X$ denotes candidate $i$'s position and $e = (e_1, \ldots, e_M)$ denotes the profile of candidacy decisions.[13] Candidates are Downsians; when choosing his position, each candidate $i$ seeks to maximize his vote share $v_i(e)$ (*vote share*

[11]Keep in mind that policy-motivation does not exclude the possibility that a Partisan is also office-motivated, i.e., cares about the policy outcome as well as the rent he can extract when holding office.

[12]Cox (1990) follows a similar approach but extends the analysis to elections for multi-member districts, including the proportional representation system. For the purpose of our survey we focus only on single office elections. Hence, it is Cox (1987) that is immediately relevant for us.

[13]That the set of feasible actions for each candidate is $X$ and not $X \cup \{\emptyset\}$ means there is no option to not be part of the race. This is a common feature of the *exogenous* candidacy models. Alternatively, one could model the set of feasible actions as $X \cup \{\emptyset\}$ but then set $\delta(\emptyset) > 0$ and $\delta(x) = 0$ for $x \in X$. This ensures that no candidate chooses action $\emptyset$.

*maximizer*) or his plurality $v_i(e) - \max_{k \neq i}\{v_k(e)\}$ (*plurality maximizer*).[14] As for the rest, the model corresponds to the unifying model described in Sect. 2 above.

Cox first studies the properties of pure-strategy Nash equilibria under the plurality rule. He shows that, in multi-candidate races (i.e., when the number of candidates $M \geq 3$), equilibria, if they exist, must be non-centrist and the degree of non-centrism must be substantial in the sense that at least one candidate must locate outside the 25–75 % quantile range of voters' ideal policies. Furthermore, Cox shows that the existence of equilibria cannot be generically guaranteed. Cox (1987, 1990) concludes that, under the plurality rule "[m]ulticandidate equilibria either do not exist, or they are decidedly non-centrist."

Cox then extends his analysis to cover the class of rank scoring rules. Under a (normalized) rank scoring rule, each voter submits a strict rank ordering over the $M$ candidates and the candidate ranked in the $j$th position is given a score $s_j \in [0, 1]$ such that $s_j \geq s_{j+1}$ for all $j = 1, \cdots, M-1$, with the normalization restrictions that $s_1 = 1$ and $s_M = 0$.[15] The election winner is the candidate whose total score is the biggest. A wide variety of different voting procedures can be considered as special cases of a rank scoring rule. For example, $s_1 = 1, s_j = 0$ for all $j \geq 2$ is equivalent to the plurality rule; $s_j = \frac{M-j}{M-1}$ is equivalent to the Borda count; $s_1 = \ldots = s_{M-1} = 1$, $s_M = 0$ is equivalent to the negative vote rule.

To compare rank scoring rules, Cox focuses on symmetric pure-strategy Nash equilibria, which he calls convergent equilibria (since, in these equilibria, all candidates adopt the same position). The extent of polarization under convergent equilibria can be parametrized by $|\alpha - 1/2|$ where $\alpha$ denotes the quantile of the distribution of voters' ideal policies at which the candidates are located. To characterize the convergent equilibria, we define *Cox's threshold of diversity* (see Myerson 1999), which is defined by

$$\tau(s, M) = \frac{1}{M}\sum_{i=1}^{M} s_i$$

Thus, Cox's threshold of diversity, $\tau(s, M)$, corresponds to the average score under rank scoring rule $s$ in a $M$-candidate race. For example, $\tau(s, M) = \frac{1}{M}$ for the plurality rule; $\tau(s, M) = \frac{1}{2}$ for the Borda count; $\tau(s, M) = \frac{M-1}{M}$ for the negative vote rule. This threshold specifies the largest possible share of the electorate, the preferences of which can be disregarded by all candidates.

---

14 Cox also considers the possibility that the candidates have a more complicated objective which he calls *complete plurality maximization*, meaning a candidate, for a given margin of victory, seeks to maximize the plurality relative to other candidate(s). Cox's results generally do not depend on which of the objectives described above are candidates' objectives, hence he collectively refers to them as "admissible candidate objectives".

15 The normalization is without loss of generality.

Cox (1987) states the following:

**Proposition 1** *Let $Q_\alpha$ denote the position of the $\alpha-$quantile of the distribution of voters' ideal policies. Then, a convergent equilibrium with $e = (Q_\alpha, \cdots, Q_\alpha)$ exists if and only if $1 - \tau(s, M) \leq \alpha \leq \tau(s, M)$.*

Proposition 1 tells us both whether or not a convergent equilibrium exists, and the degree of candidate polarization under the equilibrium. The family of rank scoring rules can be divided into three classes depending on their threshold of diversity:

1. The rules with $\tau(s, M) < 1/2$, which are called best-rewarding since they correspond to rules where scores tend to decrease faster at the top than at the bottom. Proposition 1 implies there is no convergent equilibrium under these rules. The plurality rule is best-rewarding in multi-candidate races.
2. The rules with $\tau(s, M) = 1/2$, for which there is a unique convergent equilibrium, with all candidates located at the median voter's ideal policy. The Borda count belongs to this class of rules.
3. The rules with $\tau(s, M) > 1/2$, which are called worst-punishing since they correspond to rules where scores tend to decrease faster at the bottom than at the top. Proposition 1 implies there is a continuum of convergent equilibria under these rules, with all candidates located at the same position in the interval $[Q_{1-\tau(s,c)}, Q_{\tau(s,c)}]$, which includes the median voter's ideal policy. The negative vote rule is worst-punishing in multi-candidate races.

It is interesting to observe that in two-candidate races ($M = 2$), all rank scoring rules have a threshold of diversity equal to $1/2$. Proposition 1 thus implies that under any rank scoring rule, a two-candidate race yields a unique convergent equilibrium at the median voter's ideal policy, thereby deriving the Median Voter theorem as a special case.

Cox (1987) provides a similar result for the class of non-rank scoring rules, i.e., the scoring rules where a voter can cast more than one vote but does not provide a ranking of candidates. A prime example of such a procedure is approval voting where the voter can vote for as many candidates as she wishes, and the candidate who gets the biggest vote total is elected.

# 4 Rationale Behind Endogenizing Candidacy

In this section we discuss shortcomings of Cox's (1987) approach that might be addressed by endogenizing candidacy.

## *4.1 Lack of Satisfactory Micro-Foundations*

Cox (1987) considers a set of candidates that is both exogenous and kept fixed across voting procedures. These two assumptions are arguably questionable. Indeed, it is realistic to think that the decision to contest an election, especially a political election, is highly strategic. It is reasonable as well to think that the incentives for contesting an election are specific to a voting procedure and, therefore, that the set of candidates is not fixed across voting procedures but varies from one voting procedure to another. Ignoring the differential incentives for candidates to contest an election might therefore lead to erroneous predictions. A more complete theory of comparative voting procedures would endogenize candidacy and take account of the differences in strategic candidacy behavior across voting procedures.

Dutta et al. (2001) investigates strategic candidacy behavior, underlining the importance of endogenizing candidacy when studying and comparing the theoretical properties of voting procedures. The following simple example illustrates the strategic candidacy behavior considered by Dutta, Jackson and Le Breton.

*Example 1* Consider an election with three candidates (labeled a, b and c) and three voters (labeled 1, 2 and 3). Voters' preference orderings over the set of candidates are given by

$$\begin{cases} aP_1bP_1c \\ bP_2cP_2a \\ cP_3aP_3b \end{cases}$$

where $P_n$ is voter $n$'s strict preference relation. Candidate a (resp. b and c) shares the same preferences as voter 1 (resp. 2 and 3). The election is held under the plurality rule with ties broken lexicographically (i.e., candidate a wins in a tie against any other candidate, and candidate b wins in a tie against candidate c). To further simplify the example, we assume that voters are sincere. In these circumstances each candidate receives exactly one vote (candidate a from voter 1, candidate b from voter 2, candidate c from voter 3) and candidate a is elected. However, the election of candidate a is the worst possible outcome for candidate b who prefers himself to candidate c, and candidate c to candidate a. Now, candidate b would be strictly better off withdrawing from the race since candidate c would then receive a majority of votes (the votes of voters 2 and 3) and be elected, an outcome that candidate b prefers to the election of candidate a. □

While this example is highly specific, Dutta, Jackson and Le Breton establish that the type of strategic candidacy behavior illustrated in this example applies to most standard voting procedures. Specifically, they consider voting procedures that are unanimous and candidate stable.[16] A voting procedure is unanimous if a candidate

[16]They define formally a voting procedure as a function $V : 2^{\mathcal{M}} \setminus \{\emptyset\} \times \mathcal{P} \rightarrow \mathcal{M}$, where $\mathcal{P}$ is the set of all profiles of strict preference relations in which a candidate finds himself most preferred, such that for all $\mathcal{C} \in 2^{\mathcal{M}} \setminus \{\emptyset\}$ and $P \in \mathcal{P}$:

1. $V(\mathcal{C}, P) \in \mathcal{C}$,
2. $V(\mathcal{C}, P) = V(\mathcal{C}, P')$ for all $P' \in \mathcal{P}$ such that $P_n = P'_n$ for every $n \in \mathcal{N}$, where $P_n$ is a strict preference relation for voter $n$, and

is necessarily elected when he is the most-preferred candidate of *every* voter.[17] A voting procedure is candidate stable if no candidate can affect the election in a way that is beneficial for him by withdrawing from the race.[18] Dutta et al. (2001) shows that:

**Proposition 2** *If $\mathcal{N} \cap \mathcal{M} = \emptyset$, and a voting procedure V is candidate stable and unanimous, then V is dictatorial.*[19]

Thus, when potential candidates cannot vote, *every* nondictatorial and unanimous voting procedure is subject to strategic candidacy behavior.[20] Standard voting procedures are nondictatorial and unanimous.

In light of this result, Dutta et al. (2001, 1014–1015) concludes that "[t]he results here show that the outcome of *all* nondictatorial and unanimous voting procedures will be influenced by the entry decisions of candidates. This implies that it is not valid to treat the set of candidates as fixed for any nondictatorial voting procedure. As most of what we know about voting procedures treats the set of candidates as fixed, our results suggest that these need to be revisited accounting for strategic candidacy. ... [I]n order to make meaningful comparisons across voting procedures, strategic entry and exit effects must be understood."

It is important to observe though that while Dutta, Jackson and Le Breton show that strategic candidacy behavior *may* matter when comparing the theoretical properties of voting procedures, they do not establish that it *does* actually matter. In other words, they do *not* establish that the incentives for candidates to contest an election differ across voting procedures and/or that the comparative properties of voting procedures are different when candidacy is endogenous compared to when candidacy is exogenous. The literature discussed in Sect. 5 below answers this question, showing that strategic candidacy behavior varies across voting procedures and that it does matter for the comparative properties of voting procedures.

---

3. $V(\mathcal{C}, P) = V(\mathcal{C}, P')$ for all $P' \in \mathcal{P}$ such that $P|_{\mathcal{C}} = P'|_{\mathcal{C}}$, where $P|_{\mathcal{C}}$ is the profile of strict preference relations $P$ restricted to a non-empty set of candidates $\mathcal{C}$.

In words, a voting procedure chooses a candidate from the non-empty set of candidates $\mathcal{C} \subseteq \mathcal{M}$ (condition 1), is determined only by voters' preferences (condition 2), and depends only on the preferences over the set of candidates $\mathcal{C}$ (condition 3).

[17]Formally, a voting procedure $V$ is unanimous if $V(\mathcal{C}, P) = b$ for any $\mathcal{C} \subset \mathcal{M}$, $P \in \mathcal{P}$, and $b \in \mathcal{C}$ such that $bP_n c$ for all $c \in \mathcal{C} \setminus \{b\}$ and every voter $n \in \mathcal{N}$.

[18]Formally, a voting procedure $V$ is candidate stable if $V(\mathcal{M}, P) = V(\mathcal{M} \setminus \{a\}, P)$ or $V(\mathcal{M}, P) P_a V(\mathcal{M} \setminus \{a\}, P)$ for every $a \in \mathcal{M}$ and $P \in \mathcal{P}$.

[19]Formally, a voting procedure $V$ is dictatorial if there is a voter $n \in \mathcal{N}$ such that $V(\mathcal{M}, P) = top(\mathcal{M}, P_n)$ and $V(\mathcal{M} \setminus \{a\}, P) = top(\mathcal{M} \setminus \{a\}, P_n)$ for all $P \in \mathcal{P}$ and $a \in \mathcal{M}$, where $top(\mathcal{C}, P_n)$ is voter $n$'s most preferred candidate in candidate set $\mathcal{C}$.

[20]Dutta et al. (2001) and Eraslan and McLennan (2004) examine the robustness of the above result to relaxing each of the following three assumptions: (1) the set of candidates is disjoint from the set of voters; (2) preferences are strict; and (3) ties are broken deterministically.

### 4.2 Explaining Duverger's Propositions

Duverger (1954) observes empirically that (1) the plurality rule favors a two-party system while (2) runoff and proportional representation favor a multi-party system. Riker (1982) dubs the first observation Duverger's law and the second observation Duverger's hypothesis. Maurice Duverger proposed two possible explanations for these empirical observations, a "mechanical effect" and a "psychological effect". The mechanical effect is related to the way a voting procedure mechanically converts votes into sets of election winners. The psychological effect is related to the strategic behavior of voters.

Unfortunately, Cox (1987) sidesteps explaining Duverger's propositions in two important ways. First, Cox takes the voting behavior to be sincere and the set of candidates to be exogenously given, thereby assuming away the psychological effect. In this way, he foregoes the chance of discriminating between the mechanical effect and the psychological effect as possible explanations for Duverger's propositions. Second, Cox restricts attention to symmetric pure-strategic equilibria, meaning that in equilibrium all candidates are located at the same position and tie for election. Thus, the equilibrium number of parties/candidates corresponds to the assumed number of candidates in the race. In other words, Cox assumes the party system; he does not derive it from the analysis.

Where can we go from here if we want to explain Duverger's propositions? One possible direction is to take account of strategic voting behavior. This approach is in line with explaining Duverger's propositions by the psychological effect mentioned above. This approach has been followed by, among others, Palfrey (1989), Myerson and Weber (1993), Myerson (2002) and Dellis (2013). Another possible direction, in line with the topic of this survey, is to take account of strategic candidacy behavior, by endogenizing candidacy. This approach is in the spirit of explaining Duverger's propositions by electoral barriers to exiting/entering the race. In our view this approach has the advantage over the previous one of requiring the coordination of a smaller number of players (candidates and party strategists instead of voters) who, furthermore, face higher stakes and are arguably more highly-strategic players. Finally, a third possible direction is to combine the two previous approaches, allowing both strategic voting behavior and strategic candidacy behavior. This approach has the advantage that it allows discriminating between these two types of strategic behaviors as possible explanations for Duverger's propositions.[21] We discuss in Sect. 5 below the latter two approaches.

---

[21]See Bol et al. (Forthcoming) for an attempt at using laboratory experiments to evaluate the contributions of strategic voting behavior and strategic candidacy behavior in explaining Duverger's law.

## 4.3 Allowing for Policy Motivation

Cox (1987) adopts the canonical Hotelling-Downs model in which candidates are purely office-motivated and can commit credibly to any policy position. However, it is difficult to imagine that in actual elections, candidates are not policy-motivated at all, i.e., do not care intrinsically about the policy outcome. A more satisfactory approach would take account of the policy-motivation of candidates.

However, accounting for the policy-motivation of candidates raises a dynamic consistency issue relative to candidates' commitment to policy positions. Alesina (1988) raises this issue, showing that a policy-motivated candidate has an incentive to announce any policy position that maximizes his electoral prospects and, once elected, to renege on his promise and adopt the policy he prefers rather than the policy he promised. When making their voting decision, rational voters will account for this incentive, thereby preventing a candidate from committing credibly to policy positions other than his most preferred policy. The inability of candidates to commit credibly to policy positions has received empirical support from Lee et al. (2004).

This discussion suggests that if we relax the assumption that candidates are purely office-motivated and introduce policy-motivation, then we must concomitantly relax the assumption of credible commitment. But if we consider candidates who both are policy-motivated and are unable to commit credibly to policy positions other than their most preferred policy, we must rationalize the policy preferences of the candidates; we cannot simply assume the policy preferences of candidates but must explain which types of policy preferences are represented in the set of candidates. A natural way to do so is by endogenizing candidacy. Specifically, one can consider a setting where, before the election, citizens decide strategically whether or not to contest the election. The citizen-candidate model discussed in Sect. 5 below adopts such an approach.

## 4.4 Equilibrium Characterization

As we have just discussed in the previous section, Cox (1987) adopts the Hotelling-Downs model of electoral competition. This model exhibits two important and well-known difficulties related to (pure-strategy Nash) equilibrium characterization.

1. **Equilibrium characterization in multi-candidate elections.** The simplest version of Hotelling-Downs model bypasses the difficulty of characterizing equilibria for multi-candidate elections by restricting the number of candidates to only two. In light of Duverger's law, this restriction is arguably justifiable when considering plurality rule elections. However, it is not justifiable when

comparing the theoretical properties of voting procedures. This is because all standard voting procedures are equivalent in the context of two-candidate races. In other words, assuming only two candidates would amount to eliminating all differences between voting procedures and would lead to the conclusion that all standard voting procedures exhibit the same theoretical properties, a conclusion that might not be robust to considering multi-candidate elections or to allowing them by endogenizing candidacy.

A meaningful comparison of the theoretical properties of voting procedures thus involves considering (in exogenous candidacy models) or allowing (in endogenous candidacy models) multi-candidate races. The problem though is that (pure-strategy) equilibrium characterization in the context of a Hotelling-Downs model with multiple candidates is notoriously difficult. To bypass this problem, Cox (1987) restricts attention to symmetric (pure-strategy) equilibria.[22] But this solution has several limitations on its own. One of these limitations is the incompleteness of equilibrium characterization, raising the question whether the properties identified in the context of symmetric equilibria generalize to the whole set of equilibria. Another of these limitations is that, by definition of a symmetric pure-strategy equilibrium, all candidates adopt the same position in equilibrium, which is not consistent with empirical evidence showing that candidates adopt polarized positions (e.g., Ansolabehere et al. 2001 for elections to the U.S. House of Representatives).

2. **Equilibrium existence.** A second difficulty related to equilibrium characterization in the Hotelling-Downs model concerns the issue of equilibrium existence. Even in the context of two-candidate races, pure-strategy equilibria of the Hotelling-Downs model almost never exist when the policy space is multi-dimensional (see, for instance, Duggan's 2006 survey). Moreover, even when an equilibrium exists, it is not robust to small perturbations in voters preferences. Comparisons of voting procedures using the Hotelling-Downs model have therefore been limited to settings with a unidimensional policy space. But even with a unidimensional policy space, equilibrium existence can still be an issue for some voting procedures in the context of multi-candidate races. This is the case for the plurality rule as shown, for example, in Cox (1987).

Endogenizing candidacy offers a possibility for overcoming these two difficulties. It is important to remember though that it is not the only possibility, that there are other ways to address these difficulties. Moreover, as we argue in Sect. 5 below, endogenizing candidacy is no panacea. Indeed, this approach has its own limitations, such as the possible existence of multiple equilibria.

[22] Cahan and Slinko (2012) provides an attempt at characterizing non-symmetric (pure-strategy) equilibria in scoring rule elections.

## 5 Models with Endogenous Candidacy

Models of electoral competition with endogenous candidacy differ along two main dimensions:

1. **Candidate motivation.** Models of electoral competition belong to two traditions, the Downsian tradition and the Partisan tradition. Models that belong to the *Downsian tradition* consider candidates who are purely office-motivated. Models that belong to the *Partisan tradition* consider candidates who are policy-motivated.
2. **Sequence of entry.** Some models treat all potential candidates as making their decisions simultaneously. Other models treat candidates in an asymmetric way, some candidates being already established in the political arena while others are new. Established candidates make their decisions first, while new entrants move second, after the established candidates.

Table 3 summarizes our classification of models of electoral competition with endogenous candidacy along these two dimensions.

The first family of models belongs to the Downsian tradition and treats candidates in an asymmetric way, with some established candidates facing a threat of entry by new candidates. These models look at the stability of the party structure (the number of candidates/parties contesting the election) and the effect that the threat of entry has on the positioning of established candidates/parties. Palfrey (1984) and Weber (1992, 1997) are the seminal contributions in this family. The second family of models belongs to the Downsian tradition as well but, in contrast to the first family, treats all potential candidates in the same way, with all potential candidates making their candidacy decisions simultaneously. These models are concerned with the issue of equilibrium existence. Osborne (1993) is the seminal contribution in this family. Finally, the third family of models considers simultaneous entry decisions, as the previous family of models does, but, in contrast to the other two families that belong to the Downsian tradition, this third family belongs to the Partisan tradition. These models look at the emergence of the party structure and the rationalization of candidates' policy preferences. Osborne and Slivinski (1996) and Besley and Coate (1997) are the seminal contributions in this family.

We now discuss each family of models in turn. For each family, we will outline the key features of the canonical model, discuss how it has been used for comparing the theoretical properties of voting procedures, and highlight the predictions they make on both the degree of polarization and the number of candidates contesting

**Table 3** Classification of models with endogenous candidacy

| | Downsian tradition | Partisan tradition |
|---|---|---|
| Sequential entry | Models of entry deterrence (family 1) | |
| Simultaneous entry | Hotelling-downs models with strategic candidacy (family 2) | Citizen-candidate models (family 3) |

the election. Bol et al. (2015) discusses in greater detail each family of models when applied to elections under the plurality rule.

## 5.1 Models of Entry Deterrence

The canonical model in this family is similar to the unifying model described in Sect. 2, except for the following two specificities:

- All (potential) candidates are Downsian, seeking to maximize their probability of winning the election and committing credibly to the policy they will implement if elected. A candidate enters the race if and only if he expects to be elected with a positive probability.[23]
- The defining feature of this family of models is its partitioning of the set of potential candidates $\mathcal{M}$ into two subsets. One subset contains two established candidates, which we call candidates L and R.[24] The other subset contains one potential entrant, which we call E. We write the set of potential candidates as $\mathcal{M} = \{L, R, E\}$. The two established candidates play a Cournot game vis-à-vis each other, making their candidacy decisions simultaneously and independently, while acting as Stackelberg leaders vis-à-vis the potential entrant, who makes his candidacy decision after the two established candidates have made theirs. Modelling electoral competition in this way, as a hierarchical game, captures the possibility for the two established candidates to deter successful entry by the potential entrant. Entry deterrence is the main focus of this family of models.

Thus, the policymaking process is modeled here as a four stage game. At the first stage, each established candidate $i\,(= L, R)$ makes independently his candidacy decision $x_i \in X \cup \{\emptyset\}$. At the second stage, the potential entrant E makes his candidacy decision $x_E \in X \cup \{\emptyset\}$ after having observed $x_L$ and $x_R$. At the third stage, voters observe the triplet $(x_L, x_R, x_E)$ and vote sincerely over the set of standing candidates. A voter's indifference between an established candidate and the entrant

[23] Contributions in this family differ in the assumptions they make on the entry cost and on candidates' objectives. Palfrey (1984) and Weber (1992) consider settings where there is no entry cost, with all potential candidates standing for election and seeking to maximize their vote share. Weber (1997) considers a setting in which candidates seek to maximize their vote share and a potential candidate contests the election if and only if he expects to receive at least a fraction $q \in [0, 1]$ of the votes. Finally, Callander (2005a,b) and Callander and Wilson (2007) consider settings in which candidates seek to maximize their winning probability and a potential candidate contests the election if and only if he anticipates to win with positive probability.

[24] To simplify exposition, in this survey we consider the case where there are only two established candidates. Weber (1997) and Callander (2005a,b) consider a version of the model with an arbitrary finite number $M \geq 2$ of established candidates.

is broken in favor of the established candidate. Finally, at the fourth stage, the elected candidate implements his position.[25]

In terms of comparison of voting procedures, this family of models has been used to compare the plurality rule with the runoff rules. The runoff rules involve a repeated process of candidate elimination. Some runoff rules, such as the plurality runoff rule (see footnote 4 for a description of this rule), involve repeated voting. Other runoff rules, such as the alternative vote rule, involve a single ballot-paper, but repeated counting.[26] The latter runoff rules are sometimes called instant runoff rules. Runoff rules are commonly used for political elections around the world (see Farrell 2001 for examples).

Recall from Sect. 4 that Duverger's hypothesis states that runoff rules favor a multi-party system. As illustrated in Table 1 above, France provides a good example of Duverger's hypothesis; France is holding its national elections under runoff rules and has a multi-party system. At the same time, Australia provides a counterexample to Duverger's hypothesis; Australia is holding elections to the House of Representatives under a runoff rule, but has a party system which is dominated by two main parties. In view of these observations, a theory that seeks to explain Duverger's hypothesis must be able to explain both the cases where it applies and the cases where it does not. Callander (2005b) proposes such a theory. In particular, Callander is able to explain why a two-party system (with polarized candidates) can be stable under a runoff rule. Formally,

**Proposition 3** *Consider an election under a runoff rule. For all* $x_L \in [\alpha, m]$*, where* $\alpha$ *solves* $F(\alpha) = 1 - 2F\left(\frac{\alpha+m}{2}\right)$*, a two-party equilibrium exists, in which:*

*1.* $x_L \leq m \leq x_R$ *and* $|x_L - m| = |x_R - m|$*,*
*2.* $x_E = \emptyset$*, and*
*3. candidates L and R are each elected with probability 1/2.*

Thus, Proposition 3 establishes the existence of equilibria in which 1) the two established candidates stand at two positions located symmetrically around the median voter's ideal policy $m$ (1) and tie for election (3), while the potential entrant chooses to not contest the race (2). Moreover, the threat of entry by the potential entrant puts an upper-bound on the degree of polarization between the two established candidates, so that the potential entrant, if he were to enter in-between

[25] To simplify exposition, we follow Callander (2005b) and assume that the density of voters' ideal policies $f$ is single-peaked and symmetric around the median voter's ideal policy $m$.

[26] For example, the alternative vote rule used for elections to the Australian House of Representatives requires every voter to cast a single ballot, in which she rank-orders all the candidates, from first to last. If one candidate is ranked first by a majority of voters, he is elected. Otherwise, the candidate who is ranked last on the greatest number of ballots is eliminated. Votes are then recounted using transformed ballots from which the eliminated candidate has been removed. A candidate is elected if he is ranked first on a majority of the transformed ballots. Otherwise, a second candidate is eliminated and votes are recounted using newly transformed ballots. The counting process is repeated until a candidate is elected.

the two established candidates, would finish third and be eliminated in the first round.

The key feature of runoff rules explaining the existence of such two-party polarized equilibria is that the candidate furthest from the median $m$ cannot win under a runoff rule since, even if he makes it to the last round, he will then be defeated by the other candidate still standing who, being closer to the median $m$, will be preferred by a majority of voters. This feature of runoff rules has three important implications:

1. The two established candidates must be located symmetrically around $m$; otherwise, the one established candidate who is furthest from $m$ would be elected with probability zero (being defeated in the last round, if he makes it), which would contradict that he maximizes his probability of winning.
2. Neither of the two established candidates has a *centrifugal incentive* to deviate to a position further from $m$ since, by doing so, he would be located further from $m$ than the other established candidate, implying that his probability of winning would then drop from one half to zero.
3. The threat of entry by the potential entrant eliminates the *centripetal incentives* for the established candidates to converge towards the median $m$. To see this, suppose candidate $L$ were to deviate and locate at $x_L + \epsilon \in (x_L, x_R)$ instead of $x_L$. This would induce the potential entrant to enter at a position $x \in (x_L, x_L + \epsilon)$ since he would then win outright. Indeed, as was mentioned above, $|x_R - x_L|$ is such that a candidate in-between, in this case candidate $L$, would be eliminated at the first round. Moreover, $|x - m| < |x_R - m|$ implies that the entrant would defeat candidate $R$ at the second round.

To sum up, the threat of entry can stabilize a two-party system under a runoff rule and can support polarization by eliminating the centripetal incentives for the two established candidates to converge towards the median.[27]

By contrast, under the plurality rule the threat of entry leaves the centripetal incentives unabated. In other words, the third implication above does not apply to the plurality rule. To see this, consider again the case where candidate $L$ deviates and locates at $x_L + \epsilon$ for $\epsilon > 0$ small. If the potential entrant were to enter and contest the election at $x \in (x_L, x_L + \epsilon)$, he might be able to defeat candidate $L$, but he cannot defeat candidate $R$ who would then win outright. This is because the entrant would capture votes from candidate $L$, but not from candidate $R$. Candidate $R$ would then receive (almost) half of the votes, while the other two candidates would split the other half. Candidate $L$ can thus deviate and locate closer to $m$, capture votes from candidate $R$ and still deter successful entry. Hence, under the plurality rule, the threat

[27]It is important to keep in mind that this analysis provides a *possible* explanation for the stability of a two-party system under a runoff rule. There are other possible explanations such as ethnic composition (e.g., Ordeshook and Shvetsova 1994) or the specificities of each runoff rule. Discriminating between these different possible explanations is an important question which is not addressed in this analysis.

of entry does not eliminate the centripetal incentives of the established candidates to converge towards the median.

To sum up, this family of models highlights the following key difference between the plurality rule and the runoff rules: the threat of entry eliminates the centripetal incentives for established candidates to converge towards the median and stabilize a two-party system when the election is held under a runoff rule, but not when the election is held under the plurality rule. This happens because if an established candidate deviates towards the median, the entrant can eliminate both established candidates under a runoff rule, thanks to the multiple rounds, eliminating one established candidate at a time (the deviator at the first round and the other established candidate at the second round). This is not possible under the plurality rule where there is a single round.

While Callander (2005b) highlights an interesting difference between the plurality rule and the runoff rules, the analysis exhibits several limitations relative to explaining polarization. First, there is a continuum of equilibria in runoff rule elections, including one in which the two established candidates locate at the median $m$. Hence, the model can explain indiscriminately both convergence and polarization. Second, if the model can predict polarization in runoff rule elections, it cannot predict the polarization observed under the plurality rule. This is because Callander (2005b) considers a relatively high entry cost, with a candidate entering the race only if he expects to be elected with positive probability. This high entry cost insulates the established candidates from the threat of entry in plurality rule elections as long as they have not fully converged towards the median. Weber (1997) shows that for intermediate values of the entry cost (with a candidate entering the race if and only if he expects to obtain a vote share of at least $q \in [Q, 1/2]$ for some $Q > 0$) entry deterrence generates a stable two-party system with polarized candidates under the plurality rule as well. In this case, the threat of entry on the flank is serious and generates centrifugal incentives that counterbalance the centripetal incentives, and prevent full convergence by the two established candidates even under the plurality rule. A similar prediction is obtained by Callander (2005a) in the context of multi-district elections under the plurality rule, and by Callander and Wilson (2007) in the context of single-seat elections but with voters' abstention due to alienation.

## 5.2 *Hotelling-Downs Models with Strategic Candidacy*

The canonical model in this family is similar to the model described in Sect. 2, with a finite number $M \geq 3$ of potential Downsian candidates seeking to maximize their probability of winning and committing credibly to the policy they will implement if elected. A potential candidate enters the race if and only if he expects to be elected with a positive probability. Thus, this model differs from Cox (1987) in that candidacy is here endogenous instead of exogenously given. Also, this model differs from the above model of entry deterrence in that all potential candidates are here ex ante identical; the model here makes no distinction between established candidates and potential entrant.

Contributions belonging to this family of models have compared three types of voting procedures: the plurality rule, the plurality runoff rule and the multiple vote rules.

### 5.2.1 Plurality Rule

Osborne (1993) considers the case of the plurality rule and shows that:

**Proposition 4** *Consider an election under the plurality rule. We have that:*

- *If $M = 3$, then for any distribution of voters' ideal policies $F$, the game has no pure-strategy Nash equilibrium.*
- *If $M \geq 4$, then for almost any distribution of voters' ideal policies $F$, the game has no pure-strategy Nash equilibrium.*

Thus, Proposition 4 establishes that Cox's findings for multi-candidate plurality rule elections are robust to endogenizing candidacy. Specifically, it shows that when there are multiple potential candidates, pure-strategy Nash equilibria almost never exist under the plurality rule and, if an equilibrium exists, it is non-convergent. This follows because of the restrictions that centripetal and centrifugal incentives impose on the distribution of voters' ideal policies $F$. To see this, recall that a potential candidate contests the election if and only if he expects a positive probability of winning. This means that, in equilibrium, all standing candidates tie for first place. Let $\underline{x}$ (resp. $\overline{x}$) denote the leftmost (resp. rightmost) position at which a candidate is standing.

We start by observing that there must be at least two candidates standing for election. Indeed, if no candidate is standing for election, one potential candidate could then enter the race and win outright. If there is only one candidate standing for election, a second potential candidate could then enter the race and, at a minimum, tie for first place. In both cases, the assumption that a potential candidate contests the election if he expects a positive probability of winning would be violated.

We continue by observing that $\underline{x} < m < \overline{x}$, i.e., an equilibrium, if it exists, is non-convergent. Indeed, if all candidates contesting the election were adopting the same position (i.e., $\underline{x} = \overline{x}$), then either another potential candidate could enter on their flank and win the election outright, or one of the candidates could improve his probability of winning by deviating and adopting a position directly on the left or the right. Likewise, if all candidate positions were located on the same side of the median (i.e., $m \leq \underline{x} < \overline{x}$ or $\underline{x} < \overline{x} \leq m$), then one candidate could improve his probability of winning by deviating and locating at $m$.

We further note that the centripetal incentives to converge towards the median imply that there must be two candidates standing at $\underline{x}$ and two candidates standing at $\overline{x}$. Indeed, if there was only one candidate at $\underline{x}$, this candidate could increase his winning probability by deviating and adopting a position slightly closer to the median $m$, capturing votes from his neighbor(s) while leaving unchanged the vote total of every other candidate; since our deviator was already tying for first place, he would now win outright. To eliminate these centripetal incentives, two candidates must be standing at $\underline{x}$.[28] By the same argument, two candidates must be standing at $\overline{x}$. Hence there must be at least four candidates standing, explaining why pure-strategy Nash equilibria do not exist when $M = 3$.

Finally, the left and right constituencies of $\underline{x}$ (resp. $\overline{x}$) must be of the same size, i.e., there must be as many voters on the left of $\underline{x}$ (resp. $\overline{x}$) as there are voters on the right of $\underline{x}$ (resp. $\overline{x}$) who are voting for a candidate at $\underline{x}$ (resp. $\overline{x}$); otherwise, one of the two candidates at $\underline{x}$ (resp. $\overline{x}$) would have a centripetal or centrifugal incentive to move slightly to the left or right of $\underline{x}$ (resp. $\overline{x}$), depending on whether the left constituency or the right constituency is the biggest, since he would then capture this whole constituency and win outright. As we shall see below, this restriction on left and right constituencies is key for the comparison of voting procedures. Osborne (1993) shows that this restriction imposes conditions on the distribution of voters' ideal policies $F$ that are satisfied for almost no distribution.[29]

### 5.2.2 Plurality Runoff Rule

Because they generate centripetal and centrifugal incentives that are different from those generated by the plurality rule, other voting procedures are not exposed to this equilibrium existence issue. Brusco et al. (2012) illustrates this point for the plurality runoff rule, establishing the existence of multiple pure strategy Nash equilibria under this rule. As in Callander (2005b), the distinctive feature of the plurality runoff rule compared to the plurality rule is that a candidate cannot win the election if he is located further from the median $m$ than all the other candidates. This feature eliminates the centripetal and centrifugal incentives causing the equilibrium existence problem under the plurality rule. We now explain why.

Consider the situation where all potential candidates enter the race at the median $m$. All candidates tie for first place and are equally likely to be elected.

1. This candidacy profile *does not* constitute an equilibrium under the plurality rule since one candidate would have a centrifugal incentive to move slightly to the

[28] At the same time, there cannot be more than two candidates standing at $\underline{x}$ since one of them could then win outright by deviating and adopting a position on the left flank or on the right flank of $\underline{x}$. Indeed, the deviator would then capture essentially half of the votes that were going to the candidates at $\underline{x}$, instead of receiving only one third or less of these votes. Since all candidates must be tying for first place, the deviator would then win outright, implying that the deviation is beneficial and contradicting that we are at an equilibrium.

[29] Bol et al. (2015) discusses alternative approaches that have been taken in order to ensure the existence of a pure-strategy Nash equilibrium under the plurality rule. To give an example, Feddersen et al. (1990) shows that when the voting behavior is assumed to be strategic, instead of sincere, an equilibrium exists under the plurality rule, in which all candidates standing for election adopt the position preferred by the median voter.

left or right of $m$. By doing so, the deviator would win outright since he would capture (almost) half of the votes, while the other candidates would split the other half.

2. By contrast, this candidacy profile *does* constitute an equilibrium under the plurality runoff rule since neither candidate has a centrifugal incentive to deviate and move to a position different from $m$. If a candidate were to do so, he would qualify for the second round but would then be defeated by the other standing candidate, who is located at $m$ and who is therefore preferred by a majority of voters.

Besides an equilibrium with all potential candidates standing at $m$, equilibria with candidates standing at two polarized positions do exist under the plurality runoff rule. To understand why, we now construct an equilibrium in which four candidates stand for election, two at a position $x_L$ and the other two at a position $x_R$, with $x_L$ and $x_R$ located symmetrically around $m$ and not too distant from each other. Each of the four candidates qualifies for the second round with an equal probability and is equally likely to be elected. Neither candidate has a centripetal incentive to move towards the median since he would be defeated at the first round, $x_L$ and $x_R$ being not too distant from each other. Moreover, neither candidate has a centrifugal incentive to move further away from the median since he would then be the furthest candidate from the median and would necessarily be defeated at the second round (if not already at the first round).[30] Key to observe is that in the case of the plurality runoff rule, the elimination of the centripetal and centrifugal incentives does not require that the left and right constituencies of a candidate position are of equal sizes. This contrasts with what Osborne finds for the plurality rule.

Thus, Brusco, Dziubinski and Roy essentially confirm Callander's (2005b) main conclusion, namely, that the plurality runoff rule can support a two-party system with polarized candidates. However, Brusco, Dziubinski and Roy treat all candidates identically, while Callander makes a distinction between established candidates and potential entrant. This means that Brusco, Dziubinski and Roy explain the *emergence* of a two-party system, not just its *stability* as Callander does. For the rest, Brusco et al. (2012) exhibits the same drawbacks as Callander (2005b), namely, equilibrium multiplicity and the indiscriminate explanation of both convergence and polarization, as illustrated by the two equilibria we have constructed here.

[30] Moreover, no other potential candidate wants to enter the race at another position, for the same reasons that neither of the four candidates wants to deviate. And no other potential candidate wants to enter at $x_L$ or $x_R$ since he would split the votes with the other two candidates at this position, implying these three candidates would be eliminated at the first round and the two candidates at the other position would be moving to the second round.

### 5.2.3 Multiple Vote Rules

Xefteris (2016) provides another example of voting procedures which, by generating centripetal and centrifugal incentives different from those generated by the plurality rule, are not exposed to the equilibrium existence issue identified in Osborne (1993). Specifically, Xefteris considers a whole family of voting procedures, called the multiple vote rules (also sometimes called non-rank scoring rules). Under a multiple vote rule each voter votes for (or is allowed to vote for up to) $k$ candidates and the candidate who receives the most votes is elected. Multiple vote rules differ in the (maximum) number of votes $k \in \mathbb{N}$ that a voter can cast. We call $k$-vote rule the multiple vote rule that allows voters to cast (up to) $k$ votes. Observe that this family of voting procedures includes, as a polar case, the plurality rule, which corresponds to the $k$-vote rule where $k = 1$.

Xefteris establishes that the plurality rule is the only multiple vote rule for which an equilibrium exists for almost no distribution of voters' ideal policies $F$. Formally,

**Proposition 5** *Consider the family of multiple vote rules. We have that:*

1. *If $k = 1$, then there is no equilibrium for almost every distribution of voters' ideal policies $F$.*
2. *$\Phi_k$ is non-degenerate if and only if $k \geq 2$, where $\Phi_k$ is the set of distributions of voters' ideal policies for which equilibria with two positions at which candidates are standing exist under the k-vote rule.*[31]

The first part of Proposition 5 follows straightforwardly from Osborne (1993). The second part is established by constructing equilibria with candidates standing at two positions located symmetrically around the median $m$.

The intuition underlying Proposition 5 is that the centripetal and centrifugal incentives put stronger restrictions on the left and right constituencies of a candidate position under the plurality rule than under the other multiple vote rules. This point is informally illustrated in the following example. Consider a candidacy profile with candidates standing at two positions, $x_L$ and $x_R$, that are located symmetrically around the median $m$. Suppose that at each of the two positions, there are $k + 1$ candidates standing under the $k$-vote rule. The centripetal incentives for a candidate at $x_L$ to move slightly closer to the median are eliminated if $\sigma_R \leq \sigma_L + \frac{k-1}{k}\sigma_R$, where $\sigma_L$ and $\sigma_R$ correspond to the left and right constituencies of $x_L$, respectively. This condition implies that the (approximate) vote total of the deviator would be smaller than the (approximate) vote total of each of the other candidates still standing at $x_L$. By the same argument, the centrifugal incentives for a candidate at $x_L$ to move slightly further away from the median are eliminated if $\sigma_L \leq \frac{k-1}{k}\sigma_L + \sigma_R$. Taken together, these two conditions impose the following restriction on the relationship between the size of the left constituency of $x_L$ and the size of its

[31] The equilibrium concept is here a refinement of pure-strategy Nash equilibrium. We refer the interested reader to Xefteris (2016) for more details on this and on the concept of genericity.

right constituency: $\frac{\sigma_R}{\sigma_L} \in \left[\frac{1}{k}, k\right]$. Key to observe is that this restriction is looser the bigger $k$ is. In particular, this restriction is satisfied for the plurality rule ($k = 1$) only if $\sigma_L = \sigma_R$, i.e., only if the left constituency and the right constituency are of equal sizes. As illustrated in this example, the specificity of the plurality rule is that a deviator is able to capture a whole constituency, left or right, something which is not possible under the other multiple vote rules where the other candidates still standing at the position the deviator has just deserted continue to receive votes both from the left constituency and from the right constituency.[32]

## 5.3 *Citizen-Candidate Models*

The third family of models, called the citizen-candidate models, allows for simultaneous entry by all potential candidates but assumes that the potential candidates are Partisans, i.e., that, just like the voters, potential candidates have policy preferences. Formally, potential candidates are here assumed to be voters (i.e., $\mathcal{M} \subset \mathcal{N}$), which contrasts with the previous families of models where potential candidates are Downsians, and where the set of potential candidates is disjoint from the set of voters (i.e., $\mathcal{M} \cap \mathcal{N} = \emptyset$). Moreover, given that potential candidates are here Partisans, and consistent with Alesina's (1988) argument discussed in Sect. 4.3, it is here assumed that a candidate cannot credibly commit to implementing upon being elected any policy except his ideal policy. Thus, the set of feasible actions for each potential candidate $i$ is $\mathcal{E}_i = \{x_i, \emptyset\}$, where $e_i = x_i$ if $i$ stands for election and $e_i = \emptyset$ if he does not contest the election. If potential candidate $i$ stands for election, he bears an entry cost $\delta > 0$. Voting behavior is either assumed to be sincere (as in Osborne and Slivinski 1996), or strategic (as in Besley and Coate 1997). The rest of the model is similar to the model described in Sect. 2.

Thus, this model differs from all previous models in that candidates are here Partisans instead of Downsians. Also, this model differs from Cox (1987) in that candidacy is here endogenous instead of exogenous, and it differs from the model of entry deterrence in that there is here no distinction between established candidates and potential entrant.

### 5.3.1 Citizen-Candidate Model with Sincere Voting

To clearly explain the mechanics underlying the citizen-candidate approach, we focus on a simplified variant of the citizen-candidate model *à la* Osborne and Slivinski. In particular, we assume that the rents from holding office are $\beta = 0$

[32]To be sure, this example is not rigorous; its only purpose is to illustrate the mechanism underlying the difference between the plurality rule and the other multiple vote rules. The interested reader can consult the formal proofs in Xefteris (2016).

(i.e., that candidates are purely policy-motivated) and that the density of voters' ideal policies $f$ is weakly single-peaked and symmetric around the median, $m$. We furthermore take a functional form for voters' utility function, assuming that $u^n(x) = -|x - x_n|$ (which does not affect the qualitative results).

We start by characterizing the set of equilibria under the plurality rule, partitioning the set into three subsets: (1) the subset of one-candidate equilibria, in which there is only one potential candidate standing for election; (2) the subset of two-candidate equilibria, in which two potential candidates contest the race; and (3) the subset of multi-candidate equilibria, in which more than two candidates contest the election.

**Proposition 6** *Consider an election under the plurality rule. We have that:*

1. ***One-candidate equilibria.*** *An equilibrium in which potential candidate $i \in \mathcal{M}$ runs unopposed exists if and only if each of the following two conditions holds:*
   (a) $\delta \geq |x_i - x_j|$ *for each* $j \in \mathcal{M}$ *with* $|x_j - m| < |x_i - m|$*; and*
   (b) $\delta \geq \frac{1}{2}|x_i - x_j|$ *for each* $j \in \mathcal{M}, j \neq i$*, with* $|x_j - m| = |x_i - m|$.
2. ***Two-candidate equilibria.*** *An equilibrium in which* $i, j \in \mathcal{M}$, $i \neq j$*, run against each other exists if and only if each of the following three conditions holds:*
   (a) $x_i < m < x_j$ *and* $|m - x_i| = |m - x_j|$*;*
   (b) $\frac{1}{2}|x_i - x_j| \geq \delta$*; and*
   (c) *one of the following three sets of conditions holds: (i)* $F\left(\frac{x_i+m}{2}\right) > 1/3$*; or (ii)* $F\left(\frac{x_i+m}{2}\right) = 1/3$ *and* $\delta \geq \frac{1}{3}|x_i - m|$*; or (iii)* $F\left(\frac{x_i+m}{2}\right) < 1/3$ *and* $\delta \geq |x_i - m|$.
3. ***Multi-candidate equilibria.*** *There is no equilibrium with three or more candidates running for election.* ||

An appealing attribute of the citizen-candidate model is that it is not exposed to the equilibrium existence issue arising under the plurality rule in the Hotelling-Downs family of models. However, this attribute comes at a price, namely, equilibrium multiplicity as is clear from the above proposition and as was pointed out by Roemer (2003).

We now discuss the intuition underlying the above proposition.

In one-candidate equilibria, the candidate is located at a position sufficiently close to the median $m$ so that no other potential candidate wants to enter the race. This requirement is straightforwardly satisfied for any other potential candidate whose ideal policy lies further away from the median voter's ideal policy since he would be defeated by the candidate, if he chose to enter. A potential candidate whose ideal policy lies closer to the median would be preferred by the median voter, and thus a majority of voters, and would be elected outright. Similarly, a potential candidate whose ideal policy is as far away from the median would leave the median voter indifferent and would tie for first place. In these latter two cases, the potential candidate is deterred from entering the race if his expected utility gain from implementing his ideal policy is smaller than the entry cost $\delta$ (conditions 1a and 1b).

In two-candidate equilibria, each candidate must tie for first place. Otherwise, the losing candidate would be better off not running since he would save the entry cost without changing the policy outcome. By the same logic, the two candidates must be standing at two different positions. Otherwise, one of them would be better off not running since he would save the entry cost without changing the policy outcome. Given that the two candidates must be tying for first place while standing at two different positions, their ideal policies must be located symmetrically around the median so that they split equally the votes and tie for first place (condition 2a). In a two-candidate equilibrium neither of the two candidates should be better off not running and no other potential candidate should want to enter the race. The former happens when the two candidates are sufficiently polarized, so that their expected utility gain from adopting their ideal policy exceeds the entry cost (condition 2b). The latter happens when the two candidates are close enough to each other so that a potential candidate entering in-between would be defeated (condition 2ci) or would face an entry cost that exceeds his expected utility gain from being elected and adopting his ideal policy (conditions 2cii and 2ciii). Observe that potential candidates with more extreme ideal policies are necessarily deterred from entering the race since they would split votes with the candidate on their side of the median, thereby triggering the outright election of the other, less-preferred candidate.

If a multi-candidate equilibrium were to exist, the leftmost candidate or the rightmost candidate (or both) would be better off not running since his votes would be transferred to his closest neighbor. This vote transfer would improve the electoral prospects of this neighboring candidate and worsen the electoral prospects of the other candidates. This rules out the existence of equilibria with three or more candidates running for election.

*Comparing the Plurality Rule with the Plurality Runoff Rule* An important contribution of Osborne and Slivinski (1996) is its ability to compare the equilibrium outcomes under the plurality rule with those under the plurality runoff rule (hereafter referred to as the runoff rule).

Osborne and Slivinski show that the set of one-candidate equilibria supported under the plurality and runoff rules is identical. This is so because the incentives of the running candidate and of any potential second candidate are identical under the two rules. Also, similar to the plurality rule, there are no multi-candidate equilibria under the runoff rule when the rents from holding office, $\beta$, are small.

Under the simplified variant of the model we consider in this survey, it can be shown that the set of candidate positions in any two-candidate equilibrium under the plurality rule is of the type $\{m-\varepsilon, m+\varepsilon\}$ where $\varepsilon \in [c, \overline{\varepsilon}_p]$ or $[c, \overline{\varepsilon}_p)$. The upper bound $\overline{\varepsilon}_p$ is derived by the condition that if the positions were polarized beyond this point, some potential candidate located at a position in between will want to enter and, upon entry, will win the election. Turning to the runoff rule, it is shown that the set of candidate positions in any two-candidate equilibrium is also of the type

$\{m - \varepsilon, m + \varepsilon\}$ where $\varepsilon \in [c, \overline{\varepsilon}_r]$ or $[c, \overline{\varepsilon}_r)$. The upper bound $\overline{\varepsilon}_r$ is derived by the condition that if the positions were polarized beyond this point, a candidate located at a position in between will want to enter and, upon entry, will come among the top two voter-getters and thereby advance to the second round and eventually win. It can be shown that $\overline{\varepsilon}_r \leq \overline{\varepsilon}_p$.[33] This establishes the following.

**Proposition 7** *Under the simplified variant of the citizen-candidate model à la Osborne and Slivinski (1996) with* $\beta = 0$ *and* $f$ *(weakly) single-peaked and symmetric around the median m, the set of equilibrium positions under the runoff rule supports weakly less polarization than under the plurality rule.*

While the above result provides a sharp comparison, the magnitude of the difference in the extent of polarization may not be significant; indeed, for the case of atom-less distribution it is of measure zero. Osborne and Slivinski also argue that their result is robust to small perturbations around (weakly) single-peaked and symmetric densities $f$. However, for a general shape of distributions there is no unambiguous comparison between the degrees of polarization under the two rules; the equilibrium positions under the runoff rule *may* be more polarized.

Turning to the issue of the number of candidates, we saw that under the simplified variant of the Osborne-Slivinski model developed above the number of candidates under either rule is either one or two. Hence, while the model supports Duverger's law, it fails to find support for Duverger's hypothesis. Turning to the more general case, Osborne and Slivinski show that a two candidate election is more likely under the plurality rule than under the runoff rule, and equilibria with three or four candidates are more likely to exist under the runoff rule relative to the plurality rule. However, the authors are not able to provide more general results, in particular for the case of equilibria with more than four candidates.

*Comparison with Other Voting Procedures* Dellis and Oak (2016) uses the citizen-candidate model with sincere voting to explore the comparative properties of a family of non-rank scoring rules, which they parametrize by the minimum number of votes a voter must cast $s$ $(\geq 1)$ and the maximum number of votes a voter can cast $t$ $(\geq s)$. They call this family of voting procedures the $(s, t)$ rules. The family of $(s, t)$ rules can distinguish between rules which allow partial abstention ($s < t$) and rules which do not ($s = t$); rules which allow partial abstention allow each voter not to cast all her votes, while rules which do not allow partial abstention force each voter to cast all her votes. Under each $(s, t)$ rule, the candidate receiving the biggest total score is elected. This broad family of voting procedures includes as special cases the plurality rule ($s = t = 1$), the negative vote rule ($s = t = c - 1$) and approval voting ($s = 1, t = c - 1$), where $c$ denotes the number of standing candidates.

Dellis and Oak first show that allowing citizens to cast multiple votes alters the electoral incentives of candidates. In particular, it creates an incentive for multiple

[33]Intuitively, the restriction that no potential entrant, upon entry, can be among the top two voter-getters is a stronger restriction than that such entrant be the top vote-getter.

candidates to run at the same or at similar positions; in the literature, such candidates are referred to as *clone candidates*. This possibility, of generating clone candidacies, does not occur under the plurality rule since under that rule the candidates running at the same or a similar position will split the votes and therefore reduce the chance of their ideal policy from being implemented. Secondly, the presence of clone candidates dampens the incentive for more centrist candidates to enter the race since such entrants are not assured of getting partisan votes as they are already "absorbed" by the partisan clones. Hence, the authors show that allowing multiple votes leads to multiple candidacies, thereby confirming Duverger's hypothesis and leading to greater polarization, which runs against the purported rationale provided by electoral reformists who favor allowing multiple votes. In a related paper, Dellis et al. (2016) shows that the alternative vote rule does not suffer from clone candidacy and therefore supports (weakly) less polarization relative to the plurality rule.

To sum up, the citizen-candidate model with sincere voting helps us focus on the differences in candidates' entry/exit incentives under different voting procedures. The literature finds weak support for Duverger's law as well as Duverger's hypothesis though the multiplicity of equilibria makes it difficult to make sharp predictions in this regard. In their original formulation, Duverger's propositions were explained by the strategic behavior of voters (the psychological effect discussed in Sect. 4.2). An interesting feature of the citizen-candidate model is that the mechanics of the Duvergerian outcomes is driven by strategic behavior of the *candidates rather than voters*. More interestingly, regarding polarization of candidate positions, the literature indicates that merely allowing voters to cast more votes does not lead to more centrist outcomes. Rather, whether, and to what extent, multiple votes generate clone candidacies has a bearing on the matter. Voting procedures such as the plurality runoff rule and the alternative vote rule, which do not encourage clone candidacy, support (weakly) less polarization relative to the plurality rule.

### 5.3.2 Citizen-Candidate Model with Strategic Voting

The other variant of the citizen-candidate model is due to Besley and Coate (1997). The mechanics of the model are similar to the one described above. The main difference is that the set of voters, $\mathcal{N}$, is finite and it is assumed that voting is strategic. Furthermore, it is assumed that voters use weakly undominated strategies. This refinement notwithstanding, strategic voting makes it possible to sustain equilibria driven by self-fulfilling voter coordination. To see this, consider a model similar to the one discussed in the section above but with a finite number of voters with their ideal points distributed uniformly over the interval $[0, 1]$.[34] Let's first look at the case of the two-candidate equilibria under the plurality rule. If voting were strategic, even the most extreme positions (here 0 and 1) are supported as

[34]I.e., suppose that there is one citizen at each of the locations $0, 1/N, 2/N, \cdots, 1$ for some large integer $N$.

equilibrium policy outcomes under the plurality rule. This outcome can be supported if, following entry by a centrist candidate, the voters in the set $[0, 1/2)$ would continue to vote for the candidate at 0, while the voters in the set $(1/2, 1]$ would continue to vote for 1. These voting strategies are admissible since they are weakly undominated.[35] In other words, voters are worried about *wasting their vote* on the centrist candidate whom they deem unviable, and this becomes a self-fulfilling prophecy. On the other hand, as we saw above, when voting is sincere there is an upper bound on the degree of polarization due to the constraint that if the positions of the incumbent candidates were farther apart, then a centrist candidate would enter and receive the centrist votes. In this example that we are considering, for a range of entry costs (specifically, $1/9 \leq \delta \leq 1/3$), the most extreme two-candidate equilibrium positions supported under sincere voting are $1/6$ and $5/6$.[36]

*Comparison with Other Voting Procedures* There is a set of papers that use the citizen-candidate model with strategic voting to compare different voting procedures. Dellis and Oak (2006) considers elections under approval voting. They show that approval voting generates more moderate outcomes relative to the plurality rule. The intuition is the following: under approval voting, the set of weakly undominated strategies requires that a voter must vote for his most preferred candidate and must not vote for his least preferred candidate. This in turn implies that candidate profile where two extremist candidates were running against each other may not be sustained as equilibrium due to a threat of credible entry by a centrist candidate. For instance, when voter ideal points are uniformly distributed as in the example above, if there were two candidates running at the 0 and 1 positions then a candidate at the median, situated at 1/2, upon entry will get at least 50 % of the votes (i.e., votes of the voters with ideal policy in the range $[1/4, 3/4]$) whereas the extremist candidates will get at most 50 % of the votes. Hence, the entrant will either be an outright winner, or will be in a three way tie. If the entry cost is sufficiently small (in particular, $\delta < 1/6$) the centrist candidate would enter the race.

Dellis (2009) considers the family of scoring rules, which includes the $(s, t)$–rules discussed above. Concerning the $(s, t)$-rules, it is shown that whether partial abstention is allowed (i.e., $1 \leq s < t$), or not (i.e., $s = t > 1$) matters to the outcomes. In particular, not allowing partial abstention allows supporting at least as much polarization as the plurality rule. As we saw above, due to the wasting-the-vote effect, the plurality rule supports a maximal degree of polarization when voting is strategic. However, Dellis shows that giving voters multiple votes need not reduce this effect, since more votes may be *soaked up* by multiple candidates entering at extreme positions. On the other hand, allowing partial abstention puts an upper bound on polarization for the same reason as we described above for the case of approval voting.

[35] Under the plurality rule, voting for any candidate *except the least preferred* candidate is a weakly undominated strategy.

[36] It can be verified that at $\delta \in [1/9, 1/3]$ no candidate located in between $1/6$ and $5/6$ would want to enter, nor either of the incumbent candidates would want to step down.

## 6 Discussion

We began our survey by citing two types of motivations, theoretical and empirical, for endogenizing candidacy in models of electoral competition. We now examine whether, and to what extent, the investment in endogenous candidacy has yielded returns on these fronts.

On the theoretical side, we know from the work of Dutta, Jackson and Le Breton that strategic candidacy behavior matters. But their work is silent on whether strategic candidacy behavior differs and/or has different implications across voting procedures. The models of electoral competition with endogenous candidacy show that strategic candidacy behavior does differ across voting procedures and that these differences are critical to generating different equilibrium outcomes. The differential incentives for clone candidacies, identified in citizen-candidate models, provide a clear illustration of the differences in strategic candidacy behavior and their implications. Similarly, Dellis and Oak (2006) illustrates, using a citizen-candidate model with strategic voting, differential *entry* incentives under approval voting and the plurality rule: it is shown that under the plurality rule, even if two candidates are located at extreme left and right positions, then a moderate candidate can be deterred from entering and contesting the race. This outcome is sustained by voters' self-fulfilling expectations about wasting-their-vote on such a moderate candidate. On the other hand, there is an incentive for such a candidate to enter under approval voting since such entrant would be guaranteed to gather sufficiently many centrist votes.

Continuing with the theoretical rationale for endogenous candidacy, we argued that *in principle* these models mark an improvement over exogenous candidacy since they provide a more satisfactory micro-foundation for the stability and/or emergence of different party/candidate configurations (in terms of their number and positions). However, some endogenous candidacy models, much like the exogenous candidacy models, fail to have equilibrium existence under some voting procedures. For instance, as discussed in the previous section, models in family 2 do not have equilibria under the plurality rule. While the citizen-candidate models (family 3) generically have equilibrium existence, they suffer from the problem of equilibrium multiplicity. In so far as the purpose of a model is to offer testable predictions regarding the comparison of voting procedures, neither equilibrium non-existence nor equilibrium multiplicity are desirable features.

On the empirical side, we saw that endogenous candidacy models are an improvement over the exogenous candidacy models in that they are able to provide microfoundations for Duverger's propositions and the cases where they do not apply. However, explaining some cases requires sometimes conditions that are quite restrictive.

Turning to the issue of policy polarization, one gets ambiguous results as well. Depending on the model used, one may conclude that multiple vote procedures support either more, less, or as much polarization. We saw from Cox's work that each voting procedure generates, to a varying degree, centripetal and centrifugal forces. The degree of polarization under a voting procedure depends on the relative strengths of the centripetal and centrifugal forces. While these forces work in Cox (1987) on the *intensive margin*, the models of endogenous candidacy add to the mix, or consider, another dimension, namely an *extensive margin*. Depending on the specific model, the possibility of candidate entry/exit works to accentuate or inhibit the centripetal and centrifugal forces. For instance, consider the model of entry deterrence in Callander (2005b). In this model, the standard Downsian incentives under a runoff rule exert a centripetal force, inducing each party to move closer to the median. However, the threat of entry on its flanks exerts a centrifugal force, preventing the parties from moving too close to the median. The result of these forces is the possibility of a polarized, two-party equilibrium under the runoff rule. On the other hand, using the citizen candidate model with sincere voting (e.g., Osborne and Slivinski 1996 and Dellis et al. 2016), we find that under a runoff rule it is the threat of entry by a centrist candidate that exerts a centripetal force, generating equilibria where candidates assume less polarized positions relative to the plurality rule.[37] On this count, Callander (2005b) performs better than the citizen-candidate models, making predictions that are consistent with the stylized facts presented in the introduction of this survey.

Of particular interest in this regard is Dellis and Oak (2007), which compares the plurality rule vis-à-vis approval voting using a model of endogenous candidacy which may be thought of as a hybrid of models of families 2 and 3. They assume that candidates are policy motivated (as in the citizen-candidate models of family 3), but are able to credibly commit to policy positions other than their ideal ones (as in models of family 2). They show that approval voting can lead to polarized equilibria. The intuition behind this result is that approval voting provides an incentive for clone candidacies. When two positions, distributed symmetrically around the median, are each populated by more than one candidate, a candidate is deterred from moving closer to the median out of the fear that he will lose the votes of more extreme voters. Hence, the centripetal incentives are diluted due to clone candidacy. The plurality rule, on the other hand, does not support clone candidacy and therefore generates a centripetal force leading to equilibrium outcomes at or close to the median position. Hence, this model predicts that approval voting supports a greater degree of polarization than the plurality rule does.

[37] In the citizen-candidate model, there exists only the *extensive margin* for candidate behavior, i.e., each candidate either chooses to run at his ideal policy or not be in the race. In this family of models the system *as a whole* exhibits centrifugal, or centripetal, tendencies in the sense that it induces candidates located at symmetric, and therefore mutually balancing, positions around the median to stand for election, where their positions are further apart from, or closer to, the median.

## 7 Conclusion

Models of endogenous candidacy offer a promising avenue for comparative analysis of voting procedures. In particular, they can provide a more satisfactory micro-foundation for emergence and stability of different configurations of political parties (in terms of their number and positions) observed across elections around the world. They also provide an avenue for experimental research by providing testable predictions regarding different voting procedures, including those which so far have never been used in real life elections. As yet, there is no one model of electoral competition with endogenous candidacy that can satisfactorily explain all the institutional features. As we discussed in the section above, different models provide different insights regarding candidates and voters behavior under alternative voting procedures, yet each has its own blind spots or weaknesses. As a whole, this class of models suffers from a multiplicity of equilibria. Another shortcoming of these models is their lack of a serious treatment of political parties.We briefly discuss some of the approaches to addressing the shortcoming of the existing models as well as works that lie outside the scope of the survey.

Different approaches have been taken to tackle the issue of equilibrium multiplicity (in citizen-candidate models). One approach that has yielded success, in terms of reducing equilibrium multiplicity is to introduce informational frictions (see for example, Grosser and Palfrey 2014; Messner and Polborn 2004; and Eguia 2007). Another approach consists in refining the solution concept for citizen-candidate models à la Besley and Coate (see for example, De Sinopoli and Turrini 2002; Dhillon and Lockwood 2002; and De Sinopoli 2004). However, these works typically focus on a particular voting procedure, usually the plurality rule; whether there exists a refinement that is capable of generating unique equilibrium predictions across a large class of voting procedures remains a topic for ongoing research.

The other criticism levelled at the models of electoral competition, including models with exogenous as well as endogenous candidacy, is the absence of political parties. In the literature reviewed in this paper, there was essentially no distinction between a party and a candidate. However, much of the political science literature, going back to the work of Maurice Duverger, takes the formation and evolution of political parties and issues surrounding them seriously. The need to articulate the relationship between candidates and political parties is particularly significant in the citizen-candidate models, where candidates are endowed with policy preferences. John Roemer points out forcefully the need to incorporate political parties into the analysis of electoral competition. Speaking about the citizen-candidate model, he writes: "...parties organize political competition, and discipline candidates to commit to policies which are, in general, not their ideal policies" (Roemer 2006, 423). Morelli (2004) and Levy (2004) are attempts at introducing parties in the citizen-candidate approach, where parties serve as a commitment device for candidates.

Models of endogenous candidacy, in particular the citizen-candidate models, have been used for examining issues other than the one we focused on in this survey. Closer to the topic of this survey, Morelli (2004), which we mentioned above, also compares voting procedures (using a citizen-candidate model), but with multi-member districts, comparing the plurality rule with proportional representation. Other examples are De Sinopoli and Iannantuoni (2007), which investigates the cases where Duverger's hypothesis does not apply under proportional representation, and Iaryczower and Mattozzi (2013), which compares numbers of candidates, campaign spending and polarization in proportional representation and in plurality rule elections.

Another interesting question that the endogenous candidacy approach is naturally amenable to is regarding the attributes of the candidates. For instance, what attracts people into politics—rents from office or policy considerations? With endogenous candidacy we can study endogenous emergence of Downsian or Partisan candidates rather than assuming them to be of one type or another (see for instance, Callander 2008 and Dziubinski and Roy 2013). One can also look at other attributes such as honesty or ability that get selected into political office (see for instance, Caselli and Morelli 2004; Messner and Polborn 2004; and Mattozzi and Merlo 2008). One promising area of research would be to examine how the selection of such attributes vary with the choice of the voting procedure governing elections.

## References


Alesina, A. (1988). Credibility and policy convergence in a two-party system with rational voters. *American Economic Review, 78*(4), 796–805.

Ansolabehere, S., James, S. Jr., & Charles, S. III. (2001). Candidate positioning in U.S. house elections.*American Journal of Political Science, 45*(1), 136–159.

Besley, T., & Stephen C. (1997). An economic model of representative democracy. *Quarterly Journal of Economics, 112*(1), 85–114.

Bol, D., André B., & Simon L. S.-V (forthcoming). Which matters most: Party strategic exit or voter strategic voting? A laboratory experiment. *Political Science Research and Methods*.

Bol, D., Arnaud D., & Mandar, O. (2015). *Endogenous Candidacy in Electoral Competition: A Survey on the Number of Candidates and their Polarization*. Working paper.

Brusco, S., Marcin D., & Jaideep, R. (2012). The hotelling-downs model with runoff voting. *Games and Economic Behavior, 74*(2), 447–469.

Cahan, D., & Arkadii, S. (2015). *Nonconvergent Electoral Equilibria Under Scoring Rules: Beyond Plurality*. Working paper.

Callander, S. (2005a). Electoral competition in heterogenous districts. *Journal of Political Economy, 113*(5), 1116–1145.

Callander, S. (2005b). Duverger's hypothesis, the run-off rule, and electoral competition. *Political Analysis, 13*(3), 209–232.

Callander, S. (2008). Political motivations. *Review of Economic Studies, 75*(3), 671–697.

Callander, S., & Catherine, W. (2007). Turnout, polarization, and Duverger's law. *Journal of Politics, 69*(4), 1047–1056.

Caselli, F., & Massimo, M. (2004). Bad politicians. *Journal of Public Economics, 88*(3–4), 759–782.

Cox, G. (1987). Electoral equilibrium under alternative voting institutions. *American Journal of Political Science, 31*(1), 82–108.
Cox, G. (1990). Centripetal and centrifugal incentives in electoral systems. *American Journal of Political Science, 34*(4), 903–935.
Dalton, R. J. (2008). The quantity and the quality of party systems: Party system polarization, its measurement, and its consequences. *Comparative Political Studies, 41*(7), 899–920.
De Sinopoli, F. (2004). A note on forward induction in a model of representative democracy. *Games and Economic Behavior, 46*(1), 41–54.
De Sinopoli, F., & Alessandro, T. (2002). A remark on voters' rationality in a model of representative democracy. *Journal of Public Economic Theory, 4*(2), 163–170.
De Sinopoli, F., & Giovanna, I. (2007). A spatial voting model where proportional rule leads to two-party equilibria. *International Journal of Game Theory, 35*(2), 267–286.
Dellis, A. (2009). Would letting people vote for multiple candidates yield policy moderation? *Journal of Economic Theory, 144*(2), 772–801.
Dellis, A. (2013). The two-party system under alternative voting procedures. *Social Choice and Welfare, 40*(1), 263–284.
Dellis, A., & Mandar, O. (2006). Approval voting with endogenous candidates. *Games and Economic Behavior, 54*(1), 47–76.
Dellis, A., & Mandar, O. (2007). Policy convergence under approval and plurality voting: The role of policy commitment. *Social Choice and Welfare, 29*(2), 229–245.
Dellis, A., & Mandar, O. (2016). Multiple votes, multiple candidacies and polarization. *Social Choice and Welfare, 46*(1), 1–38.
Dellis, A., Alexandre G.-B., & Mandar, O. (2016). *Policy Polarization and Strategic Candidacy in Elections Under the Alternative Vote Rule*. Working paper.
Dhillon, A., & Ben, L. (2002). Multiple equilibria in the citizen-candidate model of representative democracy. *Journal of Public Economic Theory, 4*(2), 171–184.
Downs, A. (1957). *An economic theory of democracy*. New York: Harper.
Duggan, J. (2006). Candidate objectives and electoral equilibrium. In B. Weingast, & D. Wittman (Eds.), *The Oxford handbook of political economy*. Oxford: Oxford University Press.
Dutta, B., Matthew, J., & Michel L. B. (2001). Strategic candidacy and voting procedures. *Econometrica, 69*(4), 1013–1037.
Duverger, M. (1954). *Political parties: Their organization and activity in the modern state*. New York: Wiley.
Dziubiński, Marcin, and Jaideep, R. (2013). Hotelling-downs competition with free entry when voters have an option to contest. *Journal of Public Economic Theory, 15*(4), 620–653.
Eguia, J. (2007). Citizen candidates under uncertainty. *Social Choice and Welfare, 29*(2), 317–331.
Eraslan, H., & Andrew, M. (2004). Strategic candidacy for multivalued voting procedures. *Journal of Economic Theory, 117*(1), 29–54.
Farrell, D. (2001). *Electoral systems. A comparative introduction*. New York: Palgrave.
Feddersen, T., Itai, S., & Stephen, W. (1990). Rational voting and candidate entry under plurality rule. *American Journal of Political Science, 34*(4), 1005–1016.
Grosser, J., & Thomas, P. (2014). Candidate entry and political polarization: An antimedian voter theorem. *American Journal of Political Science, 58*(1), 127–143.
Hotelling, H. (1929). Stability in competition. *Economic Journal, 39*(153), 41–57.
Iaryczower, M., & Andrea, M. (2013). On the nature of competition in alternative electoral systems. *Journal of Politics, 75*(3), 743–756.
Laakso, M., & Rein, T. (1979). Effective number of parties: A measure with application to West Europe. *Comparative Political Studies, 12*, 3–27.
Lee, D., Enrico, M., & Matthew, B. (2004). Do voters affect or elect policies? Evidence from the U.S. house. *Quarterly Journal of Economics, 119*(3), 807–859.
Levy, G. (2004). A model of political parties. *Journal of Economic Theory, 115*(2), 250–277.
Lupu, N. (2015). Party polarization and mass partisanship: A comparative perspective. *Political Behavior, 37*, 331–356.

Mattozzi, A., & Antonio, M. (2008). Political careers or career politicians? *Journal of Public Economics, 92*(3–4), 597–608.
Messner, M., & Mattias, P. (2004). Paying politicians. *Journal of Public Economics, 88*(12), 2423–2445.
Morelli, M. (2004). Party formation and policy outcomes under different electoral systems. *Review of Economic Studies, 71*(3), 829–853.
Myerson, R. (1999). Theoretical comparisons of electoral systems. *European Economic Review, 43*(4–6), 671–697.
Myerson, R. (2002). Comparison of scoring rules in poisson voting games. *Journal of Economic Theory, 103*(1), 219–251.
Myerson, R., & Robert, W. (1993). A theory of voting equilibria. *American Political Science Review, 87*(1), 102–114.
Ordeshook, P., & Olga, S. (1994). Ethnic heterogeneity, district magnitude, and the number of parties. *American Journal of Political Science, 38*(1), 100–123.
Osborne, M. (1993). Candidate positioning and entry in a political competition. *Games and Economic Behavior, 5*(1), 133–151.
Osborne, M., & Al, S. (1996). A model of political competition with citizen-candidates. *Quarterly Journal of Economics, 111*(1), 65–96.
Palfrey, T. (1984). Spatial equilibrium with entry. *Review of Economic Studies, 51*(1), 139–156.
Palfrey, T. (1989). A mathematical proof of Duverger's law. In P. Ordeshook (Ed.), *Models of strategic choice in politics*. Ann Arbor: University of Michigan Press.
Riker, W. (1982). The two-party system and Duverger's law: An essay on the history of political science. *American Political Science Review, 76*(4), 753–68.
Roemer, J. E. (2003). Indeterminacy of citizen-candidate equilibrium. Cowles foundation discussion paper no. 1410.
Roemer, J. E. (2006). Book reviews. [Review of the book *Positive Political Theory II: Strategy and Structure* by David Austen-Smith and Jeffrey S. Banks.] *Journal of Economic Literature, 44*(2), 421–425.
Sartori, G. (1976). *Parties and party systems: A framework for analysis* (Vol. 1). Cambridge: Cambridge University Press.
Volkens, A., Pola, L., Theres, M., Nicolas, M., Sven, R., & Annika, W. (2015). The manifesto data collection. Manifesto Project (MRG/CMP/MARPOR). Version 2015a. Berlin: Wissenschaftszentrum Berlin für Sozialforschung (WZB).
Weber, S. (1992). On hierarchical spatial competition. *Review of Economic Studies, 59*(2), 407–425.
Weber, S. (1997). Entry deterrence in electoral spatial competition. *Social Choice and Welfare, 15*(1), 31–56.
Xefteris, D. (2016). Stability in electoral competition: A case for multiple votes. *Journal of Economic Theory, 161*, 76–102.

# Negative Advertising During Mexico's 2012 Presidential Campaign

**Andrei Gomberg, Emilio Gutiérrez, and Zeev Thepris**

## 1 Introduction

TV and radio advertising has long been a staple of political campaigning, increasingly consuming significant financial resources. A recent item in *Wall Street Journal*[1] reports an industry forecast that during the 2016 presidential campaign in the US over USD$4.4 billion will be spend by candidates and outside groups on broadcast and cable TV advertising alone, up from the already staggering USD$3.8 billion 4 years earlier. In Mexico, with its much smaller economy, campaign spending by the three main parties in the 2006 presidential campaign amounted to over USD$300 million, much of it spent on TV ads.[2] Given the enormousness of expenditure involved, the role of advertising during political campaigns has attracted the attention of policy makers and social scientists alike.

There is, indeed, a growing literature that tries to estimate the impact of advertising on election outcomes both for developed and consolidating democracies

We express our particular gratitude to the Instituo Nacional Electoral (INE) for its assistance in locating and compiling the data. Gomberg and Gutierrez are grateful to Asociación Mexicana de Cultura for financial support.

[1]O'Connor, P. (2015, August 30). TV remains king in political ad spending. *Wall Street Journal*. Retrieved from http://www.wsj.com/articles/tv-remains-king-in-political-ad-spending-1440978256.

[2]Herrera, J. (2006, October 8). IFE: gasto de campañas en medios creció más que 100 %. *El Universal*. Retrieved from http://archivo.eluniversal.com.mx/nacion/143969.html.

A. Gomberg (✉) • E. Gutiérrez
ITAM, CIE, Mexico City, Mexico
e mail: gomberg@itam.mx

Z. Thepris
Credit Suisse, Mexico City, Mexico



M. Gallego, N. Schofield (eds.), *The Political Economy of Social Choices*, Studies in Political Economy, DOI 10.1007/978-3-319-40118-8_3

(Huber and Arceneaux 2007; Gerber et al. 2011; Da Silveira and De Mello 2011; Durante and Gutierrez 2015; Larreguy et al. 2014), much of it finding, at best, modest or short-lived impact, in obvious contrast with the willingness of campaigns to spend humongous resources on ads. One possibility is that, as many of the studies are focused on the volume of advertising or expenditure, they fail to differentiate between the differential impact of ads based on their content.

A number of theoretical models have been proposed to establish the mechanisms through which advertising affects voting behavior. DellaVigna and Gentzkow (2010), categorize these models under "belief based" and "preference based". Belief-based models' underlying assumption is that advertising contains information valued by viewers, while preference based models can predict a change in behavior regardless of the informational content of the ads. Datasets containing detailed information on the content of advertising can then be useful to shed light into which of the two groups of models best fits empirical observations. The main caveat for analysis of this kind is that one must rely on subjective assessments regarding the informational content of advertising, as it may be impossible for the researcher to identify the kind of information that is valuable to voters. However, while the informational content of a specific ad may be hard to define, there is one feature that is easily observable: whether the ad makes a direct reference to competing candidates or not. As in most cases politicians do not invest in promoting the virtues of their opponents, it is reasonably safe to assume that such mention is designed to disparage their competition: negative advertising.[3] It may be interesting, for instance, to explore whether "negative" and "positive" ads impact voters differently, implying differences in their strategic use by the campaigns.

Naturally, in part due to data availability, most of the existing literature that analyzes the effects and presence of negative advertising focuses on the US (see Lau and Rovner 2009 for a recent review of the literature). Of course, even if one abstracts from the possible differences between the US and less established democracies, the US focus tends to limit the problem to two-candidate environments, in which both negative and positive advertising by either candidate would impact the same variable: the expected vote margin between the two competitors.[4]

In contrast, multi-candidate environments allow for richer effects of advertising, and, consequently, richer set of possibilities and incentives for varying advertising content. For instance, candidates facing multiple opponents have to choose not only whether to place a positive or negative ad but also, in case the ad is negative, they need to choose which opponent to attack. As a result, the existence and content of negative advertising may vary depending on the relative expected vote share of each

[3]Our data set, of course, would allow a much richer definition of "negative campaigning". However, for the purposes of this preliminary study we decided to avoid the potential pitfalls of content analysis and stick to an easily quantifiable notion here defined. We hope that in the future more information from our data set could be exploited.

[4]Perhaps due to this, some studies have focused on the potential differential effects between positive and negative advertising on other outcomes, such as turnout (see, for example, Ansolabehere et al. 1999).

of the candidates with respect to all other competitors. This is, indeed, the focus of a study by Skaperdas and Grofman (1995) which analyses strategic incentives for negative campaigning in an election in a multi-candidate race. We believe the Mexican environment provides a particularly appropriate setting for testing the predictions of the model.

In order to either observe such effects or to test theories about them, one needs, of course, to have a detailed data set, incorporating both the quantity and content data on political advertising. In this chapter, we report a unique dataset that we constructed, which includes both the total number of ads placed by all competing political parties during Mexico's 2012 presidential campaign and the detailed information on the content of the ads aired every day during the course of the campaign by each of the competing parties. In order to illustrate how our dataset may be useful to understand candidates' choices and voters' responses to one specific feature of the advertising's content, we describe the evolution each party's negative advertising strategies (defined as ads that explicitly mention the other competing candidates or parties) over the course of the campaign, and relate it to the expected vote share in the general election for each of the competing candidates based on the available surveys. We show that parties' choice of negative advertising strategies are consistent with a model in which ads do affect voting intentions, and negative (positive) advertising affect negatively (positively) the vote share of the mentioned party, and positively (negatively) that of all other competing parties. We believe that this data set can be a starting point for further exploration on the interaction between advertising content and its impact.

The rest of the chapter is organized as follows. Section 2 describes the political and institutional context of our study. Section 3 introduces the predictions of the Skaperdas and Grofman (1995) model which we shall be testing. Section 4 describes the dataset constructed for the analysis along with descriptive statistics. Section 5 describes the evolution of negative advertising from the part of all competing candidates over the course of Mexico's 2012 presidential campaign and relates it to the model predictions. The final section concludes. Appendix contains the list of content variables in our database.

## 2 Context

Mexico is a consolidating federal democracy in which presidential elections are held every 6 years, concurrently with a congressional election (with a midterm election to the lower house of the Congress occurring halfway during the presidential term). The last presidential election, on which this study focuses, was held in 2012. In addition, in 2012, 16 of Mexico's 32 states held local (gubernatorial, legislative and/or municipal) elections concurrently with the presidential one. Crucially for our study, reforms to Mexican electoral law adopted prior to this election specified centralized allocation of political advertising through the Federal Electoral Institute (IFE), with advertising amounts regulated by law based on previous electoral results.

Furthermore, even though at the time organization of state electoral campaigns was the responsibility of state electoral institutes, the allocation of TV ads to candidates for all national and state elections was, by law, entrusted to IFE, which used randomization to allocate advertising slots to parties in proportions imposed by law (we provide a more detailed description of the environment below). Though the advertising quantity and distribution, therefore, was fixed exogenously, parties were left free to determine advertising content.

During the 2012 presidential election, three main parties disputed most of the positions at stake in both federal and local elections: the Institutional Revolutionary Party (Partido Revolucionario Institucional, PRI), The National Action Party (Partido Accion Nacional, PAN), and the Party of the Democratic Revolution (Partido de la Revolucion Democratica, PRD). At the time, there were seven political parties with a legal right to participate in federal elections. However, three of the four minor parties ran in coalition with one of these three majors. In terms of ideology, while PAN can be broadly defined as right-to-center, PRD leans to the left-to-center and PRI is generally considered as centrist. Three main presidential candidates disputed the 2012 election: Enrique Peña Nieto (EPN, running for PRI in coalition with the Partido Verde-Ecologista, PVEM), Josefina Vázquez Mota (JVM, running for PAN) and Andrés Manuel López Obrador (AMLO, running for PRD, in coalition with two other minor political parties, Partido del Trabajo, PT and Movimiento Ciudadano, MC). A fourth candidate, Gabriel Quadri de la Torre (GQT), ran for a smaller party of recent formation, New Alliance (Nueva Alianza, NA).

While the inability of surveys to correctly predict candidates' electoral performance in the 2012 election has been a source of political controversy in Mexico, there exists a general consensus with respect to the evolution of electoral preferences over the course of the campaign, that we illustrate in Fig. 1, which shows the evolution of electoral preferences for all four presidential candidates according to a daily survey conducted by a national newspaper, Milenio.

EPN was the frontrunner over the whole course of the campaign. The runner-up's position was highly contested between JVM and AMLO. JVM started the campaign with large electoral support, and lost it to AMLO who, by the end of the campaign, had a stronger electoral support than her. Finally, GQT consistently ranked as the candidate with the lowest electoral support throughout the campaign, with very low chances of winning the election. In fact, his primary objective in this campaign seems to have been to maximize his vote share, which, under Mexican law would increase his party's future perks, such as public financing and access to broadcast advertising.[5]

[5]Mexican electoral law divides these perks among the parties that achieve a certain electoral threshold (at the time it was 2 % of the total vote in the national election), with higher vote share explicitly rewarded in allocating the future resources. Parties that do not achieve the threshold, in contrast, lose their register, which implies not only the loss of public funding and advertising access, but also inability to nominate candidates in future elections. Consequently, minor parties, if they choose to run outside of the coalition, are primarily concerned with vote share maximization.

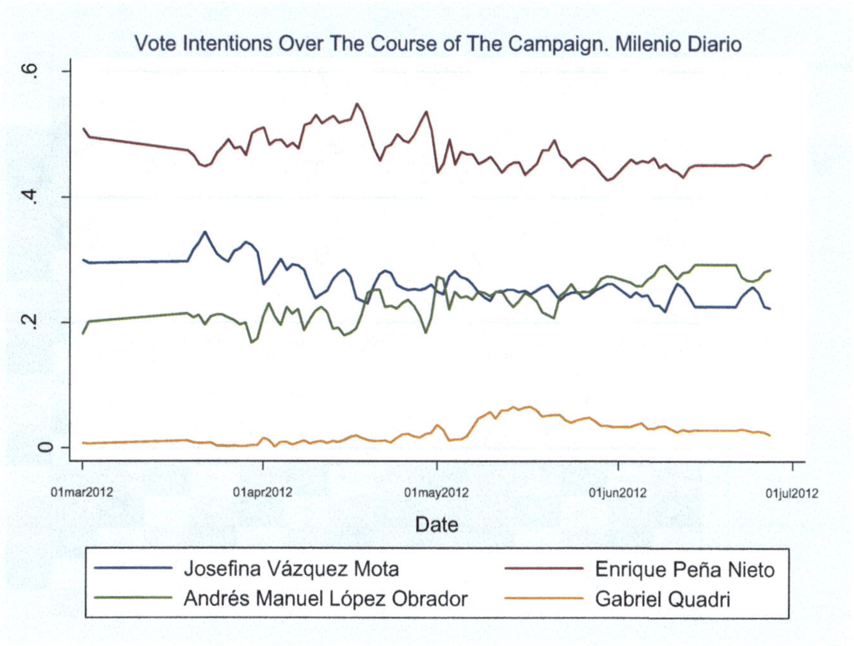


**Fig. 1** Evolution of vote intentions over the course of the campaign. Milenio Diario

As noted above, following the highly contested presidential election of 2006, in 2007 the Mexican Congress adopted an electoral reform which, among other things, drastically changed the regulation of campaign advertising. The legislation introduced prevented parties competing in any election (at the federal, state, or municipal level) from purchasing airtime on radio and TV directly from private broadcasters. Rather, it required broadcasters devote 45 min of their airtime—evenly distributed between 6:00 a.m. and midnight—to "government programs" during the 90 days of the campaign. This time was divided into 30-s slots which IFE assigned to all parties competing in the election, according to a specific rule. In particular, 30 % of all slots were evenly divided among all competing coalitions, while the 70 % of slots remaining were assigned to each coalition in proportion to its vote share in the previous election. In states where only the federal election was held, airtime was assigned to candidates in proportion to their parties' vote share in the last federal congressional elections. In states where state elections were held concurrently (16 of 32 states in Mexico), airtime was instead assigned to candidates in proportion to their parties' vote share in both the previous federal and state elections. The

| Date/Time | | MAR 30 FRI | MAR 31 SAT | APR 1 SUN | APR 2 MON | APR 3 TUE | APR 4 WED | APR 5 THU | APR 6 FRI | APR 7 SAT | APR 8 SUN | APR 9 MON | APR 10 TUE | APR 11 WED | APR 12 THU | APR 13 FRI |
|---|---|---|---|---|---|---|---|---|---|---|---|---|---|---|---|---|
| 06:00:00 - 06:59:59 | 1 | PRI | PRD | PAN | PRI | PRD | PAN | PRI | PRD | PAN | PRI | PRI | PAN | PRI | PNA | PRD |
| | 2 | PRI | PRI | PRD | PAN | PRI | PRD | PAN | PRI | PRD | PAN | PRI | PRI | PAN | PRI | PNA |
| | 3 | PAN | PRI | PRI | PRD | PAN | PRI | PRD | PAN | PRI | PRD | PAN | PRI | PRI | PAN | PRI |
| | 4 | PNA | PAN | PRI | PRI | PRD | PAN | PRI | PRD | PAN | PRI | PRD | PAN | PRI | PRI | PAN |
| | 5 | PRI | PNA | PAN | PRI | PRI | PRD | PAN | PRI | PRD | PAN | PRI | PRD | PAN | PRI | PRI |
| | 6 | PAN | PRI | PNA | PAN | PRI | PRI | PRD | PAN | PRI | PRD | PAN | PRI | PRD | PAN | PRI |
| 07:00:00 - 07:59:59 | 7 | PRD | PAN | PRI | PNA | PAN | PRI | PRI | PRD | PAN | PRI | PRD | PAN | PRI | PRD | PAN |
| | 8 | PRI | PRD | PAN | PRI | PNA | PAN | PRI | PRI | PRD | PAN | PRI | PRD | PAN | PRI | PRD |
| | 9 | PAN | PRI | PRD | PAN | PRI | PNA | PAN | PRI | PRI | PRD | PAN | PRI | PRD | PAN | PRI |
| | 10 | PRI | PAN | PRI | PRD | PAN | PRI | PNA | PAN | PRI | PRI | PRD | PAN | PRI | PRD | PAN |
| | 11 | PRI | PRI | PAN | PRI | PRD | PAN | PRI | PNA | PAN | PRI | PRI | PRD | PAN | PRI | PRD |
| | 12 | PAN | PRI | PRI | PAN | PRI | PRD | PAN | PRI | PNA | PAN | PRI | PRI | PRD | PAN | PRI |
| 08:00:00 - 08:59:59 | 13 | PRD | PAN | PRI | PRI | PAN | PRI | PRD | PAN | PRI | PNA | PAN | PRI | PRI | PRD | PAN |
| | 14 | PRI | PRD | PAN | PRI | PRI | PAN | PRI | PRD | PAN | PRI | PNA | PAN | PRI | PRI | PRD |
| | 15 | PAN | PRI | PRD | PAN | PRI | PRI | PAN | PRI | PRD | PAN | PRI | PNA | PAN | PRI | PRI |
| | 16 | PNA | PAN | PRI | PRD | PAN | PRI | PRI | PAN | PRI | PRD | PAN | PRI | PNA | PAN | PRI |
| | 17 | PRI | PNA | PAN | PRI | PRD | PAN | PRI | PRI | PAN | PRI | PRD | PAN | PRI | PNA | PAN |
| | 18 | PAN | PRI | PNA | PAN | PRI | PRD | PAN | PRI | PRI | PAN | PRI | PRD | PAN | PRI | PNA |

**Fig. 2** Assignment of ads slots. Federal election

number of ads each candidate could air in each of the broadcasters was then fixed by law. The total number of advertising slots and the time of the day during which broadcasters were supposed to air them was also fixed, and closely monitored by IFE. The specific time slots assigned each day to each of the competing parties did vary, but was assigned randomly. In particular, a lottery was used to determine how different time slots were allocated to party ads on the first day of the campaign, and ad slots are rolled over in subsequent days. It is important to note that, while the allocation of slots to parties for the states where only federal elections were held was determined by a unique national lottery, a separate lottery took place in each of the states in which state elections were held concurrently. To illustrate the outcome of a lottery, in Fig. 2 we report the allocation of time slots to parties over the first 15 campaign days for the time comprised between 6:00 and 9:00 a.m., as determined by the federal election advertising lottery.

As a consequence of this law, candidates' decisions regarding advertising were limited mainly to their content. Parties had knowledge of the specific time of day at which their ads could be placed every week, and were allowed to provide to IFE a weekly detailed list of the specific ads they would place in each of the slots assigned. IFE collected the ads provided by all of the competing parties weekly, and distributed them to all broadcasters, along with the detailed instructions regarding the time and day at which each of the ads should be placed. Compliance with these guidelines from the part of broadcasters was closely monitored by IFE, who reports compliance rates of over 98 %.

## 3 The Hypotheses to Be Tested

Skaperdas and Grofman (1995) present a model of negative advertising that, in its main features seems to be remarkably close to the institutional environment we are considering here. They consider electoral competition between J candidates, with each candidate having certain known level of voter support (perhaps reflected by polling, as in our case) and a number of undecided voters that could be affected by advertising. Crucially, candidates do not choose the overall advertising investment, which in their model is normalized to 1 (a modeling feature at serious odds with the typical environment of a US campaign, but exactly replicated in Mexican campaign law). A campaign strategy of candidate *i* consists of choosing how to allocate this total advertising budget among the mentions of herself and of other candidates (with ${y^i}_j$ denoting the amount of advertising party *i* dedicates to party *j* in its campaign and ${y^i}_1 + {y^i}_2 + \ldots + {y^i}_J = 1$). Like us, they interpret a candidate investment in her own image ${y^i}_i$ to be positive advertising, and her references to others ${y^i}_j$ as negative advertising (whenever *i* is not equal to *j*). In their model, positive advertising has the effect of increasing a candidate vote share, while the negative advertising decreases it.

Though Skaperdas and Grofman (1995) analyze the case of, at most, a three-candidate election, they separately consider the case in which one of the candidates has no possibilities of winning, but acts as a vote-share-maximizing spoiler (the role which, in this election was performed by GQT). They show that such spoiler effectively stays out of the competition between the major candidates, not affecting their negative advertising strategies. Hence, the analysis of the case of three viable candidates would directly apply to the race between EPN, AMLO and JVM.

There are four main hypotheses derived (under a number of simplifying assumption) by Skaperdas and Grofman (1995), which seem particularly relevant here. We are summarizing them in what follows.

**Hypothesis 1**: The Leading candidate engages in less negative campaigning then the followers

**Hypothesis 2**: The spoiler candidate does not engage in negative campaigning

**Hypothesis 3**: Nobody campaigns against the spoiler

**Hypothesis 4**: Candidates concentrate their negative campaigning on the stronger of their opponents.

In what follows, we will be seeking to verify these hypotheses using our data.

## 4 Data Construction

Two main datasets were exploited in this chapter. First, we took advantage of the public repository that the Federal Electoral Institute holds, with all of the TV ads aired during the 2012 campaign, by all competing candidates.[6] We then coded the information contained in each of the aired ads. To conduct the analysis and develop the first dataset used in this chapter, the content of each of the ads was categorized according to its type (i.e. Presidential, Senatorial, Local, Federal, etc.), its content (i.e. features female candidates, number of candidates featured, etc.) and other features of the message it transmitted to voters (i.e. whether it encourages the audience to vote, whether it mentions explicitly candidates' campaign proposals, whether it attacks a competing candidate and his or her party identity, etc.). In total, 63 different variables describing each of the ad's content were constructed (the full description of these variables can be found in the Appendix). Table 1 presents some descriptive statistics regarding the ads.

1116 ads with an average duration of 30 s were analyzed, amounting to approximately 9.3 h of video. EPN's campaign broadcast, in total, 521 different ads. JVM, AMLO and GQT's campaigns broadcast 214, 332, and 49 different ads, respectively. EPN's campaign broadcast, in total, 138 (383) different ads in the federal (local) campaign(s). JVM, AMLO and GQT federal (local) campaigns broadcast 105 (109), 130 (202) and 34 (15) different ads, respectively. From this set of ads used in the federal (local) campaign, 5 (3) of EPN's included a direct reference to other parties or candidates (which in this chapter we code as negative advertising). JVM's campaign broadcasted 25 (9) negative federal (local) ads, while AMLO's and GQT's broadcasted 52 (38) and 4 (0) negative ads, respectively. This already suggests that very important differences in negative advertising strategies are observed across candidates. While for JVM and AMLO a large fraction of all the ads broadcasted contained references to competing candidates or parties, only less than 2 % of EPN's ads included a reference to his competitors.

This first dataset, however, only measures the extent to which different campaign produced negative ads, and not the frequency with which each of the ads was aired. In order to better assess the importance of negative advertising for each of the

**Table 1** Total number of ads broadcast by candidate

| | Federal election | | Local elections | |
|---|---|---|---|---|
| | Total ads | Negative ads | Total ads | Negative ads |
| Enrique Peña Nieto | 138 | 5 | 383 | 3 |
| Josefina Vázquez Mota | 105 | 25 | 109 | 9 |
| Andrés Manuel López Obrador | 130 | 52 | 202 | 38 |
| Gabriel Quadri de la Torre | 34 | 4 | 15 | 0 |

[6]Which can be accessed here: http://pautas.ife.org.mx/transparencia/index.html.

**Table 2** Distribution of ads slots by candidate

| | Federal election | |
|---|---|---|
| | Percentage of slots | Percentage negative |
| Enrique Peña Nieto | 43.20 | 2.7 |
| Josefina Vázquez Mota | 26.89 | 34.0 |
| Andrés Manuel López Obrador | 21.19 | 41.4 |
| Gabriel Quadri de la Torre | 8.72 | 12.1 |

competing candidates during the 2012 and campaign, and in order to explore the evolution of their negative advertising strategies over the course of the campaign, we exploited a second source of data, also publicly available from IFE.

In particular, IFE made public the list of weekly "transmission orders" for TV networks across the country, which specifies the day and exact time at which each of the ads provided by all competing parties should be aired by each network. The dataset lists, for each of the time slots every time of the day, a unique identifier of the exact ad that should be aired. We then combined the information on the content of all of the existing ads described above with the information on the frequency with which each of those ads was aired, along with the exact date and time at which this occurred. Table 2 provides basic descriptive statistics on the allocation of negative advertising by each of the competing coalitions.

Giving equal weight to each ad slot in each of the networks, EPN's campaign was assigned, in total, 43 % of all ad slots in total over the course of the campaign. JVM, AMLO and GQT were assigned 27 %, 21 % and 9 % of all ad slots, respectively. Of all the slots assigned to him, EPN only devoted 2.7 % to negative ads. JVM, AMLO and GQT used 34, 41.4 and 12 % of the available ad slots to negative advertising. It is then clear that larger differences in negative advertising strategies from the part of competing coalitions arise when we consider the frequency with which each type of ads were aired. For both JVM and AMLO, the fraction of air time devoted to negative advertising is larger than the fraction of all ads produced that contained negative advertising strategies. For GQT, the fraction of air time devoted to negative ads is much smaller that the fraction of all ads produced by his campaign that contained negative messages.

## 5 Evolution of Negative Advertising Strategies

In this section, we refine the analysis by exploring the evolution of the presence of negative ads, and the competing candidates mentioned in them, over the course of the campaign. We present the analysis graphically, describing the negative advertising strategy of each of the competing candidates separately.

Figure 3 then shows the evolution of negative advertising strategies from the part of EPN's campaign over time. The first graph shows the fraction of all ads aired

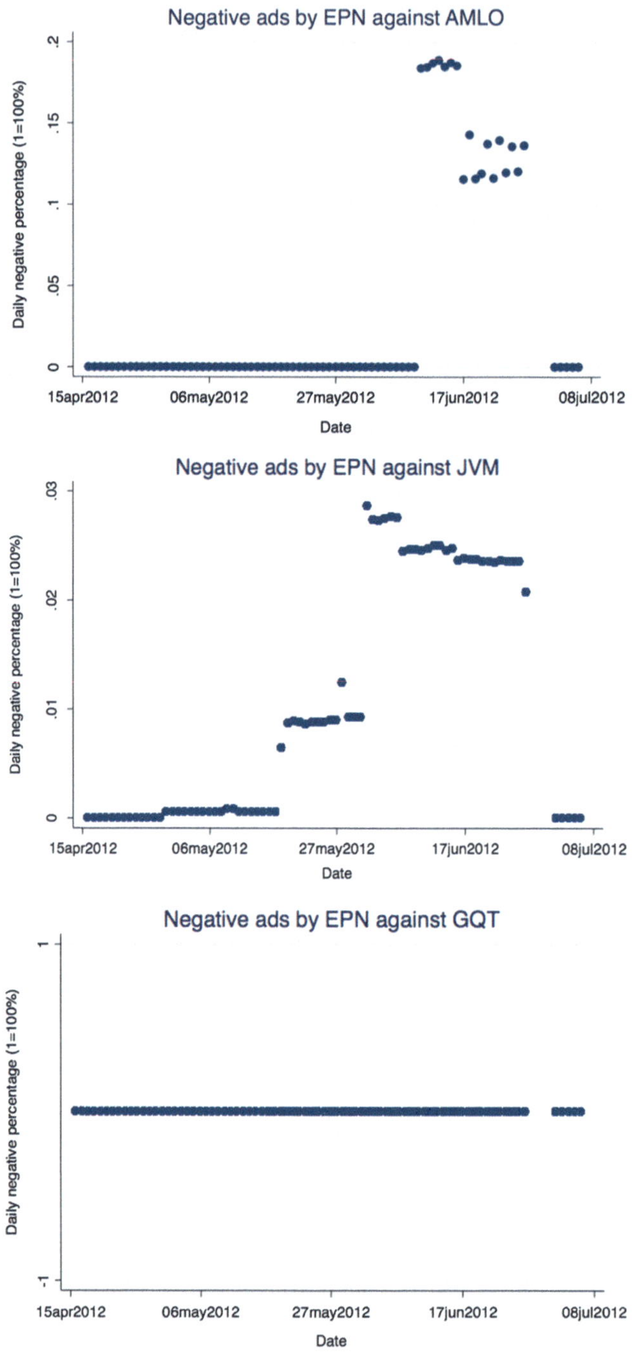


**Fig. 3** Evolution of negative advertising aired by EPN over the course of the campaign

each day of the campaign in which AMLO is mentioned. The second and third show the fraction of all ads aired each day of the campaign in which JVM and GQT are mentioned, respectively.

When looking at the evolution of the content of the ads aired by EPN over the course of the campaign, three facts are worth highlighting. First, for almost the first half of his campaign, no negative ads were aired. Second, while EPN's campaign first aired ads mentioning JVM, the intensity with which it mentioned AMLO towards the end of the campaign, when AMLO was emerging as his main competitor, was considerably larger. Finally, EPN did not air any ads mentioning GQT.

EPN's negative advertising strategies seem to match very closely the incentives that a frontrunner may face to place negative advertising during a campaign. In particular, when perceived as a clear frontrunner, he had little incentives for placing negative advertising. If it is hard to infer which of the competing candidates would benefit from the frontrunner attacking another competitor, the risk of this kind of advertising reducing his lead in preferences would deter him from attacking his competitors.

Figure 4 replicates the exercise, this time exploring the evolution of negative advertising strategies from the part of JVM's campaign. The first graph shows the fraction of all ads aired each day of the campaign in which EPN is mentioned. The second and third show the fraction of all ads aired each day of the campaign in which AMLO and GQT are mentioned, respectively.

Different facts are worth highlighting in this case. First, throughout the campaign, EPN was the main target of JVM's negative advertising efforts. Towards the end of the campaign, AMLO displaced EPN as the main target of JVM's attacks. And, finally, JVM never mentioned GQT in her ads.

Again, this seems consistent with JVM perceiving herself as the potential runner-up, having a much stronger electoral support than AMLO, and with EPN being the frontrunner. The potential return on attacking AMLO could result in an increase in EPN's lead. By attacking EPN, she had larger chances of reducing his lead without fearing a worrisome closing of her lead over AMLO. However, as the campaign advanced and her lead over AMLO disappeared, it became decreasingly profitable to attack the frontrunner, as it could benefit AMLO, and increasingly profitable to attack AMLO, to regain the position as runner-up.

Figure 5 presents the evolution of negative advertising strategies from the part of AMLO's campaign. The first graph shows the fraction of all ads aired each day of the campaign in which EPN is mentioned. The second and third show the fraction of all ads aired each day of the campaign in which JVM and GQT are mentioned, respectively.

As can be seen, AMLO's negative advertising strategies differ considerably from those of the other competing candidates. First, both EPN and JVM were the target of his negative advertising strategies throughout the whole course of the campaign. Nevertheless, while JVM was the target of a larger fraction of negative ads at the start of the campaign, EPN slowly became AMLO's most common target as the campaign evolved. Again the differences in the intensity and target choice for

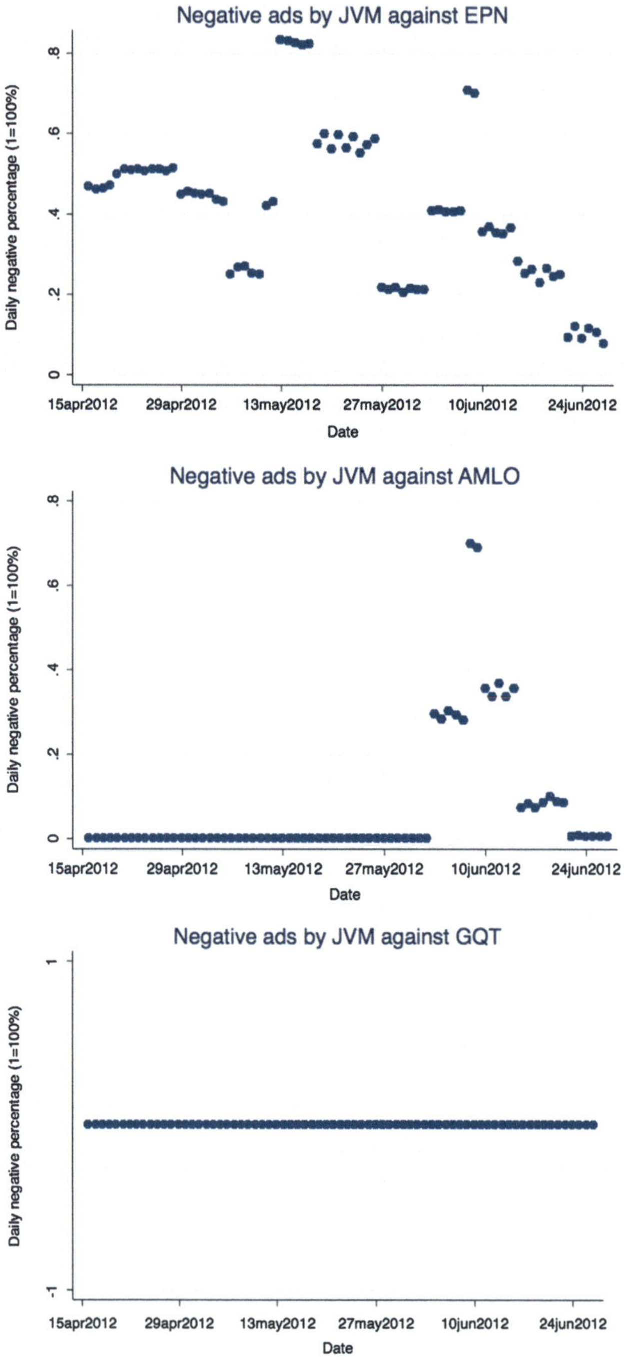


**Fig. 4** Evolution of negative advertising aired by JVM over the course of the campaign

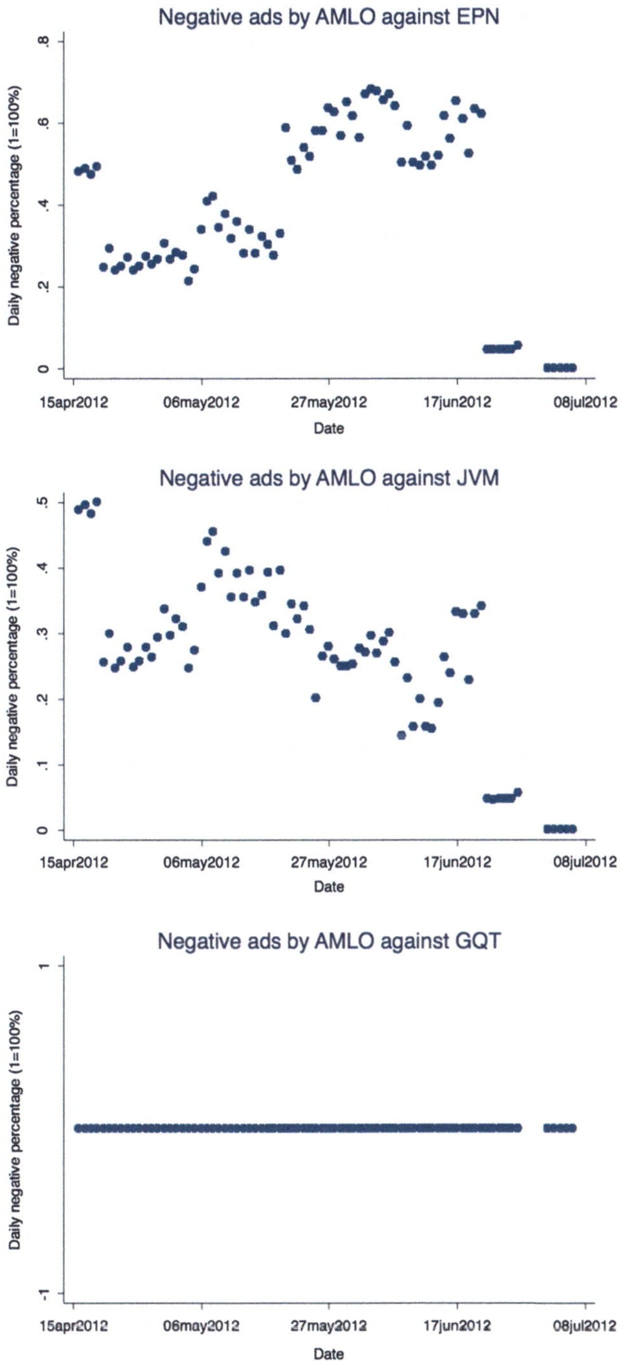


**Fig. 5** Evolution of negative advertising aired by AMLO over the course of the campaign

AMLO's advertising strategy are consistent with the scenario in which he perceived himself running behind the two over the course of the campaign. The shift towards attacking EPN more frequently as the campaign evolved is also consistent with him perceiving himself as displacing JVM to the third place.

It should not be surprising that none of the three main candidates chose GQT as the target of their attacks. Finally, Fig. 6 shows the evolution of GQT negative advertising strategies over the whole duration of the campaign. It is clear that the presence of negative advertising is low. GQT did not seem to have very strong incentives to attack his competitors, since his objectives did not involve seriously competing with them for victory.

Overall, we observe strong support for Hypotheses 1, 2 and 3 of Skaperdas and Grofman (1995) though somewhat lesser support for Hypothesis 4 (AMLO and JVM did actively campaign against each other). It could be conjectured that this is due to the perceived close contest for the second spot, with the identity of the runner-up not being always obvious in this election.

## 6 Conclusions

In this chapter we report a new data set of political advertising in 2012 federal election in Mexico, which combines publically available information on campaign spot broadcasts with the detailed coding of content of the spots involved. We use it to test basic predictions of Skaperdas and Grofman (1995) model of negative campaign advertising, as applied to the multi-candidate setting of Mexico, which we find broadly supported by evidence. While the analysis presented here is fairly simple, and uses only a small subset of the content variables available, we believe this data could be of broader importance in studying advertising campaign strategies and effectiveness in emerging democracies.

## Appendix: Ad Content Dataset—List of Variables Coded[7]

1. PAN—Whether the ad is sponsored by the party Partido Acción Nacional or not.
2. PRI—Whether the ad is sponsored by the party Partido Revolucionario Institucional or not.
3. PRD—Whether the ad is sponsored by the party Partido de la Revolución Democrática or not.
4. PT—Whether the ad is sponsored by the party Partido del Trabajo or not.

[7]The database itself may be obtained from the authors.

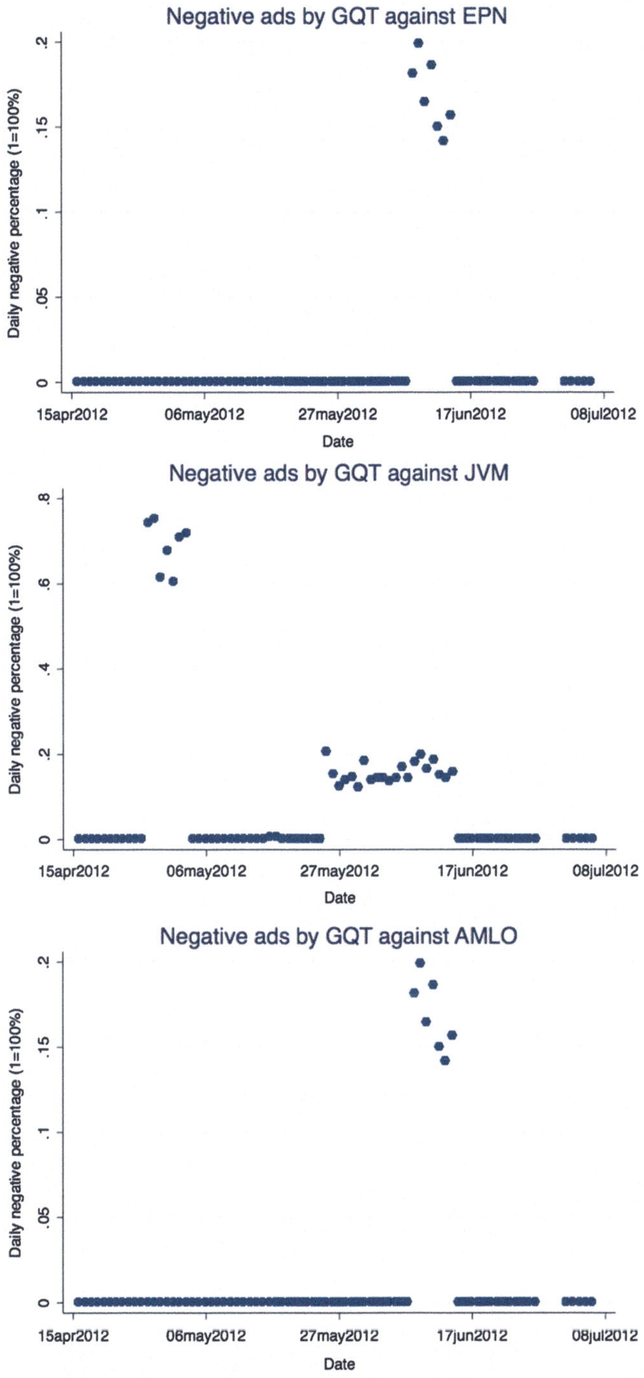


**Fig. 6** Evolution of negative advertising aired by GQT over the course of the campaign

5. PV—Whether the ad is sponsored by the party Partido Verde Ecologista de México or not.
6. MC—Whether the ad is sponsored by the party Movimiento Ciudadano or not.
7. NA—Whether the ad is sponsored by the party Nueva Alianza or not.
8. MP—Whether the ad is sponsored by the party Coalición Movimiento Progresista or not.
9. Diputado Federal—Weather the candidate is running for a position as federal representative.
10. Diputado Local—Weather the candidate is running for a position as a local representative.
11. Senator—Whether the candidate is running for the upper chamber as "Senador".
12. President—Whether the candidate is running for President of Mexico.
13. Governor—Whether the candidate is running for Governor.
14. Municipal President—Whether the candidate is running for Municipal President.
15. State—State for which the candidate is running for.
16. State Mentioned—Whether the State for which the candidate is running for is mentioned in the ad.
17. District—District for which the candidate is running for.
18. District Mentioned—Whether the District for which the candidate is running for is mentioned in the ad.
19. Whether the sponsored candidate appears in the ad.
20. Whether a candidate running for PAN appears in the ad.
21. Whether a candidate running for PRI appears in the ad.
22. Whether a candidate running for PRD appears in the ad.
23. Whether a candidate running for PT appears in the ad.
24. Whether a candidate running for PVEM appears in the ad.
25. Whether a candidate running for MC appears in the ad.
26. Whether a candidate running for NA appears in the ad.
27. Whether a candidate running for MP appears in the ad.
28. Whether the name of the party PAN is explicitly mentioned in the ad.
29. Whether the name of the party PRI is explicitly mentioned in the ad.
30. Whether the name of the party PRD is explicitly mentioned in the ad.
31. Whether the name of the party PT is explicitly mentioned in the ad.
32. Whether the name of the party PV is explicitly mentioned in the ad.
33. Whether the name of the party MC is explicitly mentioned in the ad.
34. Whether the name of the party NA is explicitly mentioned in the ad.
35. Whether the name of the party MP is explicitly mentioned in the ad.
36. Local Proposals—Whether the candidate or the ad proposes specific policies (about a specific subject, i.e. Labor Market, Energy Reform) that will affect the local community.
37. Federal Proposals—Whether the candidate or the ad proposes specific policies(about a specific subject, i.e. Labor Market, Energy Reform) that will affect the Country in general.

38. No Proposals—Whether the candidate or the add does not make any specific proposals in the ad.
39. Whether a candidate from PAN or the party itself is mentioned in a negative way (i.e. quoting bad policies or social scandals)
40. Whether a candidate from PRI or the party itself is mentioned in a negative way (i.e. quoting bad policies or social scandals)
41. Whether a candidate from PRD or the party itself is mentioned in a negative way (i.e. quoting bad policies or social scandals)
42. Whether a candidate from PT or the party itself is mentioned in a negative way (i.e. quoting bad policies or social scandals)
43. Whether a candidate from PV or the party itself is mentioned in a negative way (i.e. quoting bad policies or social scandals)
44. Whether a candidate from MC or the party itself is mentioned in a negative way (i.e. quoting bad policies or social scandals)
45. Whether a candidate from NA or the party itself is mentioned in a negative way (i.e. quoting bad policies or social scandals)
46. Whether a candidate from MP or the party itself is mentioned in a negative way (i.e. quoting bad policies or social scandals)
47. Whether the name of a candidate from PAN is explicitly mentioned in the ad.
48. Whether the name of a candidate from PRI is explicitly mentioned in the ad.
49. Whether the name of a candidate from PRD is explicitly mentioned in the ad.
50. Whether the name of a candidate from PT is explicitly mentioned in the ad.
51. Whether the name of a candidate from PV is explicitly mentioned in the ad.
52. Whether the name of a candidate from MC is explicitly mentioned in the ad.
53. Whether the name of a candidate from NA is explicitly mentioned in the ad.
54. Whether the name of a candidate from MP is explicitly mentioned in the ad.
55. Political Celebrities—Whether political celebrities appear in the ad (i.e. Carlos Salinas, Elba Esther Gordillo).
56. Other Celebrities—Whether other celebrities appear in the ad (i.e. television, radio, scientific community members).
57. In your judgment, is the primary purpose of the ad to promote a specific candidate, to attack a candidate, or to contrast the candidates:

    a. 0 'Not applicable'
    b. 1 'Attack'
    c. 2 'Contrast'
    d. 3 'Promote'
    e. 4 'Unsure or Unclear'

58. Multiple Candidates—Whether the ad promotes multiple candidates.
59. Female Candidate—Whether the ad promotes a female candidate.
60. Male Candidate—Whether the ad promotes a male candidate.
61. Whether the ad directs the viewer to take any action.

62. If 61 is yes, What is the action?
    a. 0 'Not applicable'
    b. 1 'Other'
    c. 2 'To vote for someone'
    d. 3 'To support someone'
    e. 4 'To elect or re-elect someone'
    f. 5 'To write, call, or tell someone to do something'
    g. 6 'To reject someone'
    h. 7 'To urge action or attention to a particular matter'
    i. 8 'To defeat someone'
    j. 9 'To send a message or call someone to express yourself'
    k. 10 'To vote against someone'
63. Whether the primary focus of this ad on the personal characteristics of either candidate or on policy matters
    a. 1 'Policy matters'
    b. 2 'Personal characteristics'
    c. 3 'Both'
    d. 4 'Neither'

## References

Ansolabehere, S. D., Iyengar, S., & Simon, A. (1999). Replicating experiments using aggregate and survey data: The case of negative advertising and turnout. *American Political Science Review, 93*(4), 901–909.

Da Silveira, B. S., & De Mello, J. M. P. (2011). Campaign advertising and election outcomes: Quasi-natural experiment evidence from gubernatorial elections in Brazil. *Review of Economic Studies, 78*(2), 590–612.

Della Vigna, S., & Gentzkow, M. (2010). Persuasion: Empirical evidence, with Matthew Gentzkow. *Annual Review of Economics, 2*.

Durante, R., & Gutierrez, E. (2015). Fighting crime with a little help from my friends: Political alignment, inter-jurisdictional cooperation and crime in Mexico. *CEPR discussion paper 10769*. London: Centre for Economic Policy Research. http://www.cepr.org/active/publications/discussion_papers/dp.php?dpno=10769

Gerber, A. S., Gimpel, J. G., Green, D. P., & Shaw, D. R. (2011). How large and long-lasting are the persuasive effects of televised campaign ads? Results from a randomized field experiment. *American Political Science Review, 105*(1), 135–150.

Huber, G. A., & Arceneaux, K. (2007). Identifying the persuasive effects of presidential advertising. *American Journal of Political Science, 51*(4), 957–977.

Larreguy, H., Marshall, J., & Snyder Jr., J. M. (2014). *Political advertising in consolidating democracies: Ads, clientelism, and political development in Mexico*. Working Paper.

Lau, R. R., & Rovner, I. B. (2009). Negative campaigning. *Annual Review of Political Science, 12*, 285–306.

Skaperdas, S., & Grofman, B. (1995). Modeling negative campaigning. *American Political Science Review, 89*(1), 49–61.

# Legislative Leaders as Condorcet Winners? The Case of the U.S. Congress

**Robert S. Erikson and Yair Ghitza**

Legislative scholars recognize that legislative output is affected by the legislature's institutional design. Let us assume that the goals are to avoid chaos and to enhance welfare. Toward these goals there are many variations. Strong committees (Shepsle and Weingast 1987), strong ruling parties (Cox and McCubbins 2004), institutions fostering legislative exchange (Weingast and Marshall 1988) and a strong agenda setter (Dahm and Glazer 2015) all can contribute to the interests a legislature's members (see also Krehbiel 2004). In this chapter, we explore the potential gains from electing a strong agenda setter who holds the power to steer policy toward the setter's own policy preferences. In the US context, one can loosely label this setter as a strong "speaker." Would the election of a strong speaker lead to policies that represent the views of the chamber? And would it be possible to elect a strong speaker whose views are representative of the party in an open vote, given that theoretical models of multi-dimensional voting on policy suggest the absence of a predictable equilibrium outcome and the specter of endless cycling?

Assume the institutional design of a legislature where the voting members select one of their own to serve as the Speaker who has the power to set the agenda via closed rules as items arrive in the policy hopper. On the one hand, the selection of the agenda setter provides order. The reduction of policy choices to the binary alternatives of the status quo versus the proposed new set point eliminates

To be published in *The Political Economy of Social Choices* (2016, Springer), edited by Maria Gallego and Norman Schofield. We thank Joon Yang for his research assistance.

R.S. Erikson (✉)
Columbia University, New York, NY, USA
e-mail: rse14@columbia.edu

Y. Ghitza
Catalist, Washington, DC, USA



M. Gallego, N. Schofield (eds.), *The Political Economy of Social Choices*,
Studies in Political Economy, DOI 10.1007/978-3-319-40118-8_4

indeterminacy and potential cycling of policy choices. On the other hand, there might be a risk that a legislature would elect a leader with policy preferences that differ from most of its members. Further, one must ask whether, in a necessarily multidimensional setting, the initial process of electing a leader itself would itself set off a sequence of indeterminacy, chaos, and potential cycling that electing a strong leader is designed to avoid?

To those steeped in the rational-choice literature, the answer to the latter question might seem to be obvious "yes." When policy divisions are arrayed along a single dimension, the answer is the well-known median voter solution (Black 1958). The median voter is decisive. With multiple dimensions, however, uncertainty and chaos reign. In theory, an open agenda and multiple dimensions leads to endless cycling and no clear outcome (Plott 1976). Moreover, by McKelvey's (1976) chaos theorem, if an agenda setter is imposed, the setter can guide the outcome to anywhere in policy space.

When policy choices are modeled in a policy space with multiple dimensions, it is generally true that no Condorcet winner exists; any proposed outcome (in this case the choice of Speaker and the Speaker's policy preferences) can be defeated by any some other outcome (here, another Speaker candidate from within the legislature). For electing a Speaker, would there not be endless cycling?

Agenda control is central to many models of legislative behavior. The agenda setter holds the power to influence outcomes by their setting the alternative to the status quo (Romer and Rosenthal 1978). McKelvey's (1976) chaos theory shows that by strategically moving the alternative to the evolving status quo, an agenda setter could induce a legislature to move anywhere in issue space, including those far from their Pareto set. In basic models of legislative bargaining ("divide the dollar"), the random assignment of the agenda setter plays a decisive role (Baron and Ferejohn 1989).

Real-world legislatures have the power to choose their agenda setter themselves rather than allow the choice to be arbitrarily or exogenously imposed. In theory, via majority vote, legislators can collectively vote for one of its members to dictate an agenda. For purposes of this paper, we assume that the task of the legislator is to elect a leader who will be given agenda setting power for the duration of the session.

The present paper considers the election of a legislative leader when the leader is given the power to control the agenda. Let us make two assumptions about this election of a leader. First, assume that across two or more policy dimensions the legislators share identical expectations about each others' interests. Thus, policy preferences are common knowledge with no private information. Second, assume that if elected, each potential leader has the motivation and the power to guide policy as close as possible to his or her ideal point. Assume the contest for leader consists of a tournament where each of $N$ legislator is pitted against each other in a round robin of $N*(N-1)/2$ dyadic contests where the winner of each contest is the one whose ideal point is closer to more voters in policy space. The only way this tournament could have a Condorcet winner would be if one legislator were able to defeat all others in this round robin tournament.

We model the election of an agenda setter as an open vote at the beginning of the legislative session. We contend that under plausible models of preference space, selecting an agenda setter via open voting (with pairwise voting on all possible candidates for leader) often leads to a Condorcet winner. In this tournament where every legislator competes with every other in a contest to see which one is closest to the majority of other legislators on the issues, there often is a decisive winner who defeats all comers.

There are two keys for this result to hold. First, as mentioned above, legislators share common (but not necessarily complete) expectations about each others' interests on the likely issues that will arise.[1] This may be a reasonable assumption when applied to politically sophisticated legislators. Second, also as mentioned above, the agenda setter will be sufficiently powerful and motivated so that he or she will freely pursue his or her personal policy agenda. From these two assumptions, there are no grounds for deception or bargaining. Interests are transparent, and candidates for the chosen leader will be unable to bargain regarding their policy actions. The key then is that the menu of leadership policy positions is not the entirety of continuous policy space. The policy menu is limited to the $N$ policy ideal points for the $N$ members who are candidates for the leadership position.

More formally, let there be a legislature with $N$ (typically an odd number) members. Each member has a preference profile consisting of ideal points on each of $K$ issue dimensions, where $K > 1$, with utility determined by quadratic loss. All members share a common expectation of everybody's preference profile on the $K$ dimensions, including their own. In effect, legislators share common knowledge of each others' interests. There are no games to be played; members hold no private information relevant to the leader's selection.

In a nutshell, the model is as follows: Legislators are to decide policy on a multidimensional set of issues. They seek to minimize their loss as a (possibly weighted) distance between their ideal points and policy. They share a common understanding of each others' objective functions. Their task is to select a leader to dictate the agenda. With no incentive for keeping promises and no reason for members to believe promises, the elected agenda setter will manipulate outcomes to be as close as possible to his or her ideal point.

To say that there is a likely Condorcet winner in the context of picking a dictator is to say that in a tournament of head-to-head competition, there is likely to be one dominant legislator whose policy preference is preferred by a majority to those of

[1]As described below, we use Poole and Rosenthal DW-Nominate scores on two dimensions as our data. These scores on two dimensions are estimates of legislator positions on two primary dimensions. One can observe these as expectations as how legislators will behave on the two dimensions. However, they would not be deterministic; for specific legislation that arises, the legislators take into account unobserved considerations that appear as part of an error term. An analogy is as if legislators were aware of a factor analysis which partially determined their individual ideal points, perhaps two factors defining the ideological space as observers understand it. These are the expectations. Individual preferences on any specific item will include an orthogonal error term that is not observed.

every other member. Of course, consistent with choice theory, the policy preference of the potential Condorcet dictator would not dominate all of policy space. But crucially, in the selection of an agenda setter, the choices do not include all of continuous policy space. The choices for leader positions are limited to the *N* discrete policy preferences of the *N* legislators over the *K* dimensions. In other words, although there will be a win-set in policy space that dominates the Condorcet winner's preferred position, this win-set does not necessarily include the preferences of any other member.

Generalizing from the spatial modeling literature, one might think that the presence of a Condorcet winner when picking a legislative leader is extremely unlikely. One would know that with one-dimensional policy space, a legislature would choose the median member. But with ideal points distributed multidimensionality, the answer seemingly would be an indeterminate outcome with no particular candidate for leader being able to dominate all rivals. That of course would be the answer if the feasible choices were to include the entire *K*-dimensional issue space. In general, any choice for the omnibus set point could be defeated by some other, even if it also has to defeat the status quo. But here the task is not to pick a set point from among a continuous set of alternatives in multidimensional space. Rather, the task is to pick a set point from among the *discrete* ideal points of a finite set of legislators. In short, the question is: Is there a Condorcet winner from among the set of discrete ideal points? That is, each legislator's ideal point is pitted against each others' in a tournament where their peers chose the one closest to their own position. Would this tournament produce a winner who would defeat all others?

Although we argue that our model often leads to a Condorcet winner, answering the circumstances that allow it is complicated. Estimating the requirements for and likelihood of a Condorcet winner is not a problem that yields to an obvious analytical solution. In a separate paper (Erikson and Ghitza 2016), we expand on the theory and explore the problem computationally rather than analytically. That paper simulates various legislatures with multidimensional preference structures. It then simulates the results of leadership tournaments, in each version replicating with at least one thousand simulations. That exercise concludes that at least within the confines of several models of multi-dimensional policy choice, a Condorcet winner is more likely than not. Moreover, the likelihood generally increases rather than decreases with the number of dimensions beyond one dimensions.

In the present paper, we approach the problem not by simulating outcomes with simulated data. Instead, we simulate outcomes of tournaments with real data—the ideal points of actual legislators. The actual legislatures consist of the historical data for the US House of Representatives and the US Senate of the US Congress. For each chamber and each 2-year Congress from 1789 forward, we simulate a tournament for leader based on preferences identified by Poole and Rosenthal's (1997) DW-Nominate scores. For each chamber and Congress, we ascertain the presence or absence of a Condorcet winner in the two-dimensional space presented by DW-Nominate scores. Also for each chamber and Congress since World War II, we ascertain the presence or the absence of a Condorcet winner within each political party. How often does a Condorcet winner arise? We also note the identity of tournament winners and their closeness to the center of policy space.

In our simulations using simulated data (Erikson and Ghitza 2016), the likelihood of a Condorcet winner was least with two policy dimensions. Moreover, although for higher-dimension models the likelihood of a Condorcet winner generally increased with the number of voters, this was not true for two dimensional models. Thus, the challenge is that the likelihood of a Condorcet winner is least likely for two dimensions and many members, such as for an actual congressional chamber. So our task below is to search for Condorcet winners under seemingly adverse conditions but with actual data from the real world rather than the byproduct of simulations.

Before we begin, let us present a simple set of examples from the simplest form of simulation—five voters in two dimensions.

## 1 Simple Examples

To illustrate, Fig. 1 presents four simple examples involving five voters and two dimensions. In panel A, Voter 1 is the Condorcet winner. We see this, because the "win-set" for Voter 1 (the area in issue space containing positions a majority prefers to Voter 1's position, highlighted in grey) is empty. The other four voter positions are outside the win set.

Panels B–D illustrate a case without a Condorcet winner. In Panel B we move Voter 2 from Panel A "northwest" slightly. So Voter 1 is not a Condorcet winner, but perhaps Voter 3 could be one, as he is in Voter 1's win set. However, as Panel C shows, Voter 3 is no Condorcet winner because Voter 4 is in Voter 3's win set. Finally, Panel D shows that Voter 1 is in Voter 4's win set, completing the cycle.

Clearly, none of these examples appear pathological or contrived. From this initial glimpse, sometimes a simulated set of preferences yields a Condorcet winner and sometimes it does not. In the case of five voters and two dimensions, with ideal points on the two dimensions each drawn from a normal distribution, over thousands of simulations we find a simulated Condorcet winner 90 % of the time.

Borrowing from Erikson and Ghitza (2016), we can depict the results of simulations with a "basic" model of issue preferences as the number of legislators and the number of issue dimensions expand. Figure 2 presents an informative picture based on distributions where policy dimensions are orthogonal (uncorrelated with one another, and simulated policy positions are all drawn from a normal distribution with mean zero, variance one. The figure summarizes the likelihood of a Condorcet winner for each odd number of voters, through 107, and various numbers of dimensions up through 50, each result based on thousands of one-on-one tournaments.

The first thing to notice from Fig. 2 is that under all simulated scenarios, a Condorcet winner is more likely than not. Second, the likelihood of a Condorcet winner based on the number of voters is curvilinear, generally declining beyond three voters (for whom a Condorcet winner is structurally induced to be 100 %

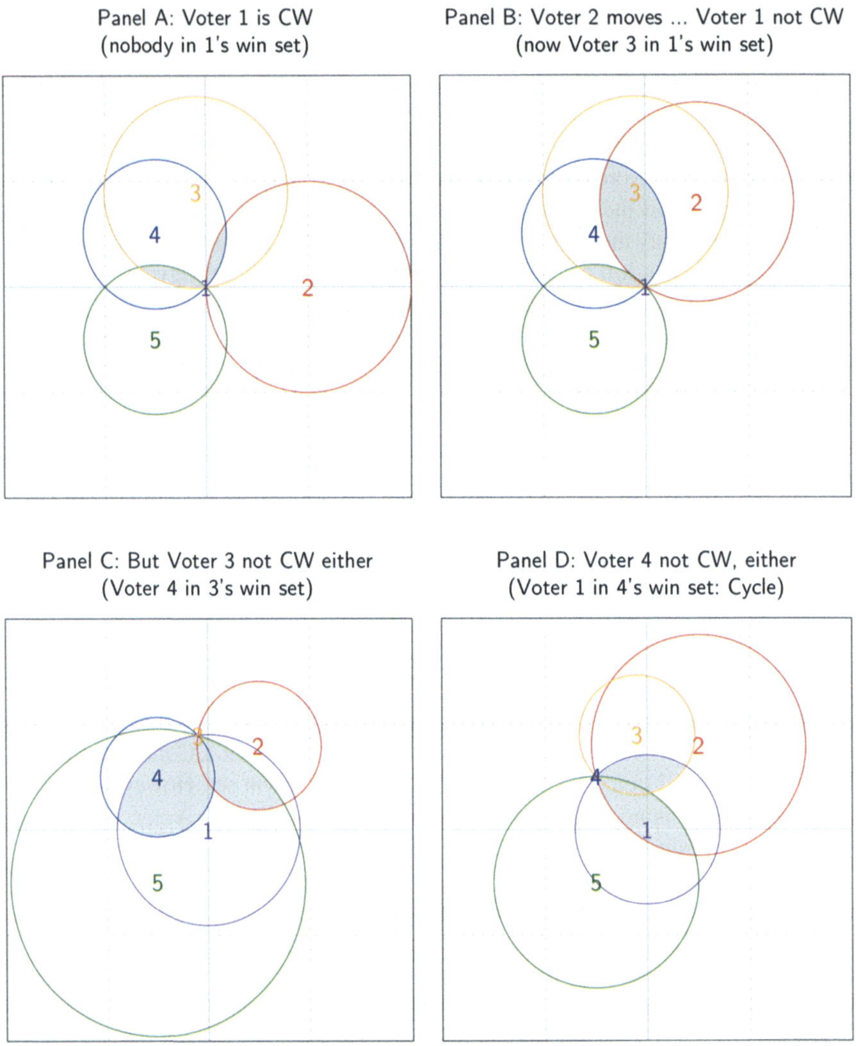


**Fig. 1** Simple example of a Condorcet winner (Panel A), and a case where there is no Condorcet winner due to cycling (Panels B–D)

likely) and then reaches a nadir after which the likelihood increases again. In the range where the number of voters approaches the number in a legislature, the likelihood of a Condorcet winner increases with legislative size. Further, for any number of legislators, and obviously excepting the one-dimensional example, the likelihood of a Condorcet winner increases with the number of dimensions. In general, the more the model gets complicated by high $N$ and high $K$, the more likely is a Condorcet winner. This may strike the reader as a counterintuitive result. These claims are true not only for the basic model depicted in Fig. 2 but also the

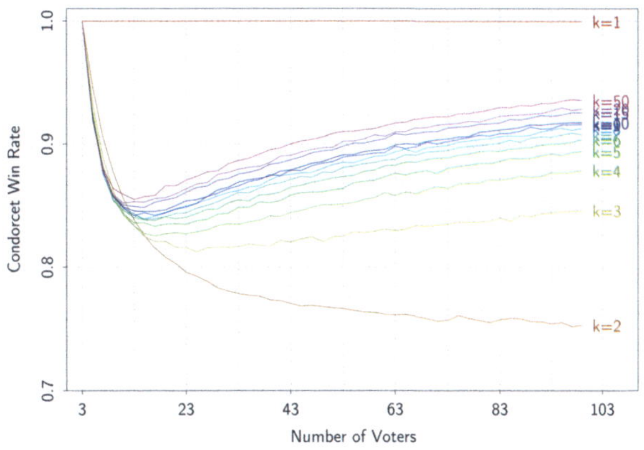


**Fig. 2** Simulated ideal points are drawn from a K-variate normal distribution. For each value of N and K, we draw 100,000 sets of simulated data. For each simulation, we run a tournament in the form of a series of pairwise contests, to determine if any legislator is a Condorcet winner

various alternative models employed in Erikson and Ghitza (2016) involving the shapes of distributions, nonorthogonality of dimensions, weighting of dimensions for salience, and the like.

One oddity from the simulations is that with exactly two dimensions, the likelihood of a Condorcet winner is not only the lowest with two dimensions but also that for only two dimensions the likelihood of a Condorcet winner decreases with the number of voters. For any number of voters, the likelihood of a Condorcet winner declines with the number of voters. At least this is true over the observed range up to 107 voters. Would, for instance, be likelihood of a Condorcet winner decline further for larger N's, such as the 435 members of the US House of Representatives?

The delicate result for two dimensions is based on contrived simulations rather than real world data. Suppose we consider observations of actual two-dimensional policy positions by actual legislators. What would be the likelihood of finding a Condorcet winner? This is the motivation for turning the analysis to the US Congress.

## 2 The Plan

Our raw data consists of DW-Nominate scores (Poole and Rosenthal 1997) on the two major dimensions for every US House member and every US Senator serving at the beginning of each Congress from the first (1789–1780) to the 113th (2013–2014).[2] The data summarize the two-dimensional voting record of the legislators. Poole and Rosenthal calculate two-dimension scores from the roll call votes, by

[2]We use the "Common Space" DW-NOMINATE Scores, removing the President from the dataset but using all other legislators in the database. We multiply the second dimension by a scaling factor of 0.4113, as is recommended on the VoteView website where the data is available (http://voteview.com/dwnomin_joint_house_and_senate.htm).

choosing a set of scores that best predict the legislators' votes on individual bills. The first dimension is generally perceived to be ideological, as a standard left-right or liberal-conservative continuum. The degree of division between the two major parties has waxed and waned historically. In the current polarized climate, there is a strong division between Democrats and Republicans.

The key to making our task possible is the presence of a second dimension. This leads to depictions of Congressional policy space as an observable scatterplot. The interpretation of policy differences on the second dimension is sometimes mysterious, but often includes policies regarding race. As the second dimension is less important than the first, Poole and Rosenthal suggest giving it less weight. They suggest specifically that with the first dimension having a standard deviation of one, the second dimension should be weighted .411 of the first. We adopt this convention for the analysis below.

For each congressional chamber for each Congress, we conduct a round-robin tournament based on DW-Nominate scores. The tournaments include $N*(N-1)/2$ mock two-candidate elections between each pair of legislators. For each contest, the winner is the one whose DW-Nominate position is closest in policy space to the most voters. Only if one legislator defeats all others by being closer to the majority of voters in every contest, is there a Condorcet winner.[3]

We conduct this exercise not only for the entire House membership and the entire Senate. We also repeat the exercise one more time for each major party in the House and Senate. The analysis of parties is limited to parties with at least ten members for the Congress.

There are several ways to evaluate the results of these fantasy tournaments. The main motivation is to observe the frequency of Condorcet winners in a recognizable set of real world settings. For instance, does the likelihood of a Condorcet winner decline with the number of voters? Can anything be learned from which tournaments lead to failures to obtain a Condorcet winner? A second motivation is one of curiosity regarding the identity of Condorcet winners, particularly within each party. Are the winners recognizable as typical of their party? Do they include actual party leaders? Could the Condorcet winners include actual leaders (Speakers of the House, majority and minority party leaders)?[4]

[3]In the world of electoral theory, elections have an odd number of voters to avoid ties. For the instances in Congress when $N$ is even rather than odd, there can be ties in the tournament. Ties are treated like losses in our analysis. To be scored as a Condorcet winner, one cannot tie any other voter in the popularity contest based on ideological proximity.

[4]By custom, US House speakers do not vote on roll calls except for breaking ties. One could, however, observe whether Speakers are centered as Condorcet winners in Congresses preceding their ascension to the speakership.

## 3 Results: The Frequency of Condorcet Winners

Our major findings can be summarized in Fig. 3, which shows the frequency of a Condorcet winner over time in both the House and Senate—for each chamber at-large and separately for each major party—today's Democrats and Republicans plus the early Whigs and Federalists. The early Democrat-Republicans [or Republicans] for the earliest Congresses are shown separately from today's Democrats.

The summary of Fig. 3 is as follows. For the full Senate, 59 % of Congresses contain a Condorcet winner, which is roughly equivalent to the frequency with the simulated data as the $N$ approaches 100. In contrast, the full House of Representatives contains a Condorcet winner in "only" 23 % of the Congress. This sharp contrast cannot be due to differences in the distributions of positions along the two dimensions, since for any Congress the distribution of DW-Nominate scores in the two dimensions generally mimic each other. Thus the difference is likely due to the fact that the House has always contained more members than the Senate throughout history.

For the parties in the House and Senate, the frequencies of a Condorcet winner are all in the middle of the range. For the Democratic and Republican parties, the frequencies with which they have Condorcet winners vary only in the narrow range from 49 to 66 %. (The shorter tenures of the other parties that are shown produce

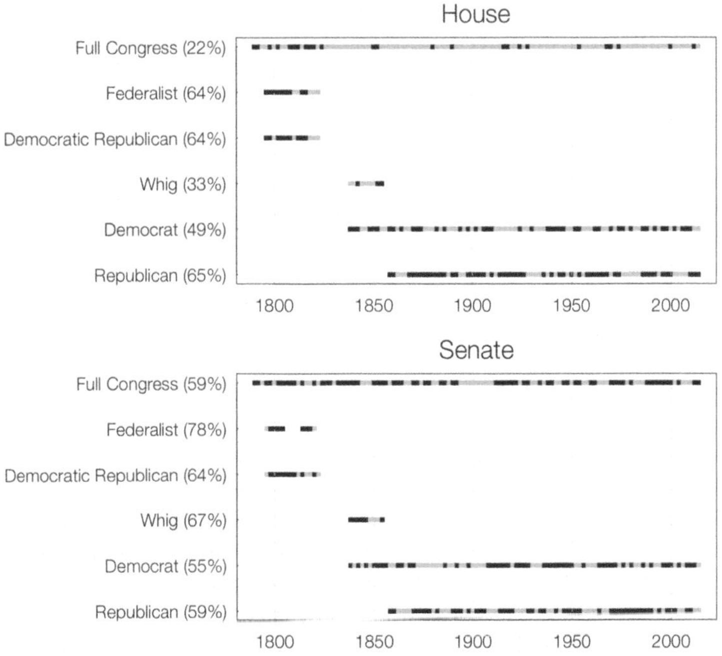


**Fig. 3** Frequencies of Condorcet winners by Congress and party over time

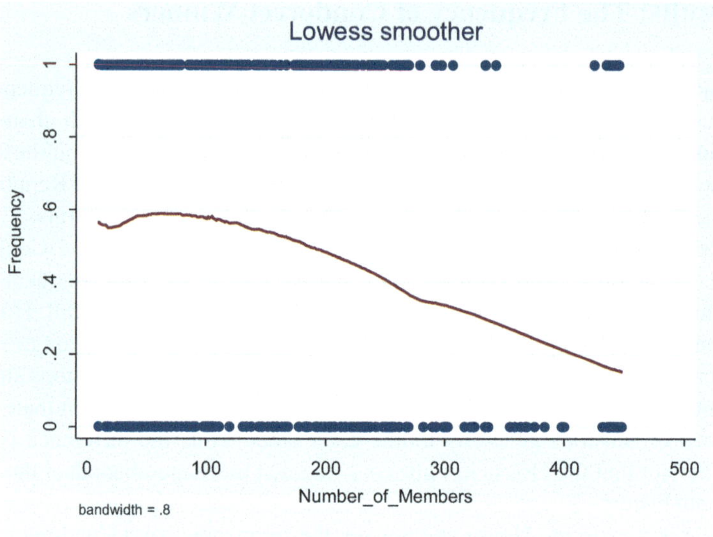


**Fig. 4** Frequency of Condorcet winners by size of membership, based on full Senate chambers, full House chambers, Democrats both in House and Senate, and Republicans both in House and Senate

more wobble.) Here, size makes no obvious difference. For instance, there is no appreciable difference between the frequencies of the majority and minority parties for the Congress. More tellingly, the House and Senate parties have nearly identical frequencies even though the House parties are about 4 times larger. Perhaps only when the *N* grows to the neighborhood of 400 does the size of the legislature appreciably matter for the possibility of a Condorcet winner.

Some purchase is provided by Figs. 4 and 5. Figure 4 pools all the data for the full House, the full Senate, and each of the Democratic and Republican parties within each chamber. With this data pooled, the case could seem compelling that the likelihood of a Condorcet winner declines with time. On the other hand, Fig. 5 suggests a different story when the different series are treated separately. The House had more Condorcet winners in its earlier days when the membership was larger. Visually the same pattern of decreased frequency with greater membership appears for both the Senate and congressional parties. However, neither trend appears to be statistically significant.

The data suggest nuance. Based on our analysis of real data, we can tentatively add confirmation to the findings from simulations that for two-dimensions (but for two dimension only—see Erikson and Ghitza 2016), the likelihood of a Condorcet winner decreases as the number of legislators (all potential leaders) increases in size. At the same time, the variation by specific institution would seem to confirm that the shapes of the distributions matter as well.

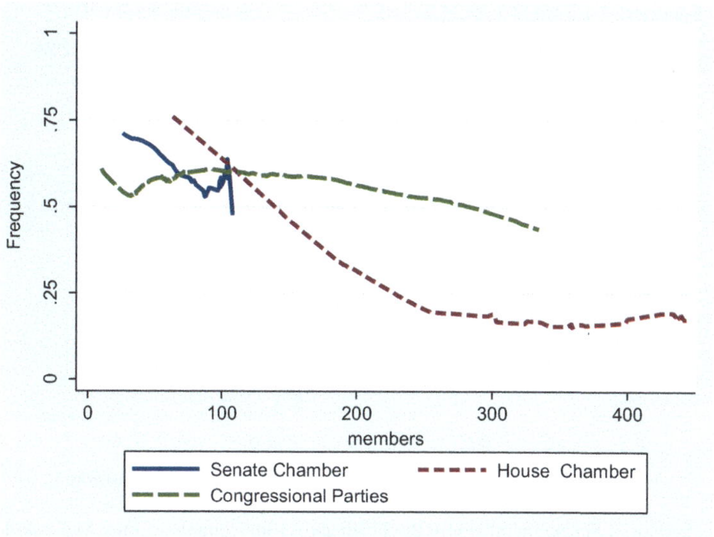


**Fig. 5** Frequency of Condorcet winners (Lowess fit) by membership size in Congress, 1st to 113th Congress. Separately for Full Senate, Full House, plus House and Senate Democratic and Republican Congressional parties (ten or more members)

## 4 Who Are the Condorcet Winners?

We can ask who are the Condorcet winners? For the recent 113th and 114th Congresses, Figs. 6 and 7 present scatterplots of DW-Nominate scores for the House and Senate, with Condorcet winners identified. For the Senate, Senator Claire McCaskill, Democrat from Missouri, was the Condorcet winner for both the 112th and 113th Congress, then under Democratic Party control. For the 113th Congress, the House had no Condorcet winner. However for the 112th Congress, Representative Don Young (Alaska) was the Condorcet winner. Note how these Condorcet winners emerge despite the very polarized setting. It is not obvious why there is no House "winner" for the 113rd Congress. The DW-Nominate scatterplot for the 113rd Congress House looks to the naked eye to be very similar to the similarly polarized distribution for the House in the 112th Congress.

Figures 8 and 9 show scatterplots of DW-Nominate scores separately for each party and chamber, eight instances in all (four party-House combinations times two Congresses), with Condorcet winners identified. Senate Democrats have a Condorcet winner only in the 112th Congress while the Republicans do so only for the 113th. In the House, Representative Goodlatt (VA) is the Condorcet winner in both elections, while the Democrats have none. This makes two recent instances when the same person is the Condorcet winner in two separate Congresses (McCaskill for Senate Democrats, Goodlatt for House Republicans). As we will see, there are many other instances of multi-Congress "winners." This is a sign that the

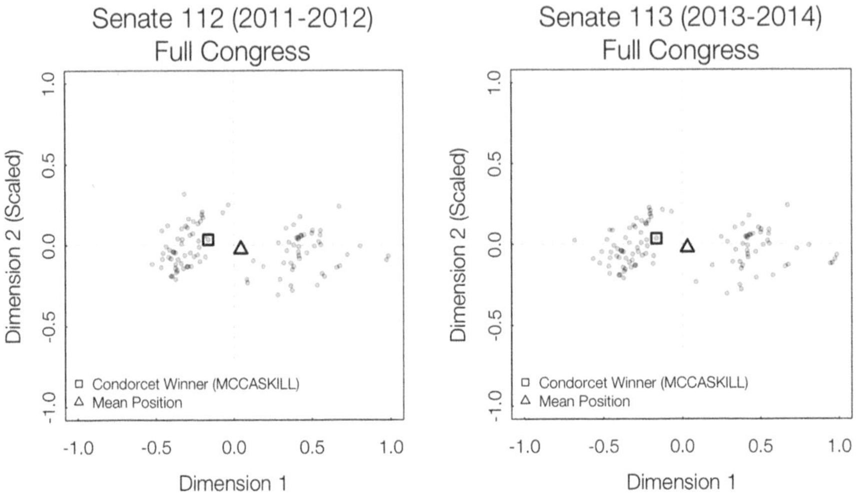


**Fig. 6** Scatterplot of Senate members for the 112th and 113th Congress. Claire McCaskill, Democrat from Missouri, was the Condorcet winner in both chambers. Mean position = intersection of means

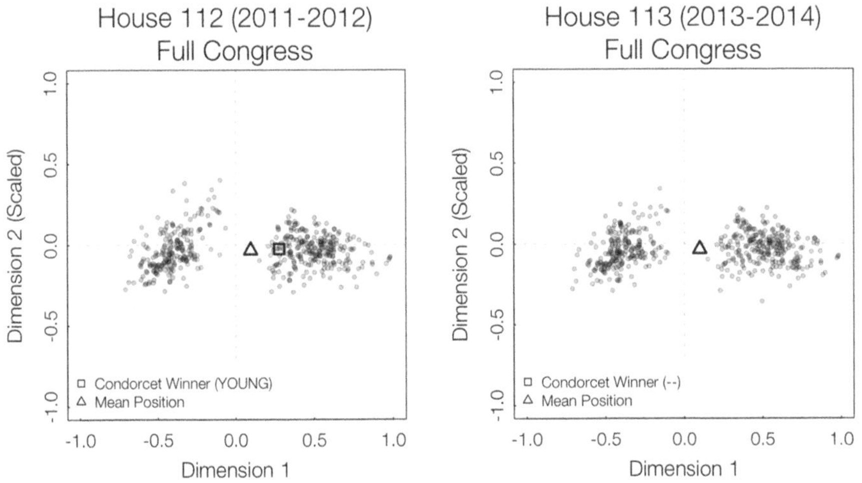


**Fig. 7** Scatterplot of House members for the 112th and 113th Congress. Republican Don Young (Alaska 1st District) was the Condorcet winner in the 112th Congress, but there was no Condorcet winner in the 113th, despite a similar set of members. Mean position = intersection of means

identity of Condorcet winners may be more than just a novelty exercise. Evidently there is some stickiness to being the center of one's congressional party.

Table 1 shows thirty Congress's worth of Condorcet winners for each House as a whole and for each party in each chamber. For the House, the "winners" are largely unremarkable and anonymous to the general public. For the Senate, however,

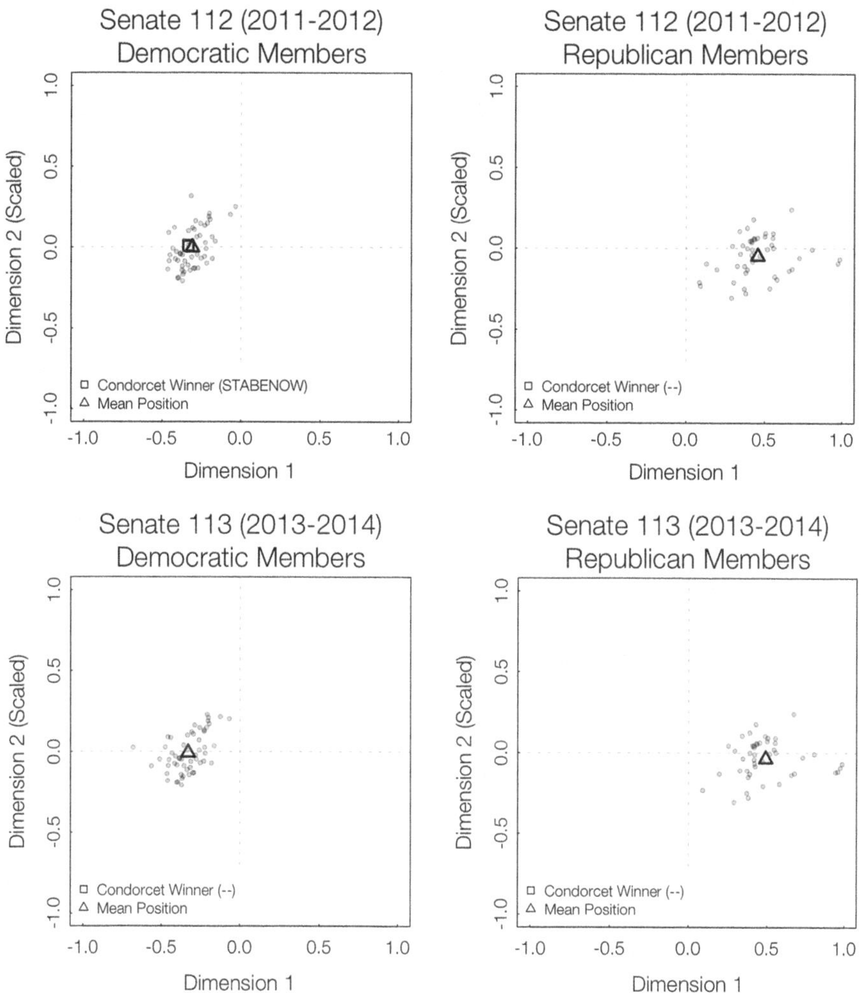


**Fig. 8** Scatterplots and Condorcet winners for Democratic and Republican Senate caucuses, shown separately. There were no party Condorcet winners for the 113th Congress. Mean position = intersection of means

Condorcet winners tend not only to be repeaters, but are also mostly well known, in the public spotlight. For example, future Vice Presidents Al Gore and Joe Biden are notable Democrats on the list. Biden was the Condorcet winner of his Senate party a record setting 5 times!

On the Republican side the familiar names Bob Dole, and Bill Frist make multiple appearances. Each once served as their party's majority or minority leader in the Senate. Dole in fact was the Condorcet winner among Republican Senators 3 times. Earlier (not shown in Table 1), Howard Baker, a future Republican leader, was

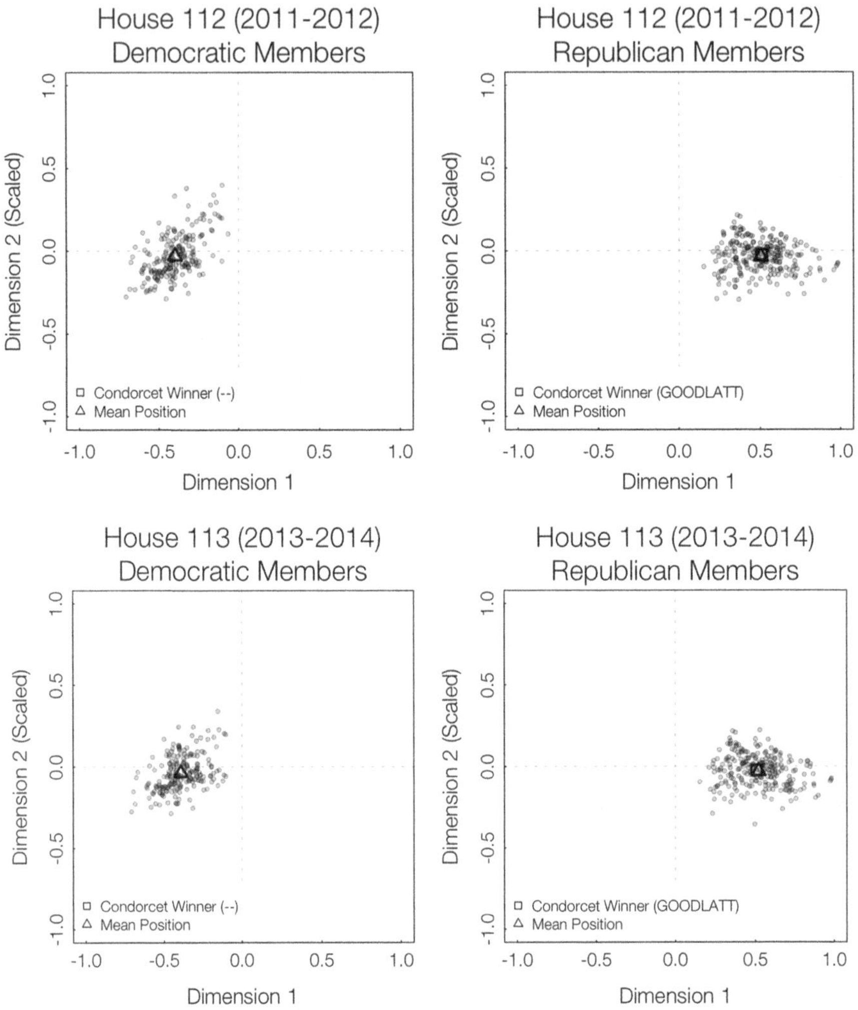


**Fig. 9** Scatterplots and Condorcet winners for Democratic and Republican House caucuses, shown separately. There was not a Condorcet winners among the Democrats. Mean position = intersection of means

also a Condorcet winner. Joining them as a Condorcet winner among Democrats is Tom Daschle—who also went on to lead his Senate partisans.

Of the eight Senators who have served as party leader between the early 1980s and 2013, four have been Condorcet winners for their party. (The four exceptions are McConnell, Byrd, Mitchell, and Reid). Moreover, Leaders Baker, Dole, Frisk, and Daschle all had been Condorcet winners within their party prior to rising to their leadership position. What should we make of this? For most but not all recent contests for party leader, the elected leader had been near the party's center

**Table 1** Condorcet winners by chamber and party, over 40 years, 1975–2014

| Congress | Full Congress | Democrat | Republican |
|---|---|---|---|
| *Senate* | | | |
| 94 (1975–1976) | McGee, WY (D) | Magnuson, WA | Roth, DE |
| 95 (1977–1978) | – | – | Roth, DE |
| 96 (1979–1980) | Baucus, MT (D) | Magnuson, WA | Roth, DE |
| 97 (1981–1982) | – | – | Simpson, WY |
| 98 (1983–1984) | – | – | Dole, KS |
| 99 (1985–1986) | – | Gore, TN | Dole, KS |
| 100 (1987–1988) | Baucus, MT (D) | – | Simpson, WY |
| 101 (1989–1990) | Robb, VA (D) | Daschle, SD | Dole, KS |
| 102 (1991–1992) | Lieberman, CT (D) | – | – |
| 103 (1993–1994) | Campbell, CO (D) | – | Murkowski, AK |
| 104 (1995–1996) | D'Amato, NY (R) | Biden, DE | – |
| 105 (1997–1998) | Smith, OR (R) | Biden, DE | Frist, TN |
| 106 (1999–2000) | Smith, OR (R) | Biden, DE | Frist, TN |
| 107 (2001–2002) | – | – | Roberts, KS |
| 108 (2003–2004) | Coleman, MN (R) | Biden, DE | – |
| 109 (2005–2006) | – | – | – |
| 110 (2007–2008) | – | Biden, DE | Isakson, GA |
| 111 (2009–2010) | – | Stabenow, MI | Isakson, GA |
| 112 (2011–2012) | McCaskill, MO (D) | Stabenow, MI | – |
| 113 (2013–2014) | McCaskill, MO (D) | – | – |
| *House* | | | |
| 94 (1975–1976) | – | Traxler, MI-8 | – |
| 95 (1977–1978) | – | – | – |
| 96 (1979–1980) | – | Cotter, CT-1 | – |
| 97 (1981–1982) | – | – | – |
| 98 (1983–1984) | – | – | – |
| 99 (1985–1986) | – | McCloskey, IN-8 | Roth, WI-8 |
| 100 (1987–1988) | – | McCloskey, IN-8 | Hopkins, KY-6 |
| 101 (1989–1990) | – | – | Hopkins, KY-6 |
| 102 (1991–1992) | – | McCloskey, IN-8 | Roth, WI-8 |
| 103 (1993–1994) | – | – | – |
| 104 (1995–1996) | – | Wynn, MD-4 | McInnis, CO-3 |
| 105 (1997–1998) | – | Lantos, CA-12 | McInnis, CO-3 |
| 106 (1999–2000) | Smith, NJ-4 (R) | – | Isakson, GA-6 |
| 107 (2001–2002) | – | Lantos, CA-12 | – |
| 108 (2003–2004) | – | – | – |
| 109 (2005–2006) | – | Lantos, CA-12 | – |
| 110 (2007–2008) | – | Wynn, MD-4 | – |
| 111 (2009–2010) | – | Miller, NC-13 | McCaul, TX-10 |
| 112 (2011–2012) | Young, AK-1 (R) | – | Goodlatt, VA-6 |
| 113 (2013–2014) | – | – | Goodlatt, VA-6 |

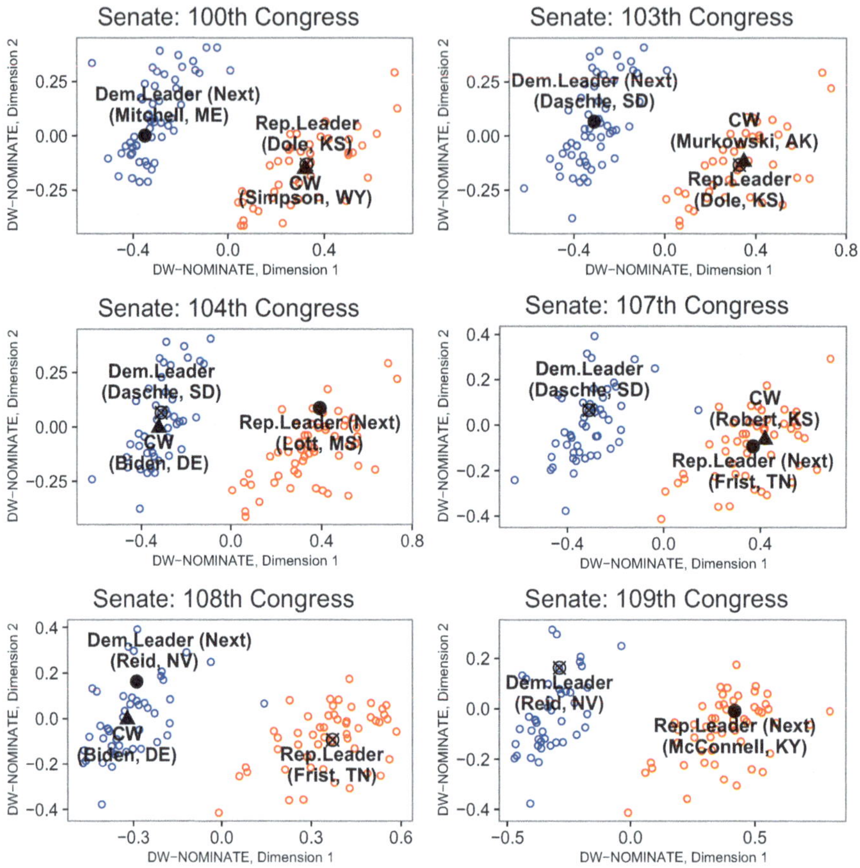


**Fig. 10** Two dimensional DW-nominate scores for the US Senate in recent Congresses immediately prior to the selection of a party leader. Shown are the positions of the future leaders in the following Congress, the leader of the other party who continues for the following Congress and the Condorcet winners (if any) for each party

in the Congress prior to their ascension to the leader position. Figure 10 depicts the locations of future Senate party leaders in the Congress directly prior to their selection.[5]

[5]In the congressional literature, most treatments of Senate parties (e.g., Campbell et al. 2002) assume that Senate parties chose caucus leaders who are close to the center of their party, which in the one-dimension case means the median position. In what is probably the most extensive examination of the ideological positioning of Senate party leaders, Hatcher (2010) finds that Senate leaders are at near the party center immediately prior to election and, if the party is in the majority, moves toward the party's extreme. One should always be aware that the ideological positioning of congressional leaders is induced by strategic considerations as well as personal ideological preferences (Roberts 2007).

One could imagine that when a party's senators vote for a new leader they engage in a mental tournament in their heads based on their understanding of each others' signature positions on two dimensions of roll calls when they pick party leaders. If the party contains a Condorcet winner, this process would lead to a decisive winner. Of course the process does not exactly work this way.[6] However, we could imagine their anticipation of tournament outcomes. Being the Condorcet winner in two dimensions has a similar stature to being the median voter in one dimension. When choosing from a fixed set of alternatives defined as voter ideal points, the Condorcet winner is as much the decisive winner as is the median voter in one dimension.

Our central point of this paper has been that even in two-dimensional settings such as distributions of DW-Nominate scores, a Condorcet winner often exists. It is remarkable that for the modern Senate, they not only exist on paper, but are often found to be the actual party leader.

## 5 Summary and Conclusions

The question has been as follows. Suppose a group of people such as a legislative chamber or a legislative party grapple with the task of electing a leader with the voters mainly concerned with the closeness of the policy result to their ideal points in policy space. This policy space has multiple dimensions. As we normally think of this problem, candidates for leaders can propose policy outcomes across the full range of continuous policy space, with the result that there is no equilibrium. When the process is modeled on paper, the result is endless cycling. However, let us impose two conditions. First, each participant shares a common knowledge of each other's policy preferences, so that there is no relevant private information. Second, the chosen leader has sufficient power to lack the incentive to bargain over policy space once in office. To simplify, think of the leader as a dictator. Since each potential leader would be a dictator once elected, their promises to promote any position other than their personal policy agenda is not credible. Given these conditions, a Condorcet winner *might* exist. In a set of paired contests with each other legislator to see who gets more votes in a contest based on policy proximity, there is considerable likelihood that there is a winner who defeats all potential challengers. In simulations of imaginary data of legislative preferences, a likely winner is found more often than not, but with the least likelihood with two dimensions and many voters (Erikson and Ghitza 2016).

[6] In all instances of our analysis, Condorcet winners are close to the center of gravity, as indicated by the intersection of means on the two dimensions. They are also close to the observed intersection of medians on the two dimensions (which of course are subject to change with the rotation of the axes). If a legislative caucus wants an automatic decision rule for deciding the leader based on ideological centrality, and they had an ideological mapping at hand such as DW-Nominate scores, the intersection of the means or the medians would be available.

This setup provides the backdrop for this chapter. This chapter explores the likelihood of a Condorcet winner in the US Congress. Congressional roll call voting is typically modeled in two dimensions with many members (currently, 100 in the Senate and 435 in the House). We consider Poole and Rosenthal's' DW-Nominate scores for the two dimensions as our data and search for Condorcet winners across congressional history. For the US House as a whole, Condorcet winners are rare, only about in one fifth of the Houses and these are mainly in the early years when the House had fewer members. For the Senate as a whole, and for each party in the House and the Senate, we find a Condorcet winner about half the time or a bit more. Thus, if congressional party caucuses were to select a winner solely based on who was closest to their views, there would be a clear winner at least half the time.

Over recent history, it turns out that half of the actual party leaders in the Senate had, been the Condorcet winner in at least one Congress prior to their rise to the leader position. This fact helps to make our findings more than a modeling curiosity. Congressional parties might not chose their leaders in the way we model the process here, but the actual selections often approximate the results of our models.

## References


Baron, D. P., & Ferejohn, J. A. (1989). Bargaining in legislatures. *American Political Science Review, 83*(4), 1181–1206.

Black, D. (1958). *The theory of committees and elections*. Cambridge: Cambridge University Press.

Campbell, A., Cox, G. W., & McCubbins, M. D. (2002). Agenda power in the U.S. Senate, 1877 to 1986. In D. W. Brady & M. D. McCubbins (Eds.), *Parties, procedure, and policy: Essays on the history of congress*. Stanford: Stanford University Press.

Cox, G. W., & McCubbins, M. D. (2004). *Setting the agenda*. Cambridge: Cambridge University Press.

Dahm, M., & Glazer, A. (2015). A carrot and stick approach to agenda setting. *Journal of Economic Behavior & Organization, 116*, 465–480.

Erikson, R. S., & Ghitza, Y. (2016). *Electing the agenda setter*. Unpublished Paper.

Hatcher, A. C. (2010). A man for all seasons: Partisan constraints on U.S. Senate majority leaders. *Party Politics, 16*, 323–346.

Krehbiel, K. (2004). Legislative organization. *Journal of Economic Perspectives, 18*(1), 113–128.

Mayhew, D. (1966). *Party loyalty among congressmen: The difference between democrats and republicans, 1947-1962*. Cambridge, MA: Harvard University Press.

McKelvey, R. (1976). Intransitives in multidimensional voting models and some implications for agenda control. *Journal of Economic Theory, 12*, 432–442.

Plott, C. (1976). Axiomatic social choice theory: An overview and interpretation. *American Journal of Political Science, XX*(3), 511–596. (Reprinted in *Social Choice Theory*, vol. I, pp. 231–316, by C. K. Rowley, Ed., April 1993, UK: Edward Elgar)

Poole, K., & Rosenthal, H. (1997). *Congress: A political-economic history of roll call voting*. Oxford: Oxford University Press.

Roberts, J. M. (2007). The statistical analysis of roll call data: A cautionary tale. *Legislative Studies Quarterly, 32*, 341–360.

Romer, T., & Rosenthal, H. (1978). Political resource allocation, controlled agendas, and the status quo. *Public Choice, 33*, 27–43.

Shepsle, K. A., & Weingast, B. R. (1987). The institutional foundations of committee power. *American Political Science Review, 81*(1), 85–104.
Weingast, B. R., & Marshall, W. J. (1988). The industrial organization of congress; or, why legislatures, like firms, are not organized as markets. *Journal of Political Economy, 96*, 132–163.

# The Duverger-Demsetz Perspective on Electoral Competitiveness and Fragmentation: With Application to the Canadian Parliamentary System, 1867–2011

J. Stephen Ferris, Stanley L. Winer, and Bernard Grofman

**JEL Classification** D7, D4

## 1 Introduction

In this paper we combine consideration of Duverger's Law (1954) with Demsetz's (1968) theory of natural monopoly to provide a novel perspective on the measurement of electoral competitiveness in a single member district, plurality rule electoral system. In the Duverger-Demsetz view, as we shall refer to it, an increase in the effective number of parties, measured using the inverse of the Hirschman-Herfindahl index of the concentration of candidate vote shares, above the long run level of 2 predicted by the 'Law' for single member plurality rule electoral systems, signals a departure from equilibrium and a *decline* in the degree of electoral competition. This runs contrary to the view, sometimes expressed in empirical studies of elections and public policy, that more candidates or parties, each with smaller vote shares reflects a more competitive environment. Consider, for example, the following recent quote

J.S. Ferris
Department of Economics, Carleton University, 1125 Colonel By Drive, Ottawa, ON, Canada K1S 5B6

S.L. Winer (✉)
School of Public Policy and Administration and Department of Economics, Carleton University, 1125 Colonel By Drive, Ottawa, ON, Canada K1S5B6
e-mail: stanley.winer@carleton.ca

B. Grofman
Department of Political Science, University of California, Irvine 3151 Social Sciences Plaza, Irvine, CA 92697–5100, USA



M. Gallego, N. Schofield (eds.), *The Political Economy of Social Choices*, Studies in Political Economy, DOI 10.1007/978-3-319-40118-8_5

that embraces this perspective while also touching upon other ways of thinking about competition:

> Electoral contestation may be defined as the degree of election-based competition in a political unit. Where contestation is minimal there is little organized opposition, and the incumbent party captures most of the votes and seats. Where contestation flourishes there are more competitors than available seats, a tight race for votes and seats, and frequent turnover in control. Contestation implies ex ante electoral uncertainty (Gerring et al. 2015, p. 574).

The suggestion here that greater fragmentation of vote shares signals more competition is analogous to the conclusion usually reached about an industry when the concentration of output across firms declines. But an electoral system is not a private goods market.[1]

To fix ideas as we proceed, we illustrate the concepts and associated indexes of competitiveness discussed for the history of the Canadian parliamentary system. In doing so we find qualified support for the Duverger-Demsetz perspective on the measurement of electoral competitiveness. This support appears in the inverse co-movement uncovered between fragmentation and a new index of the contestability of elections, as we measure these dimensions of the Canadian electoral system using the history of regular national elections from the first election in 1867 to the forty-first election in 2011.

Strictly speaking, our analysis of the conceptual issues and its empirical application to Canada applies only to a single member district, plurality rule (SMP) electoral system. However, the argument may also apply to winner take all systems that use forms of voting in which there is only one round, such as the alternative vote system. Towards the end of the paper we consider whether the argument can be applied to systems of proportional representation.

We begin in Sect. 2 of the paper by considering in some detail how increased fragmentation may be mistaken for a signal of greater competitiveness when competition is associated with electoral uncertainty and, as a practical matter, indexed by the closeness of electoral contests. This is followed in Sect. 3 by development of the Duverger-Demsetz perspective on electoral competitiveness, in which party governance in an SMP system is interpreted as being analogous to the management of a natural monopoly. Section 4 presents evidence consistent with the Duverger-Demsetz view using data from the history of Canadian parliamentary

[1]Empirical work that uses the effective number of parties (ENP) to measure competitiveness or a closely related index of party or vote fragmentation, such as $1 - 1/\mathrm{ENP}$ or one minus the winner's vote share $v_1$, includes: Ashworth et al. (2014); Alfano and Baraldi (2015)—who use a normalized Hirschman-Herfindahl (HH) index; Gerring et al. (2015)—who use $1 - v_1$ which is practically similar to the HH index of party fragmentation [for Canada from 1867 to 2011, the correlation of $1 - v_1$ with ENP is about 0.89 and with Rae's (1968) measure of fragmentation $1 - \mathrm{HH}$ it is about 0.95]. See also Ghosh (2010) for India, among others. We also note other work that associates party fragmentation with 'weak government' and increased public expenditure, a view that is complementary with, but distinct from the Duverger-Demsetz view that we shall develop in what follows. Examples of this literature include Roubini and Sachs (1989), Ricciuti (2004), Chhibber and Nooruddin (2004), Borge (2005) and Geominne et al. (2008).

elections. Here we consider how a new index of the contestability of elections relates to fragmentation of the electoral system. We show that as the effective number of parties rises above 2 and fragmentation increases, competitiveness as judged by our contestability index *declines*. Section 5 discusses the problems of extending the analysis to proportional systems, and Sect. 6 concludes.

Before continuing, it is important to note that our basic interest in what follows is not with an explanation of the effective number of parties or with the testing of Duverger's Law. We are interested in what is meant by political competition and how it can be measured. Competitiveness, as distinct from a given state of perfect or imperfect competition, is undoubtedly complex and difficult to study. In addition to (1) the entry and exit of candidates and parties and (2) the rivalry between them in an election, political competition in an electoral system also includes (3) competition among parties in the legislature between elections, (4) competition among governments and bureaus, and (5) relationships among these forces. In this paper we focus on particular aspects of the first and second dimensions of the process.

## 2 From the Hirschman-Herfindahl Index of Market Concentration to Uncertainty and Closeness in Elections

A key aspect of economic competition that lies behind the often used Hirschman-Herfindahl (HH) index of economic competitiveness (Herfindahl 1950; Hirschman 1945) concerns the ability of firms to affect market price. To the extent that individual firms are unable to influence market prices, the firm has no market power and the industry is said to be highly or perfectly competitive. This feature of a competitive market is usually translated into an index of competitiveness through the logic that if there are more firms, each of which supplies a smaller share of market demand, the ability of any individual firm to influence the market price will be reduced. The HH index is designed to encapsulate this logic and is defined as the sum of the squared market shares of the firms in an industry. That is,

$$HH_j = \sum_{n=1}^{N} s_{nj}^2, \tag{1}$$

where $s_{ij}$ is the output share of the ith of N firms in industry j. The HH index will equal one if one firm supplies the entire market and will approach zero as the number of firms increase and each firm's market share declines.

The Hirschman-Herfindahl measure of concentration has crossed over into political science as the effective number of candidates or parties (ENP), defined as one over an HH index constructed using candidate or party vote (or seat) shares

(Laakso and Taagepera 1979). The national average ENP in election $t$ with $j = 1, 2, \ldots, J$ constituencies is defined as

$$ENP_t = \sum_{j=1}^{J} \left(ENP_{jt}\right) / J, \tag{2}$$

where $ENP_{jt} = 1/\sum_{i=1}^{I} v_{ijt}^2$ is the effective number of candidates at the constituency level and $v_{ijt}$ is the vote share of candidate $i$ in constituency $j$ in election $t$. Alternatively, we could employ the vote or seat shares of political parties at the national level to define a national party-based analogue to (2). In either case ENP will equal 2 if two candidates or parties equally share the vote and will rise as the number of candidates or parties increases and their individual vote shares decline.[2]

A closely related measure of the fragmentation or fractionalization of the electorate was proposed by Rae (1968). His measure, *Fragmentation,* is defined as 1 – HH or, equivalently, 1 – 1/ENP, with ENP as in (2) above. At the constituency level, this index can be thought of as a measure of the probability that two randomly chosen individuals will not share the same partisan association. It approaches 1 as voters become more fractionalized.

Both ENP and *Fragmentation,* defined as national averages over constituency level values based on vote shares, are shown in Fig. 1 below for Canadian national elections 1 through 41 (1867–2011). (We shall ignore the middle line in the figure for now.) Both indexes indicate increasing fragmentation over the entire history, though *Fragmentation* appears to flatten out after the 25th election.[3] Here and in subsequent figures, elections during world wars and the great depression (1917–1945) are shaded, and note is also made of the extraordinary 35th election in 1993 when the incumbent party in power (the Conservatives) lost (to the Liberals), falling from 169 seats out of 295 to just 2.

[2]For extensive discussion of ENP see Taagepera (2007) and Borooah (2002). Gaines and Taagepera (2013) consider some of the problems associated with the use of ENP to measure the number of parties.

[3]Many early Canadian elections featured acclamations, particularly in elections 1 (1867: 46/181), 2(1872:51/200), 3(1874:54/206) and 13(1917:31/235). In these cases, we set $v_1 = 1$ in a constituency with an acclamation in the case of ENP defined for constituencies. The presence of 2-seat constituencies (123 before the 28th election in 1968) makes little difference to the averages over all constituencies that form the basis for the measures discussed in this section. It is interesting to note in this respect, however, that for one-seat constituencies, the mean for all elections of the constituency level ENP = 2.4, while for the two-seat constituencies, mean ENP = 4.0. We also note that there are 10,247 regular individual constituency elections in elections 1 through 41. The maximum number of candidates in any one constituency election from 1867 to 2011 (general elections 1–41) is 13. The maximum number of parties in any one election (taking self-named parties as a party without judgement of its success), independent candidates and candidates of unknown affiliation is 27, in the 19th election in 1940. On the party system in Canada up to 1908, see recently Godbout and Hoyland (2013). ON ENP in Canada, see Johnston and Cutler (2009).

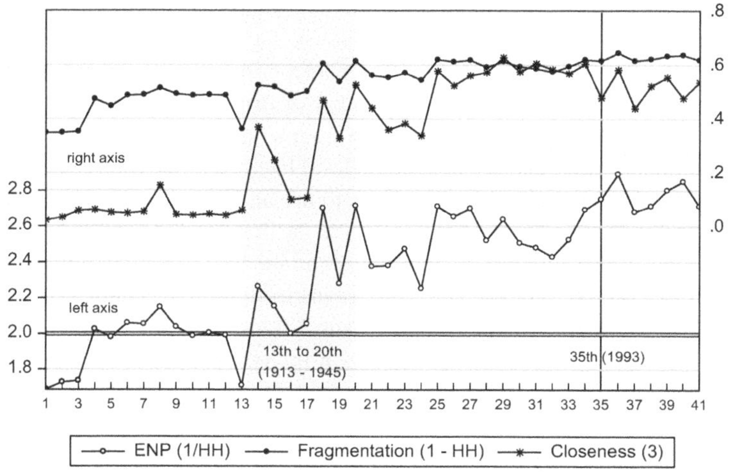


**Fig. 1** ENP, fragmentation and closeness (3), Canadian General Elections 1–41, 1867–2011

To a considerable extent the use of the HH index in political science derives from the desire to test the predictions by Duverger (1954) about the effective number of candidates and parties that will arise in a long run electoral equilibrium.[4] Duverger argues that in a single member district, plurality rule electoral system, the number of political parties at the district or constituency level tends towards 2 in the long run.[5] Cox (1997, p. 271) interprets this as an upper bound on what he refers to as the carrying capacity of the electoral system. In a majoritarian parliamentary system, in Duverger's view, factions are forced together into two parties *before* the election by the winner take all aspect of the electoral system. This contrasts with the formation of coalition government *after* the election in a system of proportional representation. Cox (1997, p. 30) attributes the Law to elites—opinion leaders, contributors, party officials, etc—who do not want to waste their influence on hopeless candidates, and to strategic choices made by individual voters for the same reason, with uncertainty in the process introduced by the problems for elites and voters of coordinating to decide who is, and who is not, a serious candidate.

It should be emphasized that Duverger's Law is in the first instance a statement about two party competition at the district or constituency level. Even if the Law holds there, the two parties competing at the local level may differ across regions, thus leading to more than two at the center. (See for example Riker 1982; Gaines 1999; Chhibber and Kolman 2004; Grofman et al. 2009, in addition to Cox and Taagepera). To go from localized two-party competition to national

[4]Cox (1997) and Taagepera (2007) provide extensive discussions of Duverger's work and references to the associated literature.

[5]The Law is not a point prediction, but a statement that there cannot typically be more than 2 candidates (Cox 1997, p. 271). So in the long run there could be less than 2.

competition between the *same* two parties requires additional assumptions. Despite this qualification however, Duverger's Law is far stronger than any result in economics concerning the number of firms in a competitive market equilibrium. In a perfectly competitive market, the number of firms is indeterminate. We shall return to the differences between indexes of fragmentation and competition defined at the constituency level and at the national level later. Here we wish to explain why it is tempting, though probably misleading, to use an HH-related index such as the average constituency value of ENP to measure the degree of electoral competition.

How could a rise in ENP or in *Fragmentation* be associated with increased electoral competitiveness despite the absence in elections of an analogue to a market price that can be manipulated by participants?[6] One influential argument is that fragmentation of vote shares may serve as a proxy for political competitiveness when competition is associated with electoral uncertainty. The reasoning begins with the view that as a practical matter, electoral uncertainty can be metered by the closeness of elections in terms of candidate or party vote shares. Then, because greater fragmentation of the party system often leads to splitting of the vote among the contenders for office, it usually results in closer and thus more uncertain and competitive election contests.

The idea that an election is competitive when its outcome is highly uncertain or 'too close to call' is a sensible one that is widely employed. See, for example, Franklin (2004), Blais and Lago (2009), Grofman and Selb (2009) and Besley et al. (2010) as well as the earlier U.S. literature on 'vanishing marginals' originating with Mayhew (1974). Whether fragmentation of vote shares is a useful proxy for electoral uncertainty via its correlation with closeness of electoral outcomes is a separate empirical matter. In the remainder of this section, we consider how concentration of vote shares at the constituency level and indexes of closeness are actually related for the history of Canadian general elections.[7]

## 2.1 *Electoral Uncertainty as Closeness in Elections and Its Relationship to Fragmentation*

One measure of the closeness of an election at the constituency or district level that appears in the literature is that proposed by Endersby et al. (2002). Their index of the closeness of the election in constituency $j$ in election $t$, $CL_{jt}(K)$, is

$$CL_{jt}(K) = K^K \prod_{i=1}^{K} v_{ijt}, \tag{3}$$

[6]See for example, Drazen and Eslava (2010), Aidt and Eterovic (2011), Aidt and Mooney (2014).

[7]There is perhaps some danger that we are setting up a 'straw person' in the rest of this section. However even if generalizing from the particular overstates our case, this exercise leads in interesting directions.

where *Π* denotes the product of terms following, $v_{ijt}$ is the vote-share of candidate *i* in jurisdiction *j in election t*, and *K* is taken by Endersby et al. to be equal to the integer value of the effective number of parties in the constituency, $ENP_{jt}$. $CL_{jt} = 0$ if there is an acclamation ($v_{1jt} = 1$); and it is $=1$ if *K* candidates have equal vote shares.

In our implementation of this index we set $K = 3$ because, historically, the sum of the first three vote shares (in elections 1–41) constitutes on average 0. 97 of the vote and has never been less than 0.90 of the vote with a small standard deviation of 0.026. As with ENP, this measure is aggregated up to the national level by averaging across constituencies.

It is important to note that because CL is designed to measure closeness, it must differ from ENP to some extent. For example, ENP treats the outcome (.5, .5; ENP = 2) as inherently different than (.33, .33, .33; ENP = 3), while both cases can be said to be examples of equally close or highly competitive elections and are treated as such by the CL index. Even so, it may still be that ENP as a measure of fragmentation may serve as a rough proxy for closeness defined by (3).

To see if that is so, Fig. 1 above also provides the national average values of CL($K = 3$) for Canadian parliamentary elections running from Confederation in 1867 until the 41st election in 2011 along with the corresponding national averages over constituencies of ENP and *Fragmentation*. Inspection of that figure, together with the correlations provided in Table 1, makes it apparent that fragmentation may indeed serve as a proxy for this measure of closeness in the Canadian case, provided we ignore the dissimilarities across indexes of short run election to election fluctuations.[8]

A second, somewhat more sophisticated measure of the closeness of elections turns out to be much less complementary to the use of fragmentation as a proxy. Following Mayhew (1974) and many others, this alternative measure of closeness embodies the idea that a close or competitive election is one in which the winning vote margin, ($v_1 - v_2$), is 'small'. What small means in this context is not unambiguous. However, any attempt to define it must take into account the potential for voting patterns to change, since a relatively small margin can be quite 'safe' (a concept which will play a key role in measuring electoral contestability later on) if the party's vote in that riding varies little across elections, while even a large margin may be unsafe in a constituency with many voters that switch their vote from one election to the next.[9]

[8]The downward trend in Closeness(3) after about the 30th election is not picked up by the fragmentation indexes.

[9]The idea of adjusting vote margins for volatility is analogous to standardizing scores by dividing the differences by a standard deviation. In the usual difference of means test, whether a difference is large or small is defined in terms of the normalized value of that difference in standard error units, with the standard error of the mean simply a specialized version of a standard deviation. In this way, any conclusion about the existence of a "meaningful" difference will reflect the level of uncertainty as to whether any observed difference might be due to chance alone.

**Table 1** Correlations

| | ENP | ENP12party | Fragmentation | Closeness(3) | $V_1 - V_2$ | $1 - v_1$ | $(v_1 - v_2)$/volatility | PS | AMS |
|---|---|---|---|---|---|---|---|---|---|
| ENP | 1.00 | 0.88 | 0.98 | 0.91 | 0.02 | 0.87 | −0.49 | −0.40 | −0.37 |
| ENP12party | 0.88 | 1.00 | 0.82 | 0.76 | 0.25 | 0.65 | −0.43 | −0.30 | −0.27 |
| Fragmentation | 0.98 | 0.82 | 1.00 | 0.91 | −0.11 | 0.92 | −0.44 | −0.39 | −0.39 |
| Closeness(3) | 0.91 | 0.76 | 0.91 | 1.00 | 0.18 | 0.73 | −0.57 | −0.56 | −0.44 |
| $V_1 - V_2$ | 0.02 | 0.25 | −0.11 | 0.18 | 1.00 | −0.47 | −0.27 | −0.36 | −0.18 |
| $1 - V_1$ | 0.87 | 0.65 | 0.92 | 0.73 | −0.47 | 1.00 | −0.30 | −0.18 | −0.26 |
| $(v_1 - v_2)$/volatility | −0.49 | −0.43 | −0.44 | −0.57 | −0.27 | −0.30 | 1.00 | 0.53 | 0.22 |
| PS | −0.40 | −0.30 | −0.39 | −0.56 | −0.36 | −0.18 | 0.53 | 1.00 | 0.27 |
| AMS | −0.37 | −0.27 | −0.39 | −0.44 | −0.18 | −0.26 | 0.22 | 0.27 | 1.00 |

*Definitions*:
*ENP* ENP calculated over candidates at the constituency level using vote shares (maximum of 13 candidates in any one constituency)
*ENP12party* ENP12party with ENP calculated at the national level using party vote shares for 12 parties—see the Appendix for definition of parties
*Fragmentation* $1 - HH = 1 - 1/ENP$, where HH is the Hirschman-Herfindahl index defined using vote shares
$v_i$ vote share of the candidate in the ith place
*Closeness(3)* an index of the closeness of candidates' vote shares $v_i$, assuming $ENP = 3$, as in Endersby et al. (2002)
$(v_1 - v_2)$*/volatility* the winner's vote margin $v_1 - v_2$ at the constituency level relative to historical volatility for that constituency
*PS* the Przeworski-Sprague (1971) volatility adjusted vote margin, defined by constituency and party
*AMS* an asymmetry adjusted measure of marginal seats, using an historical volatility and a one standard deviation test to define when an incumbent's seat is safe

Thus to reflect the relevant margin facing candidates, constituency vote margins must at least be adjusted for an estimate of the potential for voters to switch between parties from one election to the next. This point has been recognized for some time, highlighted in such works as Przeworski and Sprague (1971) whose index of closeness in elections is implemented immediately below, in Elkins (1974) who also discusses elections in Canada, and in Bartolini and Mair (1990), among others. It is important to note that the CL index is not adjusted for vote volatility, while in the Canadian case the adjustment matters significantly; the correlation of the average unadjusted margin, $(v_1 - v_2)$, with the volatility adjusted margin, $(v_1 - v_2)$ / *volatility*, over elections 1 through 41 (1867–2011) is − 0.27. (The calculation of *volatility* here is not without its own complications, and is discussed at length shortly.) That is, the correlation is *negative* as well as low. Hence in the Canadian case, the simple vote margin, which is widely used as a measure of closeness in elections, is unreliable as a measure of volatility adjusted closeness. We suspect that this problem with the unadjusted vote margin arises in many other cases.

The Przeworski and Sprague (1971) version of the volatility adjusted vote margin—hereafter, the PS index—is an especially interesting example of the class of measures of electoral uncertainty that incorporates vote volatility. As well as allowing for volatility, the PS index has embedded in it a specific view of the objectives of the losing candidates: namely that the primary objective of every candidate is to overcome his or her vote deficit vis a vis the incumbent.

To construct the PS volatility adjusted vote margin index for Canada, the lagged vote deficit faced by each party or candidate $p$ in constituency $j$ at election time $t$, $\left(v_{1jt-1} - v_{pjt-2}\right)$ is adjusted for the potential volatility of the vote to form the ratio $h^t_{pj}$:

$$h^t_{pj} = \frac{\left(v_{1jt-1} - v_{pjt-1}\right)}{Volatility_{st-1,i}}, \tag{4}$$

where volatility in the denominator is calculated across superconstituencies as $volatility_{st-1} = \sum_{p=1}^{12} \left|v_{pst-1} - v_{pst-2}\right| / 2$ and where for the incumbent, $p = 1$ and $h = 0$. Note that the ex ante-ex post issue always faced when using actual election outcomes is explicitly dealt with in (4) by using a lag of one election in both numerator and denominator.

Two particular issues that arise in calculating volatility in the denominator of (4) should be noted: (1) To allow for changes in constituency boundaries as constituencies are added and/or redrawn over time, it is necessary to derive vote and volatility measures for what we call superconstituencies. These are small aggregations of individual constituencies defined on an unchanging geographical basis. (We define 80 of these). We then assign to each party in each constituency in each election the average constituency level vote of that party in the superconstituency. This allows us to measure changes in votes across elections for each party, and thus to derive volatility measures despite the continual redistricting that has occurred over the decades; (2) A second issue that must be faced is how to define political parties. We also require that a 'party' win at least one seat in at least two general elections to be considered as such; thus the data are defined for 11 political parties (see the Appendix) plus an Other or residual category. Exploration of the consequences of using other definitions are left for the future, as are the effect of using alternative definitions of the superconstituencies.[10]

The following index for party $p$ in constituency $j$ is then calculated as:

$$c_{pj}^{t} = \begin{cases} 1 \ \text{if} \ 0 \leq h_{pj}^{t-1} \leq 1 \\ \frac{1}{h_j^{t-1}} \ \text{if} \ h_{pj}^{t-1} > 1 \end{cases}. \tag{5}$$

For a highly competitive party, $c = 1$ because the distance to go to become the winner is less than the floating vote or portion of the electorate that switched parties last time. Otherwise, the index is less than one and falling as the margin to be overcome by a party grows relative to volatility.

Aggregating across all the parties in each constituency $j$ using as weights the vote share of the party in the constituency gives:

$$C_j^t = \sum_{p=1}^{P} c_{pj}^t \ v_{pjt-1}. \tag{6}$$

$C_j = 0$ indicates no competition in the constituency and, accordingly, for ridings where there was an acclamations we set $C_j = 0$. National average competitiveness *across all constituencies* for each election, the completed PS index, is then given by the national weighted average

$$PS_{NJ}^t = \sum_{j=1}^{J} C_j^t \ adj\_vw_{jt}, \tag{7}$$

where $adj\_vw_{jt}$ is the adjusted (for acclamations) vote weight of each constituency in the national election.[11]

Figure 2 below shows the PS index along with ENP and *Fragmentation*. Each of the indexes exhibits a concave shape before the outbreak of the first world war, likely reflecting an increasing degree of competition as the number of acclamations declined sharply after the 3rd election, and then a declining degree of competition as the party system developed. But both ENP and *Fragmentation* show upward trends over the remainder of electoral history, while the PS index remains more or less

[10]After 1945, the issue of how to define a party is not problematical if one sticks to analyzing the major parties in Parliament plus a residual. Earlier decades are a different matter.

[11]The adjusted vote weight attributes an average vote to constituencies where there was an acclamation, and then adjusts vote share weights of all constituencies accordingly.

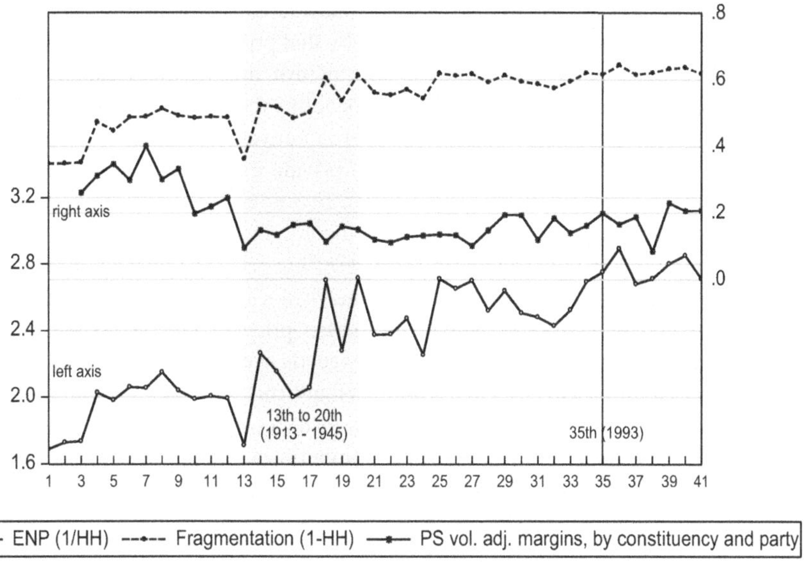


**Fig. 2** ENP, fragmentation and the PS volatility-adjusted margins, Canadian General Elections 1(3)–41, 1867(1874)–2011

flat. This impression is confirmed by the correlations in Table 1: over elections 1 through 41, both indexes of fragmentation are *negatively* correlated with the PS index of volatility-adjusted vote margins at about −0.4.

For the Canadian case then, fragmentation does not serve as a good proxy for electoral competitiveness when it is measured with an index that is designed to reflect the average closeness of individual electoral contests, taking vote-volatility into account. We hypothesize that for single member district, plurality rule electoral systems, this conclusion holds more generally.

## 3 The Duverger-Demsetz Perspective on Electoral Competition

There is a conceptual as well as an empirical basis for rejecting the view that rising fragmentation signals the greater competitiveness of elections. To develop this argument, we turn first to consider Demsetz's (1968) view of natural monopoly. The discussion here begins with the first theorem of welfare economics linking competition with economic efficiency before turning to Demsetz's contribution and its applicability to Duverger's Law and SMP elections. The implication of Demsetz's contribution for party governance is operationalized via the notion of a contestable election, an idea developed more fully in the economic context by Baumol (1982) and Baumol et al. (1982).

In any economic product market, social welfare is maximized when the difference between the total social benefit created by that product and the total social cost of producing that product is maximized. This, in turn, implies that production should be increased as long as the marginal social benefit exceeds its marginal social cost, and when the two are equalized, the market is conventionally described as being efficient. When the product in question is a private good (i.e., a good that cannot be consumed by more than one individual), two conditions are often invoked to ensure efficiency. First, the firms producing the good under increasing cost must be individually too small to influence the market selling price so that each firm becomes a price taker. Under these circumstances the firm's incentive to maximize profit means that each will produce where the market price equals its private marginal cost, and realize profits if the market price exceeds average cost. Second, there can be no barriers to new firm entry. This implies that firms will enter the market as long as profit can be made which in turn raises industry output, lowers the market price and reduces incumbent profit. In this way competition among established firms and potential entrants guarantees that only the lowest cost firms will survive and that all such firms will equate price to private marginal cost. It follows that if private and social costs are identical, competition among firms in the presence of these two conditions—price taking and the absence of barriers to entry—are sufficient for market output to be efficient and for social welfare to be maximized.

The sufficient conditions described above include two important caveats: first, that cost conditions allow atomistic firms to be the low cost option and, second, that the goods produced are not public goods that are nonrival in consumption. In most industries, however, firm-level fixed costs are present. This means that firms are typically not atomistic in size and to the extent that time and space allow some degree of market segmentation, firms will retain some degree of market power and control over price in the short run. To the extent that barriers to entry exist, such market power can persist over the longer run. In either case the ability to raise the selling price without losing all market share leads the profit maximizing firm to reduce its output and raise its price above marginal cost. The degree to which price diverges from marginal cost depends upon the degree of effective competition that arises from the firms' rivals. All other things equal, the larger the market share held by any firm, the more market power it has and the less competitive will be that industry.

Recognition that the conditions for perfect competition do not exist has led economists and policy makers to look for ways to assess how far any particular industry departs from perfect competition. Here the inability to observe directly either marginal social cost (as opposed to private average cost) or the level of economic (as opposed to accounting) profit has required the development of alternative measures to proxy the degree of competition. This has been done through observable market shares. Hence it is argued that in private markets, more firms with smaller market shares will have less market power which, in turn, will result in a smaller divergence between price and marginal cost. As noted earlier, the HH index is designed to reflect just such a tendency.

While the use of the HH index has been important in areas such as competition policy, the second caveat to the sufficient conditions discussed above means that the competitive implications of this index will not apply in a market with public good characteristics. That is, unlike a market for private goods, net social value is not maximized by equalizing the market price for each consumer and the marginal cost of each producer. Rather social welfare is maximized when marginal social cost is equated to *the sum of* the (potentially different) marginal social benefits of each individual. Such markets—light from a lighthouse, knowledge generated by a new idea, a television program for communal viewing—are often described as being natural monopolies where concurrent consumption and cost minimization combine to imply a single producer. In such cases, competition among consumers and competition from incumbent producers and potential entrants cannot be relied upon to induce efficiency.[12] Because the absence of effective competition allows the monopolist to reduce output and raise price, many economists have advocated regulation. Hence in many communities public utilities are granted a monopoly right to produce in return for a commitment to satisfy market demand at regulated prices. The latter, in turn, are designed to allow the utility only normal profits. To the extent that the regulator can determine the appropriate set of market prices, greater efficiencies can be realized.

In a provocative article entitled "Why regulate utilities?", Demsetz (1968) argued that the fact that there can be only one efficient producer does not preclude competition from being used to improve upon the welfare generated within a natural monopoly. By the splitting of two usually conjoined rights, the right to own industry assets, and the right to determine the use of these assets, competition among potential managers over the dimensions of industry output and the prices at which output is marketed can be used to better approximate an efficient solution. That is, encouraging competition among potential managers over promised levels of industry output and the prices to be set can be used to achieve better market outcomes. In essence, the competitive process will end up revealing the insider information that would be needed by regulators to set the appropriate output and pricing terms. Market competition can in this way be refigured to meet the challenge of a single producer of a public good and to overcome the information problems facing public regulators.

Suppose then that Duverger's Law is true and that 2 is the long run competitive equilibrium in an SMP system such as Canada's. The interpretation of Duverger's Law as a long run competitive equilibrium fits easily into Demsetz's (1968) re-interpretation of competitive equilibrium in economic markets served

[12]If concurrent consumers cannot be excluded from consuming, competition among consumers for the lowest (zero) price will result in insufficient revenue being generated to support the efficient level of production. On the other hand, if concurrent consumers can be excluded, efficiency could be achieved by a producer setting the Lindahl (individual) prices needed to realize a level of output at which the sum of the individual marginal values equals marginal social cost. However in the absence of competing alternatives, the sole producer will exploit its market power, raise the set of Lindahl prices and under produce relative to the potentially efficient level.

by a natural monopolist. That is, because governance of the political process has the characteristic of a public good—the policies and programs instituted by the governing party are consumed by all constituents concurrently—the governing party can be seen as analogous to the manager of a public utility. Because providing governance in a collective has the same cost structure as a natural monopoly—a single governing party/management team is the low cost service provider—social benefit is maximized when there is only one manager or governance provider. To avoid the reduction in service and higher cost that comes from the incentives facing the monopoly provider, competition must exist over the right to provide that service. This competition is provided through free and fair elections. However, for such competition to enhance welfare, there must exist not only competing sets of promised policy alternatives, but also a credible alternative manager that can step in and perform should the promised level of performance be reneged upon or not offered.

In this view, contestability in the sense developed by Demsetz and by Baumol and his co-authors—understood as the ability to credibly replace the incumbent producer—is the key mechanism by which the benefits of competition can be realized effectively by the community. In the political arena, competition in an election arises through the set of policies that competing parties view as better reflecting the wishes of the electorate. However the public good characteristic of governance means that effective competition comes not from the combined set of policy alternatives on offer, but from the set that can be provided by the credible alternative which must include the likelihood that that particular policy set will be implemented. Here the instability of minor parties in SMP systems highlighted by Duverger becomes critical. The incentive not to waste one's vote by supporting a nonviable alternative implies that the greater is the degree of party fragmentation, the less effective will second or third placed parties be as a constraint on the performance of the governing party. Because greater fragmentation means that each of the opposition parties is less likely to win a majority of seats, and since coalitions are difficult to arrange and maintain over time in SMP systems, each of these parties becomes less credible as a threat to the incumbent government. In such a fragmented party system, the pressure on the governing party to make and keep election promises is thus diminished. In short, from the Duverger-Demsetz perspective, a rise in ENP above 2 signals a decline in effective electoral competitiveness.[13]

[13]There is an additional, conceptually distinct source of inefficiency that may worsen with fragmentation. This stems from the possibility that as the number of parties increases, each party is forced by the division of the electorate to focus its electoral promises on a narrower segment of the electorate, thus moving the public sector towards special interest politics and away from concerns over the provision of general public services. See Lizzeri and Persico (2005) for an interesting exploration of this view. Chhibber and Nooruddin (2004) propose and positively test a similar hypothesis for Indian states. Bueno de Mesquita et al. (2001) present essentially the same view. See also the additional literature cited in the first footnote concerning the weak government hypothesis. A reasonable conjecture is that this source of inefficiency may be a problem in all SMP systems with weak national parties.

In the next section, we attempt to apply the Duverger-Demsetz view to Canada by measuring the contestability of elections. But before we do so, it is interesting to consider how the idea of competitiveness as electoral uncertainty fits with this approach. Is a contestable election also highly uncertain? If we are concerned with the consequences of competition, uncertainty in itself is not a necessary ingredient. That is, in the absence of performance differences across contenting parties, contestability will restrain the options of the incumbent such that replacement would arise only when the incumbent party behaves 'badly' or miscalculates the nature and distribution of voter preferences. On the other hand, if a candidate or party is superior in terms of performance, we may observe long periods of one-party dominance even in a highly contestable system, a point also made by Buchler (2011). Thus in a framework in which contestability is the center of attention, one party dominance and the absence of electoral uncertainty are not reliable indicators of a lack of competition.[14]

## 4 Analyzing the Canadian Electoral System from the Duverger-Demsetz Perspective

From the Duverger-Demsetz perspective, what matters for competitiveness is whether or not the governing party faces the threat of replacement by an alternative when it doesn't provide what voters wish. The key requirement is that the threat of replacement must be real, which requires the alternative to be credible. When the incumbent can be replaced easily by a credible alternative we may say that the electoral system is highly contestable. In this section we provide some empirical support for the Duverger-Demsetz perspective using the history of the Canadian parliamentary system to measure the contestability of elections at the national level. We show that contestability has tended to be greater when the vote is *less* fragmented.

As is well known, a good electoral strategy in a Westminster system like Canada's is to target marginal constituencies: districts especially susceptible to changing hands in an election (see, for example, Hartle 1985; Persson and Tabellini 2000 chapter 8, among others). This suggests that the proportion of marginal constituencies would be a good indicator of the contestability of an election. If every constituency is perfectly safe for its incumbent regardless of what the challenger may do, there is no competition and the position of the incumbent party in power in

[14]We use the word 'reliable' here because we are aware that the matter is not straightforward. If the survival of the incumbent was *always* assured (i.e., absolutely certain), there is no political competition. Even if the incumbent is superior, preservation of competition as a principle of governance may require throwing out such an incumbent from time to time, thus introducing uncertainty into the process.

Ottawa is clearly not contestable. On the other hand, if every seat is marginal, 'every seat is a battleground', as Bodet (2014) puts it.

Marginal seats as a measure contestability can be improved upon by incorporating a measure of the asymmetry of safe seats among parties, on the grounds that a party holding relatively more of the safe seats has an important advantage over its opposition. This is because it is able to focus its resources on constituencies that are thought to be marginal to a greater extent than its opposition. In what follows, we construct an asymmetry adjusted marginal seat index and then consider how it is related to fragmentation.

To operationalize the idea that the contestability of an election depends on the asymmetry adjusted proportion of marginal seats in an election, we must first define what marginal means. Hartle (1985) suggests that a marginal constituency is one from which economic rents cannot be taken and redistributed to other places without serious risk of electoral defeat. This is attractive as a definition of electoral marginality, but impossible to apply without the ability to measure the distribution of rents across constituencies, data which are as yet unavailable.

Previous work in Canada on marginal or safe seats includes Lovink (1973) and most recently Bodet (2014). Both of these interesting studies use data for small samples of Canada's electoral history. Bodet defines a safe seat as essentially one that lies in the upper tail of the distribution of vote margins and uses a one standard deviation above the mean based on the distribution of vote margins in the previous election (and some ancillary criteria) as his cutoff. Winning margins larger than that cutoff are considered sufficiently large to provide a substantial cushion of safeness to the incumbent party in that constituency. We also employ the one standard deviation standard in this initial exploration.

To measure a safe, or alternatively, marginal seat by party, we consider the volatility adjusted, winning vote margin for the candidate of each incumbent party $p$ (which won at time $t-1$) in constituency $j$ within superconstituency $s$ in election $t$, defined as

$$IPmargin1_{pjst} = \frac{\left(v_{1pjst-1} - v_{2jst-1}\right)}{Volatility_{st-1}}. \tag{8}$$

If this *IPmargin* falls in the upper tail of the distribution of all such margins for all parties for the previous three elections—e.g., it is more than one standard deviation above the mean—the constituency is judged to be safe for that party. Incumbent margins for the next election are constructed in the same way by adding the next election outcome and dropping the oldest to form the relevant test distribution. Note that only past election outcomes are used to judge safeness, except when a constituency does not have an incumbent (for *any* reason) in which case it is considered to be marginal or not safe. Once again the superconstituency, defined over an unchanging geographical area, is an important feature of this construction via the measurement of volatility. This is because the distributions of *IPmargins* require measures from four consecutive elections, a long period of time within which many constituencies are born, die and change boundaries.

This algorithm, applied to all constituencies in each election, leads to the number of seats that are considered to be safe in each election, $\psi_t$. We then compute the proportion of marginal seats (in the total to be elected) in each election, $MS_t$,

$$MS_t = 1 - \psi_t, \tag{9}$$

as a measure of the competitiveness of the election as a whole. In this case, $MS = 1$ indicates that 100 % of the seats in Parliament are marginal.

As noted earlier, the proportion of marginal or safe seats does not in itself provide a good index of contestability. How safe seats are distributed across parties, and in particular, whether or not there is an asymmetry in their distribution, also matter. Regardless of the number of marginal seats in total, an equal distribution of safe seats across the major parties will result in a highly contestable election compared to a situation in which the same number of safe seats are held predominately by just one party. To acknowledge the importance of the distribution of safe seats to a judgment about the contestability of an election, we adjust the proportion of marginal seats $MS_t$ by the degree of asymmetry in safe seats among the parties to produce a better measure of competitiveness at the national level.

To capture the notion of asymmetry, we borrow an idea used by Gaines and Taagepera (2013) in a somewhat different context to define the Euclidean deviation from a three party equal sharing of safe seats:

$$\phi 3_t = \sqrt{3/2} * \sqrt{\left(1/3 - S_{p1t}\right)^2 + \left(1/3 - S_{p2t}\right)^2 + \left(1/3 - S_{p3t}\right)^2} \tag{10}$$

where $S_{pkt}$ = the seat shares in Parliament of the party in *kth* place in terms of seats. Then $\phi 3_t = 0$ if the safe seats are symmetrically distributed; and $\phi 3_t = 1$ if one party has all the safe seats. In Canada's case the third 'party' is a residual consisting of all other parties except the two major parties in Canada, the Liberals and Conservatives (which are broadly defined as part of our 12 party aggregation).

An asymmetry index based on the proportion of marginal seats that acknowledges the asymmetry in their distribution may then be defined as:

$$MSadj_t = MS_t * (1 - \phi 3_t). \tag{11}$$

$MSadj_t$ (the proportion of marginal seats adjusted for asymmetry) = $MS_t$ if safe seats are symmetrically distributed among the parties, and is 0 if one party has all the safe seats.

There is one further adjustment to make before the contestability index is finished. The formulation in (11) may overweigh asymmetry. For example, if there are only 3 safe seats in 300 held by only 1 party, $MSadj = 0$. To correct this problem, we first adjust safe seats for asymmetry in their distribution:

$$AS_t = \{\psi_t \phi 3_t\}.$$

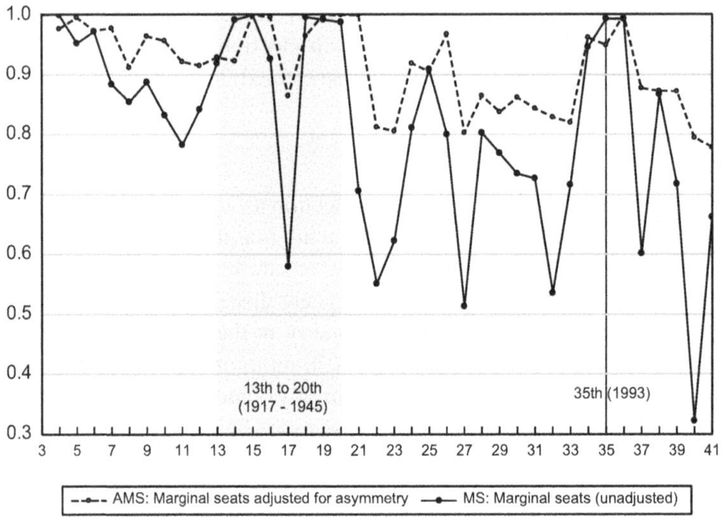


**Fig. 3** Asymmetry adjusted marginal seats vs marginal seats Canadian General Elections 4–41, 1878–2011. One standard deviation criterion

$AS = 0$ if safe seats are symmetrically distributed and $\text{AS} = \psi_t$ (the proportion of safe seats) if one party has all the safe seats. We then use $AS_t$ to define an adjusted asymmetry index of marginal seats:

$$AMS_t = 1 - \{\psi_t \phi 3_t\}. \quad (12)$$

This is our metric of electoral contestability. $AMS = 1$ if all safe seats are symmetrically distributed, and $AMS = 1 - \psi_t$ (the proportion of marginal seats) if one party has all the safe seats.

In Fig. 3 we show the AMS index for Canada in comparison to the corresponding symmetry unadjusted series MS. It can be seen that in seven or eight elections with a relatively small number of marginal seats, the asymmetry adjusted index is much higher than the unadjusted one. This pattern indicates that in these elections there is an important degree of symmetry in the distribution of safe seats even though their absolute number may be small, and illustrates the necessity of integrating the symmetry of the distribution of safe seats into the index of electoral contestability (12).

## 4.1 *Fragmentation and Contestability in the History of Canadian General Elections*

We can now consider the relationship between fragmentation and our measure of electoral contestability in a manner that sheds light on the Duverger-Demsetz perspective. To do so, it is instructive to begin by looking at a scatter diagram that relates both the AMS and PS indexes to the ENP Ratio = 2/ENP, with ENP define defined as a national average over constituency level values based on vote shares of candidates. This will be followed by regressions that confirm what a visual inspection of the data appears to indicate.

Using ENP in the form of a ratio is convenient transformation because in the Duverger-Demsetz view, the ENP Ratio will equal 1 in the long run of an electoral system that is highly contestable and will decline as the number of parties increases above 2. Recall that it is also the case that the AMS and PS indexes take a value of 1 when reflecting the highest degree of competitiveness and decline in value as competitiveness decreases. It follows that if increases in each of these indexes measure greater competiveness, all three measures should be positively related.

As the regression lines on the scatter diagram of Fig. 4 suggest, the ENP Ratio is positively related to both competitive measures, implying that fragmentation has a *negative* association with competitiveness. In the upper portion of the figure it can be see that as ENP Ratio rises towards 1, so does contestability as indicated by a rise in the values of the asymmetry adjusted marginal seat index AMS. Here, then, is qualified support for the Duverger-Demsetz view. We say 'qualified support'

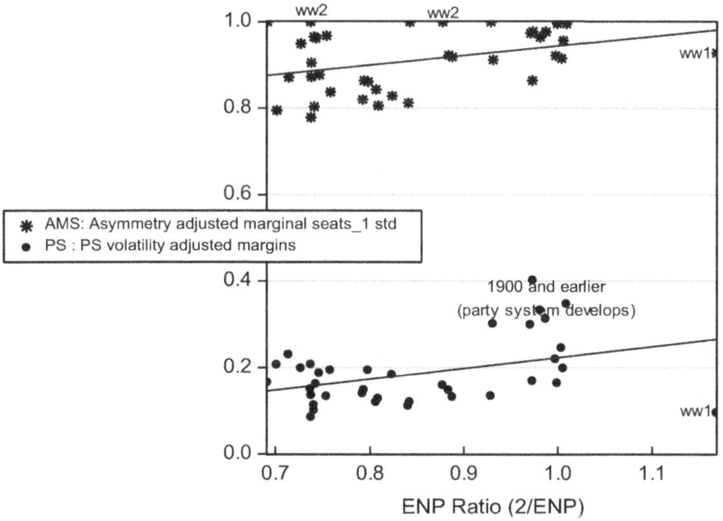


**Fig. 4** AMS and PS indexes versus the ENP Ratio, Canadian General Elections 4–41, 1878–2011

because of the potential sensitivity of our conclusion to alternative assumptions that could be made in the construction of our index that have yet to be explored.

We can also see in the figure that the PS index of volatility adjusted vote margins is also positively related to the ENP Ratio, indicating that fragmentation and our preferred measure of competitiveness as closeness are negatively related. This is further evidence that in the Canadian case at least, fragmentation is not positively correlated with competitiveness.

Closer inspection of Fig. 4 indicates that there are some episodes in which observations tend to cluster off of the regression lines. These clusters suggest controlling factors that can be used in regressions to remove anomalies that distract from the underlying relationships. Two groupings of elections are of interest in this respect: first, elections during the world wars, which likely saw the introduction of elements that would not be present under normal circumstances; and, second, elections in the period before 1900 when the party system in Canada was maturing.

The relationship between fragmentation, represented by the ENP Ratio, and the competitiveness indexes AMS and PS, as well as the role of the dummy variables for war and the early years of the party system, are explored in the regressions presented in Table 2 on the following page.

The general appearance of positive, significant coefficients on the AMS and PS indexes across the first three columns representing different versions of the equation for 2/ENP confirm what visual inspection of the scatter diagram in Fig. 4 suggests, namely that more fragmentation is associated with less contestability (see the coefficient on AMS), and less competition defined as closeness (and uncertainty) of electoral contests.

By using the ENP Ratio as the dependent variable in Table 2, we do not mean to imply that the equations in the table represent a causal relationship running *from* the AMS or PS indexes *to* ENP. Rather the regression equations represent a long run equilibrium relationship in which the number of parties, their vote shares and degree of competitiveness are all simultaneously determined. Accordingly, the dynamic least squares (DOLS) estimation in columns four and five of the table treat the model tabulated as a cointegrating relation, and allow for the possibility that standard errors of the OLS regressions may be biased by correlations across time arising among the three variables ENP, AMS and PS.[15]

Consider the DOLS results in the fourth or second to last column in Table 2. The stationarity of the residuals of this model and the fact that the coefficient estimates on AMS and PS retain their sign and significance compared to the OLS results further supports the view that contestability as well as electoral uncertainty are on average both *negatively* related to party fragmentation over the history of parliamentary elections in Canada.

[15]It does this by adding leads and lags of all three variables into the equation, so that, in principle, the calculated residuals are orthogonal to the entire process despite the mutual interdependence of the three variables.

**Table 2** The relationship between the ENP ratios and competitiveness indexes

| | ENP Ratio | | | ENP12party Ratio | |
|---|---|---|---|---|---|
| | OLS | OLS | OLS | DOLS (long run) | DOLS (long run) |
| AMS | 0.71*** (2.91) | 0.47** (2.04) | 0.46** (1.92) | 1.41*** (3.09) | 1.54** (2.43) |
| PS | | 0.64*** (3.07) | 0.55* (1.78) | 0.83* (1.77) | 0.78 (1.20) |
| ww1 | 0.31*** (3.05) | 0.37*** (4.03) | 0.37*** (3.91) | 0.20** (2.20) | −.0001 (0.001) |
| ww2 | −0.10 (1.38) | −0.06 (0.85) | −0.06 (0.80) | | |
| party_formation | | | 0.03 (0.40) | | |
| Constant | 0.20 | 0.29 | 0.32 | −0.58 | −0.82 |
| $R^2$ (adjusted) | 0.31 | 0.44 | 0.43 | 0.75 | 0.64 |
| Prob (F-statistic) | 0.001 | 0.000 | 0.000 | | |
| n | 38 | 38 | 38 | 31 | 31 |
| ADF (MacKinnon 1996, tau test) | | | −4.31* (SIC; constant and trend) | −6.56*** (SIC; no constant, no trend) | −2.79 (SIC; no constant, no trend)−7.52*** (with const. and trend) |

*, **, *** significant at 10 %, 5 % and 1 % respectively. Absolute value of t-statistic in brackets

*Notes*:

1. Dependent variables: ENP Ratio = 2/ENP with ENP calculated at the constituency level, OR ENP12party Ratio = 2/ENP12party with ENP calculated at the national level using party vote shares for 12 parties—see the Appendix for the definition of parties
2. *AMS* the asymmetry adjusted marginal seat index of electoral contestability, *PS* the Przeworski-Sprague index of electoral competition as closeness or electoral uncertainty, *ww1* (*ww2*) dummy variable equal to 1 if an election was held during world war one (two), 0 otherwise, *party_formation* dummy variable equal to 1 if the election is in 1900 or before, 0 otherwise
3. *OLS* ordinary least squares, *DOLS* dynamic least squares, maximum three lags and leads, three selected, *AIC* criterion. Only the long run DOLS equation is shown. All estimation using other default options in Eviews 9

Before turning to consider the extension of our ideas to proportional representation systems, it is useful to consider how the statistical relationship between fragmentation and contestability carries over to a situation in which fragmentation is measured at the national party level, as indicated by ENP measured using vote shares of (12) parties at the national level, labeled ENP12party in Tables 1 and 2. As Table 1 shows, in Canada this measure of the effective number of parties is highly correlated with ENP defined over candidates at the constituency level, at 0.88, and is generally larger that the constituency based ENP.[16]

We expect that a high degree of contestability at the national level will put pressure on third and fourth place parties in the constituencies. In extreme cases, it is conceivable that one party will dominate in a particular region, and another party in another, with contestability remaining high at the center where there are then two (or even more) major opposing parties, each of which has many candidates that are ‘safe’ in their own region of dominance. While our measure of contestability—the AMS index—can handle such situations, ENP defined as an average over constituencies may not. In the example outlined, average fragmentation at the constituency level will be low, while at the national level it remains more robust.

To allow for such situations, we also include the fifth column in Table 2, where the left side variable is now ENP12party Ratio, which is 2 divided by ENP defined at the national level using vote shares of 12 parties that have existed over Canadian parliamentary history.[17] Despite the complications of going from the constituency to the national level, we see that the statistically significant inverse relationship between fragmentation and the contestability index AMS still remains in the DOLS estimates, though not with quite the same statistical strength. The PS index of closeness or electoral uncertainty at the constituency level is now insignificant, perhaps reflecting situations in which contestability remains at the national level even though there are parties that have carved out for themselves safe seats that are regionally concentrated.

[16]In Canada from 1867 to about the start of the first world war, the number of parties as reflected by ENP12party declined steadily to about 2 at the outbreak of the war. Thereafter, both ENP12party and ENP defined as an average over the constituencies began to rise, with ENP12party being uniformly higher than ENP. Since Duverger's Law is a long run result, it is not clear whether or not either enp index is systematically greater than 2 in the long run. Investigation of that issue requires a dynamic empirical model of enp, which to our knowledge has not yet been constructed for Canada or elsewhere.

[17]Again, see the Appendix for the definition of party used here.

## 5 Does the Duverger-Demsetz Perspective Extend to Proportional Systems?

To complete our analysis of electoral competition from the Duverger-Demsetz perspective, we consider whether our ideas about the importance and role of contestability in SMP electoral systems can be extended to proportional representation in multi seat elections (PR)? Here we are of two minds.

On the one hand, if there is a single party in government, the exact argument given earlier for why two parties produce a highly contestable electoral system in the plurality setting will also apply to the PR setting. As we have argued earlier, the greater is the degree of party fragmentation, the less effective will be the second or third placed parties as a constraint on the performance of the governing party. Moreover, the work that has been done on extensions of Duverger's Law for proportional representation systems is based on the argument that the carrying capacity of a PR system (defined in terms of ENP at the district level) will be a function of the district magnitude, M, in that system, i.e., the number of seats being contested in a given district (or the size of median district overall). The carrying capacity of the system is either expected to have an upper bound of $M + 1$ (Cox 1997) and will thus be above 2 for $M > 1$, or expected to be, on average, the square root of $M + 1$ (Taagepera and Shugart 1989), which is above 2 for $M > 4$. Because values of $M > 1$ mean that, for PR systems, we expect to see more than 2 winning parties, this means that the likelihood of there being a single party majority tends to diminish with M (Rae 1967).

Since coalitions are likely to emerge in PR systems and because coalitions are difficult to arrange and maintain, there will sometimes be minority governments. It may therefore appear to be the case that a PR system is more contestable since it is easier to displace a minority government. However, the opposition is also likely to be fragmented under PR for the same reasons, regardless of whether there is a minority or a majority in government, and the opposition coalition, if there is one, will also tend to be difficult to maintain over time. Hence, a rise in *ENP* above 2 can also signal a decline in effective electoral competitiveness in a PR system.

From the Duverger-Demsetz perspective, the 'best' situation would be one where there is a strong coalition in government to provide public services, and a strong coalition in opposition threatening to replace it, the same situation that leads to a high degree of contestability (and efficiency) in an SMP system. In general, then, fragmentation of the party system under PR is not conducive to contestability because of the costs of forming and maintaining party coalitions, and the best outcome under PR mimics that for the SMP system. Hence we arrive at essentially the same assessment as for SMP, though by a different route.[18]

---

[18]There is an additional, conceptually distinct source of inefficiency that may worsen with fragmentation under PR. This stems from the possibility that each party in a PR system focuses its electoral promises on a narrower segment of the electorate than does a party under SMP. If so, the public sector under PR will be driven more by the demands of special interests and pay less

On the other hand, if we approach competition in terms of the measures used by Blais and Lago (2009) or Grofman and Selb (2009), we can think of competition increasing with M, because the *threshold of exclusion*—the largest vote share that a party can achieve and still be denied even a single seat—declines with M for all PR electoral rules. Thus entry of new parties is generally easier under PR than in an SMP system, and entry is another important dimension of competitiveness. Moreover, while the check on the behavior of the governing party generated by a truly viable single competitor, emphasized as the root of electoral contestability in a plurality system, does not apply in the PR context, what may apply is a growing multiplicity of viable alternatives to the present governing coalition that include some but not all elements of that coalition joined to other parties not in the present coalition.

By throwing up what is likely to be a wider range of alternatives, a fact that is sometimes taken to be a major failing of PR systems, namely the relative fragility of multiparty coalitions in terms of durability, may be given a positive interpretation from the economic perspective we have offered in this paper. To put this another way, we may say that the analysis of PR and, by implication, of SMP is not complete without considering the entry dimension of electoral competition, a dimension that is not identical to contestability.

## 6 Conclusions and Suggestions for Further Work

We have considered a number of distinctive ways to think about electoral competition in SMP systems, focusing on the issue of whether fragmentation or concentration of the electoral landscape in terms of vote shares, electoral uncertainty as measured by the closeness of contests, and the overall contestability of elections as indexed by asymmetry between major parties in the number of marginal seats stand as equivalent or even complementary indicators of greater electoral competition. The Duverger-Demsetz perspective, which emphasizes the contestability of elections, suggests these are not equivalents.

For SMP systems in particular, logic suggests that contestability will diminish with party fragmentation—in other words, that an increase in the effective number of parties (or in other related measures of fragmentation) is associated with reduced

---

attention to the provision of public services, compared to an SMP system which effectively blocks some interests that are not regionally concentrated. (For example, the Greens in Canada may have 5 % of the vote in every constituency, but they elect only one member of parliament from a place known for voters who have a strong taste for the environment). On the other hand, some argue that candidates who must appeal to voters within a small geographic area and who can differentiate themselves from their competitors by making promises for narrowly targeted pork barrel items are more likely to arise in a SMP system than in a PR system (Carey and Shugart 1995; Persson and Tabellini 2000). We cannot resolve this debate over the role of electoral systems in the link between fragmentation, special interests and inefficiency here.

electoral competition. Evidence that this view has merit was provided by showing that the effective number of parties and a new index of electoral contestability—the asymmetry adjusted index of marginal constituencies—are inversely related for the history of the Canadian parliamentary system. Robustness of the empirical work to alternative assumptions, for example about the exact way to formulate expectations of electoral success using ex post electoral data and the standards used to measure the safeness of seats, remains to be studied.

In looking to future research, there is the challenge of setting the Duverger-Demsetz perspective (or any other perspective on electoral competition) in a wider context in which various dimensions of electoral competition, including competition in legislatures between elections, are all considered together. The theory and measurement of electoral contestability in PR systems, considered only briefly here, also remains to be explored.

**Acknowledgements** This work is partly supported by a Social Sciences and Humanities Research Council of Canada grant to Winer and Ferris and by the Canada Research Chairs program. Grofman's participation is partly funded by the Jack W. Peltason Chair of Democracy Studies at the University of California, Irvine. We gratefully acknowledge the research assistance of Haizhen Mou, Alexandre Couture-Gagnon, Derek Olmstead, Sarah Mohan, Beatriz Peraza and Samira Hasanzadeh on related projects. Part of the material in this paper stems from a more general unpublished discussion of the measurement of electoral competition that has benefited from presentation at the Public Economic Theory Conference in Seattle, July 2014, at the University of Montreal in October 2015, at U.C. Irvine in February 2015, at The Political Economy of Social Choices, BIRS Workshop, Oaxaca, México, July 2015 and the IIPF Congress in Dublin, August 2015. We thank André Blais, Bill Cross, Momy Dahan, Marc Duhamel, Richard Johnston, Rein Taagepera, Carole Uhlaner, Marcel Voia and seminar participants for helpful remarks. Errors and omissions remain the responsibility of the authors. The paper is a part of a larger project on the meaning, measurement and consequences for public policy of electoral competitiveness in mature democracies including Ferris et al. (2008) and Winer et al. (2014).

## A.1 Appendix: The Data, Definition of Variables, and Measuring Vote Volatility Using Superconstituencies

### *A.1.1 Data*

Data on votes by constituency, by candidate and by party for regular parliamentary general elections 1–41 were collected for each election from series supplied by Elections Canada. This data is available online through the Parliament of Canada website at: http://www.parl.gc.ca/About/Parliament/FederalRidingsHistory/HFER.asp.

The 12 party classification employed throughout the paper is based on three criteria: A party exists as such if it gained at least 4 % of the popular vote in at least one election and contested at least 1 % of the seats in at least one election—there are 23 parties satisfying these two criteria—*plus* it must have won at least 1 seat in at least two elections. There are 11 parties satisfying all these criteria over the history of the modern state: Liberal, Conservative, Labour, the National Party, the Bloc Quebecois, Social Credit, Reform-Alliance, the CCF_NDP, Raillement Creditiste, the Progressives, and the United Farmers of Alberta, with a residual category denoted as 'Other'. Liberal and Conservative include small groups that voted with the major party at various times as is the usual custom.

### A.1.2 Variables

***ENP*** = ENP calculated over candidates at the constituency level using candidate vote shares (max. 13 candidates in any one constituency).

***ENP12party*** = ENP calculated over 12 parties (11 plus Other) at the national level using party vote shares.

***ENP Ratio, ENP12party Ratio*** = 2 divided by the corresponding ENP number.

***Fragmentation*** = 1 – HH = 1 – 1/ENP, where HH is the Hirschman-Herfindahl index defined using vote shares.

$v_i$ = vote share of the candidate in the ith place.

***Closeness(3)*** = an index of the closeness of candidates' vote shares $v_i$, assuming ENP = 3, as in Endersby et al. (2002).

***($v_1 - v_2$) / volatility*** = the winner's vote margin $v_1 - v_2$ at the constituency level relative to historical volatility for that constituency.

***PS vol-adj. margins*** = the Przeworski-Sprague (1971) volatility adjusted vote margins by constituency by party.

***AMS_1std*** = an asymmetry adjusted measure of marginal seats, using an historical volatility and a 1 standard deviation test to define when an incumbent's seat is safe.

***ww1*** = 1 for election number 13 (1917); 0 otherwise.

***ww2*** = 1 for election numbers 19 and 20 (1940 and 1945); otherwise 0.

***party_formation*** = 1 for elections between 1 and 9 (1867 until 1900); 0 otherwise.

### A.1.3 Volatility

Adjusting vote margins for volatility is not easy to do over long periods of time because of redistricting. For a country like Canada that has had consistent growth in the number and frequent changes in the size of individual constituencies, new ridings appear in many elections. Without a past, a constituency can have no history of vote variability and cannot be included in the construction of a volatility adjusted vote margin. To circumvent the loss of information on winning margins through growth and redistricting, we construct a large number of regional super-constituencies—80 in total—based on geographic regions that persist throughout Canada's election history and that can be used to establish small area vote volatility

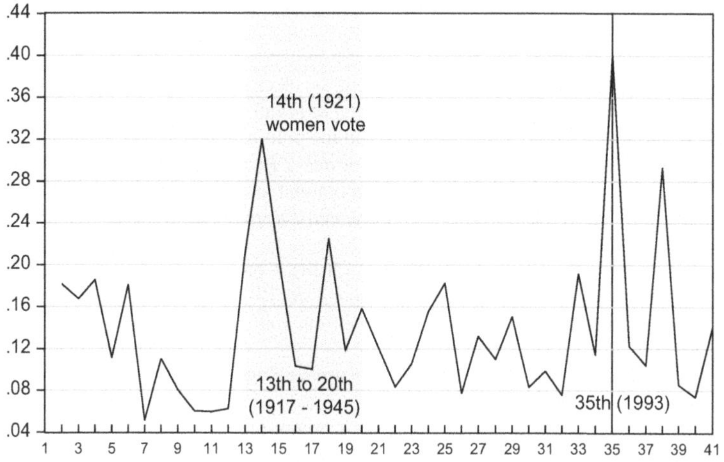


**Fig. 5** Historical volatility of party vote shares using superconstituencies Canadian General Elections 2–41

in the period when a new constituency is created or an old one is reshaped. The number and name of individual constituencies in a superconstituency may change over time, but its geographical boundaries remain fixed. To give one example, the area around Ottawa was used as the base for one of Ontario's 29 superconstituencies. Electorally it consisted of 1 riding in 1867 and rose to include 7 ridings by 2011.[19]

Aggregate volatility is then computed as follows: Average vote shares by party over constituencies within a superconstituency for each election are computed. For each superconstituency in each election, the absolute value of the changes in these (party-specific) average vote shares across adjacent elections is computed, summed and divided by 2. Each of these superconstituency specific differences in vote shares is then weighted by the relative number of constituencies inside each superconstituency, and summed to derive an aggregate volatility number for each election.

Volatility so computed is shown in Fig. 5 for the 2nd to 41st election (1869–2011) in Canada. The peaks in the 14th and 35th elections are noticeable. Whether there is a trend in volatility or not is difficult to determine.

[19]Note that the use of one past period to construct our volatility measure means that the index can begin only in the second election. This also implies the unavoidable loss of some information when new provinces are added to the country, such as Newfoundland's entry into Canada in 1949.

## References


Aidt, T., & Eterovic, D. S. (2011). Political competition, electoral participation and public finance in 20th century Latin America. *European Journal of Political Economy, 27*, 181–200.

Aidt, T., & Mooney, G. (2014). Voting suffrage and the political budget cycle: Evidence from the London Metropolitan Boroughs 1902–1937. *Journal of Public Economics, 112*, 53–71.

Alfano, M. R., & Baraldi, A. L. (2015). Is there an optimal level of political competition in terms of economic growth? Evidence from Italy. *European Journal of Law and Economics, 39*, 263–285.

Ashworth, J., Geys, B., Heyndels, B., & Wille, F. (2014). Competition in the political arena and local government performance. *Applied Economics, 46*(19–21), 2264–2276.

Bartolini, S., & Mair, P. (1990). *Identity, competition and electoral availability*. Cambridge University Press: Cambridge

Baumol, W. J. (1982). Contestable markets: An uprising in the theory of industry structure. *American Economic Review, 72*(1), 1–15.

Baumol, W. J., Panzar, J. C., & Willig, R. D. (1982). *Contestable markets and the theory of industry structure*. New York: Harcourt Brace Jovanovich.

Besley, T., Persson, T., & Sturm, D. M. (2010). Political competition, policy and growth: Theory and evidence from the United States. *Review of Economic Studies, 77*, 1329–1352.

Blais, A., & Lago, I. (2009). A general measure of district competitiveness. *Electoral Studies, 28*, 94–100.

Bodet, M. A. (2014). Strongholds and battlegrounds: Measuring party support stability in Canada. *Canadian Journal of Political Science, 46*(3), 575–596.

Borge, L.-E. (2005). Strong politicians, small deficits: Evidence from Norwegian local governments. *European Journal of Political Economy, 21*(2), 325–344.

Borooah, V. K. (2002). A general measure of the effective number of parties in a political system. In F. Cabrillo (Ed.), *Constitutional economics and public institutions* (pp. 146–159). Edward Elgar.

Buchler, J. (2011). *Hiring and firing public officials: Rethinking the purpose of elections*. Oxford University Press: New York.

Bueno de Mesquita, B., Morrow, J. D., & Siverson, R. M. (2001). Political competition and economic growth. *Journal of Democracy, 12*(1), 58–72.

Carey, J. M., & Shugart, M. A. (1995). Incentives to cultivate a personal vote: A rank ordering of electoral formulas. *Electoral Studies, 14*(4), 417–439.

Chhibber, P., & Kollman, K. (2004). *The formation of national party systems: Federalism and party competition in Canada, Great Britain, India and the United States*. Princeton: Princeton University Press.

Chhibber, P., & Nooruddin, I. (2004). Do party systems count? The number of parties and government performance in the Indian states. *Comparative Political Studies, 37*(2), 152–187.

Cox, G. W. (1997). *Making votes count: Strategic coordination in the world's electoral systems*. Cambridge University Press: New York.

Demsetz, H. (1968). Why regulate utilities? *Journal of Law and Economics, 11*(1), 55–65.

Drazen, A., & Eslava, M. (2010). Electoral manipulation via voter-friendly spending: Theory and evidence. *Journal of Development Economics, 92*(1), 39–52.

Duverger, M. (1954). *Political parties: Their organization and activity in the modern state*. London: Methuen.

Elkins, D. J. (1974). The measurement of party competition. *American Political Science Review, 68*(2), 682–700.

Endersby, J. W., Galatas, S. E., & Rackaway, C. B. (2002). Closeness counts in Canada: Voter participation in the 1993 and 1997 Federal Elections. *Journal of Politics, 64*(2), 610–631.

Ferris, J. S., Park, S. B., & Winer, S. L. (2008). Studying the role of political competition in the evolution of government size over long horizons. *Public Choice, 137*, 369–401.

Franklin, M. N. (2004). *Voter turnout and the dynamics of electoral competition in established democracies since 1945*. Cambridge University Press: New York.
Gaines, B. J. (1999). Duverger's Law and the meaning of Canadian exceptionalism. *Comparative Political Studies, 32*(7), 835–861.
Gaines, B. J., & Taagepera, R. (2013). How to operationalize two-partyness. *Journal of Elections, Public Opinion and Parties, 10*, 1–18.
Geominne, S., Geys, B., & Smolders, C. (2008). Political fragmentation and projected tax revenues: Evidence from Flemish Municipalities. *International Tax and Public Finance, 15*, 297–315.
Gerring, J., Palmer, M., Teorell, J., & Zarecki, D. (2015). Demography and democracy: A global, district-level analysis of electoral contestation. *American Political Science Review, 109*(3), 574–591.
Ghosh, S. (2010). Does political competition matter for economic performance? Evidence from sub-national data. *Political Studies, 58*, 1030–1048.
Godbout, J. F., & Hoyland, B. (2013). The emergence of parties in the Canadian House of Commons (1867–1908). *Canadian Journal of Political Science, 46*(4), 773–797.
Grofman, B., Blais, A., & Bowler, S. (2009). *Duverger's law of plurality voting* (Studies in public choice, Vol. 13). New York: Springer.
Grofman, B., & Selb, P. (2009). A fully general index of political competition. *Electoral Studies, 28*, 291–296.
Hartle, D. (1985). *Achieving electoral success in a multi-constituency system through discriminatory parties*. Unpublished, Economic Council of Canada.
Herfindahl, A. O. (1950). *Concentration in the U.S. Steel Industry*. Unpublished doctoral dissertation, Columbia University, New York.
Hirschman, A. O. (1945). *National power and the structure of foreign trade*. Berkeley: University of California Press.
Hirschman, A. O. (1964). The paternity of an index. *American Economic Review, 54*(5), 761–762.
Johnston, R., & Cutler, F. (2009). The puzzle of local three-party competition. In B. Grofman, A. Blais, & S. Bowler (Eds.), *Duverger's Law and plurality voting* (pp. 83–96). Springer.
Laakso, M., & Taagepera, R. (1979). Effective number of parties: A measure with application to West Europe. *Comparative Political Studies, 12*, 3–27.
Lizzeri, A., & Persico, N. (2005). A drawback of electoral competition. *Journal of the European Economic Association, 3*(6), 1318–1348.
Lovink, J. (1973). Is Canadian politics too competitive? *Canadian Journal of Political Science, 6*, 342–379.
MacKinnon, J. G. (1996). Numerical distribution functions for unit root and cointegration tests. *Journal of Applied Econometrics, 11*, 601–618.
Mayhew, D. R. (1974). Congressional elections: The case of the vanishing marginals. *Polity, 6*(3), 295–317.
Pedersen, M. (1979). The dynamics of European party systems: Changing patterns of electoral volatility. *European Journal of Political Research, 7*, 1–26.
Persson, T., & Tabellini, G. (2000). *Political economics: Explaining economic policy*. MIT Press: Cambridge.
Przeworski, A., & Sprague, J. (1971). Concepts in search of an explicit formulation: A study in measurement. *Midwest Journal of Political Science, 15*(2), 183–218.
Rae, D. (1967). *The political consequences of electoral laws*. Yale University Press.
Rae, D. (1968). A note on the fractionalization of some European party systems. *Comparative Political Studies, 1*, 413–418.
Ricciuti, R. (2004). Political fragmentation and fiscal outcomes. *Public Choice, 118*(3–4), 365–388.
Riker, W. H. (1982). The two-party system and Duverger's Law: An essay on the history of political science. *American Political Science Review, 76*(4), 753–766.
Roubini, N., & Sachs, G. (1989). Political and economic determinants of budget deficits in the industrial democracies. *European Economic Review, 33*, 903–938.

Taagepera, R. (2007). *Predicting party sizes: The logic of simple electoral systems*. Oxford: Oxford University Press.
Taagepera, R., & Shugart, M. (1989). *Seats and votes: The effects and determinants of electoral systems*. New Haven, CT: Yale University Press.
Winer, S. L., Kenny, L. W., & Grofman, B. (2014). Explaining variation in the competitiveness of U.S. Senate elections, 1922–2004. *Public Choice*. doi:10.1007/s11127-014-0176-0.

# Modelling the Effect of Campaign Advertising on US Presidential Elections

**Maria Gallego and Norman Schofield**

## 1 Introduction

The collection of articles included in Hendricks and Kaid's (2011) edited volume *Techno Politics in Presidential Campaigning: New Voices, New Technologies and New Voters*, argue that the 2008 US presidential election was a landmark campaign because it elected the first African-American President and brought about dramatic changes in the way presidential campaigns would be conducted from 2008 onwards. We argue that the emergence of smart phones, social media[1] coupled with large data sets on voters' personal characteristics and public and private pre-election polls has had a profound effect on political campaign advertising and on US Presidential elections as new media and new technologies gave voice to new candidates and voters. The congruence of these factors has revolutionized the way electoral campaigns are conducted. In particular, candidates now tailor their campaign message to voters taking their distinguishing sociodemographic characteristics and in their political

[1]Barak Obama was he first candidate to use social media in his 2008 bid for the US presidency. Obama ran a very successful fund raising campaign with a portion of his funds having been donated by voters contributing small amounts through social media. Obama also used social media to contact voters directly during the campaign.

M. Gallego (✉)
Department of Economics, Lazaridis School of Business and Economics,
Wilfrid Laurier University, 75 University Avenue West, Waterloo, ON, Canada N2L 3C5
e-mail: mgallego@wlu.ca

N. Schofield
Department of Political Science, Washington University in St. Louis, Campus Box 1027,
One Brookings Drive, St. Louis, MO 63130-4899, USA
e-mail: schofield.norman@gmail.com



M. Gallego, N. Schofield (eds.), *The Political Economy of Social Choices*,
Studies in Political Economy, DOI 10.1007/978-3-319-40118-8_6

preferences into account. Campaign strategists are now able to identify the core supporters of each candidate and undecided voters.

Ansolabehere et al. (2003) examine campaign contributions in Congressional and Presidential elections and note that candidates, parties and organizations raised and spent about $3 billion in the 1999–2000 election cycle. Since the federal government, at that time, spent about $2 trillion, the prize from influencing politics was of considerable value. They argue that the reason so "little" is spent campaigning is that contributions are a consumption, rather than an investment, good. However, they note that the electoral motive is not insignificant as the marginal impact of $100,000 spent in a House race, ceteris paribus, is about 1 % gain in vote. Thus, highlighting the powerful influence of ad campaigns on elections.

Moreover, by being able to directly communicate with potential supporters, candidates have greater control over their message than when using mass media (TV and radio). This highlights that models of US presidential elections must now take into account that candidates can directly communicate with voters. Currently, political campaigns use different means of communicating with voters. Not only have candidates used email, pod casts, RRS feeds and cell phone texting to communicate with voters but more recently new technologies such as blogs, YouTube, Twitter, Facebook have allowed candidates to sent messages directly to voters' smart phones and social accounts. The model in this paper assumes that candidates directly communicate with voters through smart phones and social media and that voters have preferences over candidates' message frequency.

Voters' judgments about candidates' competence are modelled by the notion of *valence*. The formal model can then be linked to Madison's understanding of the nature of the choice of Chief Magistrate (Madison 1787/1999). Schofield (2002) suggests that Madison's argument on the "extended Republic" may have been influenced by Condorcet's "Jury Theorem" (Condorcet 1785/1984) as it is based on the notion of electoral judgment rather than preference. Recent models involving valence contribute to a Madisonian conception of elections in representative democracies as methods of aggregation of both preferences and judgments.[2]

The standard spatial model assumes that it is only candidate *positions* that matter to voters. Within the context of the spatial model, controversy has arisen over whether rational candidates will converge to an electoral center, as suggested by Downs (1957) or whether elections are fundamentally unstable, as argued by Riker (1980, 1982, 1986).[3] However, as Stokes (1963, 1992) emphasizes, the non-policy evaluations, or *valences*, of candidates by the electorate are just as important as their policy preferences. Clarke et al. (2009) compare a "Downsian" or spatial model of the 2000 US presidential election with a valence model of the same election, based on voters' perceptions of candidates' traits and find that "the two models have approximately equal explanatory power." As Sanders et al. (2011) comment, valence theory is based on the assumption that "voters maximize their utilities by

---

[2]See Ansolabehere et al. (2003), Groseclose (2001), Aragones and Palfrey (2002), Schofield (2006), Zakharov (2009).

[3]See Mckelvey (1976), Schofield (1978), Saari (1997), Austen-Smith and Banks (1999).

choosing the party that is best able to deliver policy success." Valence measures the *bias* in favor of one or other of the party leaders (Mckelvey and Patty 2006). From our work, we argue that neither the Downsian convergence result nor the instability results gives an accurate picture of democratic elections. Instead, both position and valence matter in a fundamental way.

In this paper, we also argue that advertising campaigns are crucial during elections. Whereas a candidate's policy refers to the pure public good the candidate promises to implement if elected; the sole purpose of the advertising portion of the campaign is to convince individual voters that they matter to candidates so as to give voters a further impetus to vote for them. In our model, advertising campaigns give candidates another dimension in which to compete in the election and gives candidates more degrees of freedom during the electoral campaign. The fundamental difference between policy and advertising campaigns is that while advertising campaigns are a tool used by candidates during the election, it is candidate's policy that gets implemented once elected. Advertising campaigns, as a second campaign instrument, allows candidates the ability to adjust their advertisements and policy platforms to gain additional votes.

We extend Schofield's (2007) electoral model to allow candidates to use two instruments—the policy platform and advertising (ad) campaign—to maximize vote shares. Candidates use these two instruments to convince voters to vote for them. Thus, candidates' electoral campaign consists of their policies and ad campaigns. We assume that the effectiveness of the advertising campaigns varies across voters as voters not only have policy preferences but also have preferences over the ad campaign messages send by candidates. In our model, voters are characterised by their preferred policy and campaign tolerance level,[4] with campaign tolerances varying across voters to capture differences voters' desire to be contacted by candidates.

The voting literature has also identified that voters assess candidates in various ways. Voters make decisions examining how close each candidate's policy is from theirs and take into account candidates' traits (popularity, charisma, age, gender, etc.), the *traits* valence; as well as their beliefs on candidates' ability to govern, the *competency valence*. It is also well known that voters with certain sociodemographic characteristics are also more prone to voting for one of the candidates,[5] the *sociodemographic valence*. Recent research has found evidence that differences in voting preferences depend not only on candidates' policy positions but also on the leaders traits and the sociodemographic and competency valences. Empirical findings show that the sociodemographic and competence valences are a major determinant of the electoral outcome (Gallego and Schofield 2013, 2016; Schofield et al. 2011a,b). There is evidence that voters' choice of candidate is influenced by

[4]A voter's campaign tolerance is given by the ideal messages a voter wants to receive from candidates.

[5]For example, African-American voters are much more likely to vote for the Democratic candidate than to vote for the Republican candidate. Thus, Democratic candidates have a higher average sociodemographic valence among African-American voters than do Republican candidates.

voters' non-policy evaluation of candidates and by the distance between candidates' policy and voters' ideal policy (Schofield and Gallego 2011). We incorporate the sociodemographic, traits and competence valences into voters' utility functions in our model.

We assume that voters' sociodemographic, traits and competence valences vary around their mean values with the idiosyncratic component drawn from Type-I Extreme value Distributions. Since candidates do not observe voters' idiosyncratic valence components but know the distribution of these components, candidates view voters' utility as stochastic and cannot perfectly anticipate how each individual votes but can estimate their expected vote shares. The model examines how voters' preferences and characteristics affect candidates' policy and ad campaigns and how candidates' choices affect their electoral prospects.

Our results show that if candidates adopt the same electoral campaign, i.e., the same policy and ad campaign, they adopt the electoral mean[6] as their campaign strategy. We derive the conditions under which candidates converge to the electoral mean. The electoral mean $(z_m, a_m)$ is a strong (weak) local Nash equilibrium (LNE) of the election if candidates' expected vote share at $(z_m, a_m)$, $V_j^m$, is high enough, i.e., higher that the sufficient (necessary) pivotal vote share, $\mathcal{V}_{sp}^m$ ($\mathcal{V}_{np}^m$). When candidate $j$'s expected vote share at $(z_m, a_m)$ is less than the necessary pivotal vote share, i.e., if $V_j^m < \mathcal{V}_{np}^m$, $j$'s vote share is at a minimum or at a saddle point and $j$ wants to move away from the electoral mean, that is, to change its policy or advertising campaign or both, to increase its vote share. In this case, other candidates may also find it in their interest to move away from the electoral mean. The electoral mean is not a LNE of the election.

Suppose the electoral mean is a local Nash equilibrium (LNE) of the election. In equilibrium, candidates take voters' policy and campaign tolerances levels into account. The policy and advertising campaigns are the weighted average of voters' ideal policy and campaign tolerance levels. Even though we assume that the one-person-one-vote principle applies, candidates weight voters differently in their electoral campaign. In particular, candidates give little or no weight to their core supporters or to voters who care not at all for them, giving maximal weight to undecided voters, i.e., to voters who vote with probability close to a half for the candidate. Thus, candidates' equilibrium campaigns reflect the policies and the campaign tolerance levels of undecided voters.

Our comparative statics results show that when voters give more importance to the policy or the advertising dimensions, that the sufficient pivotal vote share increases thus making it more difficult for the electoral mean to be a strick LNE of the election. Whether the necessary pivotal vote share increases or decreases depends on the relative variance of voters' ideals in one dimension compared to the aggregate variance of voters' preferences around the electoral mean.

Recent literature focuses on the effects of campaign expenditure on US elections. Meirowitz (2008) develops a model in which candidates select their campaign effort

[6]The electoral mean is defined by the mean of voters' ideal policies and campaign tolerance levels.

(time, energy and money) to win elections and shows that marginal asymmetries in costs or technology can explain incumbency advantage ignoring the policy side of the election. Herrera et al. (2008) find that greater volatility in voters' policy preferences forces the two parties to spend more on the election. Ashworth and Bueno de Mesquita (2009) assume that candidates buy valence to increase their electoral chances. In our model, asymmetries between candidates are generated by the effectiveness of the policy and ad campaigns on each voter; moreover, the volatility of voters' policy and campaign preferences affect the pivotal vote shares and this affects the LNE of the election and thus the electoral outcome. The ad messages sent by candidates influence voters' choices but this effect varies across voters and so cannot be considered a valence.

Section 2 describes the preliminaries of the model. The stochastic electoral model of the election is presented in Sect. 3. Candidates' best response functions are derived in Sect. 5, a discussion of the common electoral campaign is presented in Sect. 6 with the conditions under which there is convergence to the electoral mean derived in Sect. 7. Comparative statics—Sect. 8—shows how the parameters of the model affect convergence to the electoral mean. Final comments are given in Sect. 9. The derivation of the second order conditions for convergence and of the comparative statics are given in the Appendix.

## 2 Preliminaries

We develop a stochastic electoral model in which at least two candidates compete in the US presidential election (e.g., Obama, McCain and Nader ran in 2000 election) and model an election where candidates' *electoral campaigns* consist of their policy platform and advertising (or ad) campaign. Since voters perfectly observe candidates' policy announcements, the advertising potion of the campaign serves *only* to convince voters that they matter to candidates and gives candidates another instrument with which to compete during the election. The ad portion of the campaign consists of messages[7] candidates send to voters. We assume that as messages increase so do the campaign advertising costs regardless of the effectiveness of the ad campaign.[8] We study candidates' electoral campaign game in response to the anticipated electoral outcome.

Let $C = \{1, ..., j, ..., c\}$ be the set of candidates competing in the election. Prior to the election, candidates simultaneously announce their electoral campaign.

[7]Messages are measured in continuous, rather than discrete, terms so that messages of different lengths can be compared. If message 1 is twice as long as message 2, then message 1 is assigned a number twice as large as message 2. In this paper, we assume messages are one-dimensional.

[8]The other implicit assumption here is that candidates have sufficient funds to finance their advertising campaigns. We leave it to another paper to examine the effect that activists and their donations have on electoral campaigns.

Candidate $j$'s electoral campaign, for all $j \in \mathcal{C}$, consists of a policy, $z_j \in \mathcal{Z} \subseteq \mathbb{R}$, and ad messages, $a_j \in \mathcal{A} \subseteq \mathbb{R}$. Denote by $\mathcal{K}_j = \mathcal{Z} \times \mathcal{A}$ each candidate' electoral campaign space, and by $\mathcal{K} = \prod_{j \in C} \mathcal{K}_j$ the campaign space of all candidates.

Let **z** and **a** be the electoral campaign profile of all candidates, i.e.,

$$\mathbf{z} \equiv (z_1, \ldots, z_c) \quad \text{and} \quad \mathbf{a} \equiv (a_1, \ldots, a_c). \tag{1}$$

The vector $\mathbf{z}_{-j}$ (respectively $\mathbf{a}_{-j}$) denotes the profile where all policies and ad messages *except j's policy* (*ad message*) are held constant. Let $n$ and $\mathcal{N} = \{1, .., i, .., n\}$ represent the number and the set of voters in the country.

The utility voters derive from each candidate depends on candidates' policies and advertising campaigns. Voter $i$'s utility is given by the vector of utilities, one for each candidate, i.e.,

$$u_i(\mathbf{z}, \mathbf{a}) = (u_{i1}(z_1, a_1), .., u_{ij}(z_j, a_j), \ldots, u_{ic}(z_c, a_c))$$

where the utility $i$ derives from candidate $j$, $u_{ij}(z_j, a_j)$, is given by

$$\begin{aligned} u_{ij}(z_j, a_j) &= -\beta(x_i - z_j)^2 - e(t_i - a_j)^2 + (\alpha_j + \xi_{ij}) + (\tau_j + \varsigma_{ij}) + (\lambda_j + \epsilon_{ij}) \\ &= u^*_{ij}(z_j, a_j) + \xi_{ij} + \varsigma_{ij} + \epsilon_{ij}. \end{aligned} \tag{2}$$

Voter $i$ is characterised by her ideal policy $x_i \in \mathcal{Z}$ and by the messages she ideally wants to receive from any candidate, her *campaign tolerance level* $t_i \in \mathcal{A}$ with the $(x_i, t_i)$ pair drawn from a joint distribution $\Gamma_x \times \Gamma_a$. Note that the dependence of $u_{ij}$ on $(x_i, t_i)$ is taken as understood and that $u^*_{ij}$ in (2) measures the *observable* component of $u_{ij}$. The coefficients $\beta$ and $e$ in (2) are positive, the same for all voters, common knowledge and measure the importance voters give to $j$'s policy and ad campaign. Voters policy preferences, $-\beta(x_i - z_j)^2$ in (2), says that the farther $j$'s policy $z_j$ is from $i$'s ideal policy $x_i$, the lower the utility $i$ derives from candidate $j$.

With policies being perfectly observable, the only reason candidates advertise in this model is to give voters a *further* impetus to vote for them and do so through ad messages. Since candidates directly contact voters, we assume voters have preferences over how often they want to be contacted by candidates, i.e., by their campaign tolerance level. Voters judge candidates' messages by examining the message frequency compared to their campaign tolerance level and since from the point of view of each voter there can be too much or too little advertising, we model voters as having quadratic preferences over ad campaigns. The *effectiveness of j's ad campaign on voter i*, $e(t_i - a_j)^2$ in (2), depends on the importance voters give to candidates' ad campaigns, $e$, *and* on the messages $j$ sends, $a_j$, relative to $i$'s campaign tolerance $t_i$.[9] When $a_j > t_i$, $i$ believes $j$ engaged in too much advertising,

[9]Since advertising costs rise with the message frequency, $i$'s campaign tolerance indirectly measure $i$'s preferred campaign spending.

an irritant to voter $i$ that leads to campaign fatigue[10] and if $a_j < t_i$, $i$ believes that $j$'s ad campaign reflects $j$'s lack of concern for voters.

Voters sociodemographic characteristics affect their choice of candidate.[11] Voter $i$'s *sociodemographic valence* for candidate $j$ is modelled by $\alpha_j + \xi_{ij}$ in (2). Candidate $j$'s *mean sociodemographic valence*, $\alpha_j$, captures the idea that voters with similar sociodemographic characteristics (gender, age, class, education, financial situation, etc.) share a common evaluation or bias for $j$. Voters are identified by a $d$-vector $\boldsymbol{\eta}_i$ denoting their individual sociodemographic characteristics with mean $\boldsymbol{\eta} = \frac{1}{n}\sum_{i\in\mathcal{N}} \boldsymbol{\eta}_i$. The importance sociodemographic characteristics have on the utility $i$'s derives from $j$ are modelled by a $d$-vector $\boldsymbol{\gamma}_j$ *common* to all voters. The composition $\alpha_j = \{(\boldsymbol{\gamma}_j \cdot \boldsymbol{\eta})\}$ measures the *mean* sociodemographic valence for $j$. The effect that $i$'s sociodemographic characteristics have on $i$'s choice of candidate has an idiosyncratic component, $\xi_{ij}$, that varies around $\alpha_j$ according to a Type-I extreme value distribution, $E(0, \frac{\pi}{6})$, with mean zero and variance $\frac{\pi}{6}$.

The effect of candidate $j$'s *traits* (race, gender, age, charisma, experience in public life, etc.) on voter $i$'s choice is modelled by $\tau_j + \varsigma_{ij}$ in (2). The commonly known *mean trait valence* for $j$, $\tau_j$, measures the mean influence $j$'s traits have on voters' choice with the effect on voter $i$ also depending an idiosyncratic component, $\varsigma_{ij}$. Candidates are identified by a $t$-vector $\mathbf{t}_j$ capturing their individual traits. The *average* importance voters give to these traits is given by a $t$-vector $\boldsymbol{\varpi}_j$ *common* to all voters. The composition $\tau_j = \{(\mathbf{t}_j \cdot \boldsymbol{\varpi}_j)\}$ measures the mean effect that $j$'s traits have on voters. We assume that voters' beliefs on the effect that candidates traits have on their utility are *not* observable. Voter $i$'s private component, $\varsigma_{ij}$, varies around $\tau_j$ and is drawn from a Type-I extreme value distribution, $T(0, \frac{\pi}{6})$, with zero mean and variance $\frac{\pi}{6}$.

The term $\lambda_j + \epsilon_{ij}$ in (2) represents $i$'s belief on candidate $j$'s competence. The *mean competence valence*, $\lambda_j$, measures the *common belief* voters have on $j$'s ability to govern. The private portion of $i$'s competence signal, $\epsilon_{ij}$, varies around $\lambda_j$ and is drawn from a Type-I extreme value distribution, $F(0, \frac{\pi}{6})$ with mean zero and variance $\frac{\pi}{6}$.

We assume that the three idiosyncratic components in voters' utility are drawn independent of each other. Moreover, Train (2003) has shown that the sum of independent Type-I Extreme Value Distributions all with the same mean and variance has a Type I Extreme Value Distribution. Thus, since $\xi_{ij} \sim E(0, \frac{\pi}{6})$, $\varsigma_{ij} \sim T(0, \frac{\pi}{6})$ and $\epsilon_{ij} \sim F(0, \frac{\pi}{6})$ are Type I Extreme Value Distributions then their sum $\varkappa_{ij} = \xi_{ij} + \varsigma_{ij} + \epsilon_{ij} \sim \mathcal{X}(0, \frac{\pi}{6})$ also has a Type I Extreme value Distribution.

[10]As happens when voters are too frequently contacted by robo-calls or Twitts. Voters' vary in the frequency with which they want candidates to contact them, including the number of visits they want to have by campaign volunteers. Some gladly talk to volunteers or read and re-Twitt the messages they receive, and others just close their doors on volunteers or deleting the messages they receive.

[11]See footnote 5 on how African-Americans vote in the US.

Under these circumstances, voter $i$'s utility in (2) can be re-written as follows:

$$\begin{aligned} u_{ij}(z_j, a_j) &= -\beta(x_i - z_j)^2 - e(t_i - a_j)^2 + \alpha_j + \tau_j + \lambda_j + \varkappa_{ij} \\ &= u^*_{ij}(z_j, a_j) + \varkappa_{ij}. \end{aligned} \quad (3)$$

Since the idiosyncratic sociodemographic, traits and competence components of voters' utilities are unobserved by candidates and become known to voters after candidates choose their campaign strategies and candidates know that these components are drawn from Type I Extreme Value Distributions, voters' utilities are stochastic. Thus, given policies and advertising levels $(\mathbf{z}, \mathbf{a})$ and the distributions $E$, $T$, $F$ and $\mathcal{X}$, the probability that voter $i$ chooses candidate $j$ is given by

$$\rho_{ij}(\mathbf{z}, \mathbf{a}) = \Pr[u_{ij}(z_j, a_j) > u_{ih}(z_h, a_h), \text{ for all } h \neq j \in \mathcal{C}]$$

where Pr is the probability operator generated by the distribution assumptions on $E$, $T$, $F$ and $\mathcal{X}$ and where $u_{ij}(z_j, a_j)$ is given by (3). The probability that $i$ votes for $j$ is given by the probability that $u_{ij}(z_j, a_j) > u_{ih}(z_h, a_h)$ for all $j, h \in \mathcal{C}$, i.e., given by the probability that $i$ gets a higher utility from $j$ than from any other candidate.

Since the idiosyncratic sociodemographic, traits and competence components and their sum follow Type-I extreme value distributions, the probability that $i$ votes for $j$ has a logit specification,

$$\rho_{ij} \equiv \rho_{ij}(\mathbf{z}, \mathbf{a}) \equiv \frac{\exp[u^*_{ij}(z_j, a_j)]}{\sum_{h=1}^{c} \exp[u^*_{ih}(z_h, a_h)]} = \left[\sum_{h=1}^{c} \exp[u^*_{ih}(z_h, a_h) - u^*_{ij}(z_j, a_j)]\right]^{-1} \quad (4)$$

for all $i \in \mathcal{N}$ and $j \in \mathcal{C}$ with the dependence of $\rho_{ij}$ on $(\mathbf{z}, \mathbf{a})$ sometimes omitted.

Candidates' objective is to maximize their vote shares. Since candidates do not observe the idiosyncratic components of voters' utility, they cannot perfectly anticipate how voters will vote but can estimate their *expected* vote shares. Candidate $j$'s *expected vote share* is the average of the voting probabilities of all voters in the country, i.e.,

$$V_j(\mathbf{z}, \mathbf{a}) \equiv \frac{1}{n} \sum_{i \in \mathcal{N}} \rho_{ij}(\mathbf{z}, \mathbf{a}) \quad (5)$$

where $\rho_{ij}$ is given by (4) with the expected vote shares adding to 1, $\sum_{j \in C} V_j(\mathbf{z}, \mathbf{a}) = 1$.

Candidate $j$'s chooses its policy and ad campaign to maximize its expected vote share, i.e.,

$$\max_{z_j \in \mathcal{Z}, a_j \in \mathcal{A}} V_j(\mathbf{z}, \mathbf{a}) \quad \text{for all} \quad j \in \mathcal{C}.$$

## 3 The Stochastic Electoral Model

The timing of events in this stochastic election model is as follow:

(1) Candidates simultaneously announce their policy platforms and advertising campaigns.
(2) After observing platforms and advertising levels, each voter observes the idiosyncratic components of her utility function.
(3) The election takes place.
(4) The President elect implements the announced policy platform.

We now formally define the election model.

**Definition 1** The stochastic election model can be represented by a game in normal form $\Delta\langle \mathcal{C}, \mathcal{K}, \Gamma_x \times \Gamma_a, \mathbf{V}\rangle$, where

1. **Players**: $\mathcal{C} = \{1, .., j, .., c\}$ is the set of candidates competing in the election.
2. **Strategies**: $\mathcal{K}_j = \mathcal{Z} \times \mathcal{A}$ denotes the electoral campaign space of candidate $j$ with $\mathcal{Z}$ denoting the policy space and $\mathcal{A}$ the ad messaging space, and where the electoral campaign space of all candidates is given by $\mathcal{K} = \Pi_{j\in C}(\mathcal{K}_j)$.
3. **Voters**: Voters are characterized by their ideal policy and campaign tolerance level, $(x_i, t_i) \in \mathcal{Z} \times \mathcal{A}$ for $i \in \mathcal{N}$ and by the mean sociodemographic, traits and competence valences, $(\alpha_j, \tau_j, \lambda_j)$. Voters' ideal policies and campaign tolerance levels are drawn from the joint distribution $\Gamma_x \times \Gamma_a$. The utility voter $i$ derives from candidate $j$ is given by (2) and subject to three random shocks generated by Type-I Extreme Value Distributions, $E(0, \frac{\pi}{6})$, $T(0, \frac{\pi}{6})$ and $F(0, \frac{\pi}{6})$ whose sum $\mathcal{X}(0, \frac{\pi}{6})$ also follows a Type-I Extreme value distribution.
4. **Candidates' Payoff Functions**: For any candidate $j \in \mathcal{C}$, $j$'s expected vote share function is given by $V_j : \mathcal{K}_j \rightarrow [0, 1]$ in (5). Let $\mathbf{V}(\mathbf{z}, \mathbf{a}) = ((V_j(\mathbf{z}, \mathbf{a})$ for all $j \in \mathcal{C})$ represent the payoff function profile of all candidates.

Let us find the *local Nash equilibria* (LNE) of this stochastic election model.

## 4 The Equilibrium of the Stochastic Election Model

The electoral campaign profile $(\mathbf{z}^*, \mathbf{a}^*)$ is a local Nash equilibrium (LNE), if when candidates choose their policies in $\mathbf{z}^*$ and ad campaigns in $\mathbf{a}^*$, each candidate is at a local maximum in its expected vote share functions. That is, the vector $(\mathbf{z}^*, \mathbf{a}^*)$ is such that no candidate may either shift its policy *or* advertising level by a *small* amount—ceteris paribus—to increase its vote share. Formally,

**Definition 2** A strick (weak) local Nash equilibrium of the stochastic electoral campaign model $\Delta\langle \mathcal{C}, \mathcal{K}, \Gamma_x \times \Gamma_a, \mathbf{V}\rangle$ is a vector of candidate policies and advertising campaign levels, $(\mathbf{z}^*, \mathbf{a}^*)$ such that for each candidate $j \in \mathcal{C}$:

1. There exists a small neighborhood of $z_j^*$, $B(z_j^*) \subset \mathcal{Z}$, such that

$$V_j(\mathbf{z}^*, \mathbf{a}^*) > (\geq) V_j(z_j', \mathbf{z}_{-j}^*, \mathbf{a}^*) \quad \text{for all} \quad z_j' \in B(z_j^*) - \{z_j^*\}.$$

2. There exists a small neighborhood of $a_j^*$, $B(a_j^*) \subset \mathcal{A}$, such that

$$V_j(\mathbf{z}^*, \mathbf{a}^*) > (\geq) V_j(\mathbf{z}^*, a_j', \mathbf{a}_{-j}^*) \quad \text{for all} \quad a_j' \in B(a_j^*) - \{a_j^*\}.$$

Since the observable component of voter $i$'s utility for candidate $j$, $u_{ij}^*(z_j, a_j)$ in (2) for all $i \in \mathcal{N}$, and the distributions of voters' ideal policies and campaign tolerance levels, $\Gamma_x \times \Gamma_a$, are continuously differentiable, then so are $\rho_{ij}$ in (4) and $V_j(\mathbf{z}, \mathbf{a})$ in (5). We can then estimate how $j$'s expected vote share changes if $j$ *marginally* adjusts its policy *or* ad campaign, ceteris paribus.

*Remark 1* If in Definition 2 we can substitute $\mathcal{Z}$ for $B(z_j^*)$ and $\mathcal{A}$ for $B(a_j^*)$, then a local Nash equilibrium is also a pure strategy Nash equilibrium (PNE) of the election.

*Remark 2* In political models, the critical equilibrium may be characterized by positive eigenvalues for the Hessian of one of the political parties. As a consequence the expected vote share functions of such a candidate fails pseudo-concavity. Therefore, none of the usual fixed point arguments can be used to assert existence of a "global" pure Nash equilibrium (PNE). For this reason, we use the concept of a "critical Nash equilibrium" (CNE), namely a vector of strategies which satisfies the first-order condition for a local maximum of candidates' expected vote share functions. Standard arguments based on the index, together with transversality arguments can be used to show that a CNE will exist and that, generically, it will be isolated. A local Nash equilibrium (LNE) satisfies the first-order condition, together with the second-order condition that the Hessians of all candidates are negative (semi-) definite at the CNE. Clearly, the set of LNE will contain the PNE.

The parameters $(\beta, e, \alpha_j, \tau_j, \lambda_j)$ for all $j \in \mathcal{C}$ in voter $i$'s utility function in (2) as well as the distributions $\Gamma_x$ and $\Gamma_a$ of voters' ideal policies and campaign tolerance levels, $x_i$ and $t_i$, are exogenously given in the model. Using these parameters and the distribution of voters' preferences, we can relate any vector of candidate policies and advertising levels, $(\mathbf{z}, \mathbf{a})$, to a vector of expected vote share functions, i.e., we can estimate the vector of expected voters shares $\mathbf{V}(\mathbf{z}, \mathbf{a})$ given by

$$\mathbf{V}(\mathbf{z}, \mathbf{a}) = (V_j(\mathbf{z}, \mathbf{a}) \text{ for all } j \in \mathcal{C}).$$

Using simulations we can determine if one of the LNE is a Nash equilibrium of the election.

Candidate $j$'s expected vote share function is at a local maximum if the following *first* and *second* order conditions are satisfied. Candidate $j$'s *first order necessary conditions (FONC)* determine $j$'s *best response* policy and advertising functions.

There are then $2 \times p$ best response functions and their associated $2 \times p$ *critical values*,[12] one for each policy and ad campaign for each candidate.

We then examine whether at these critical values the expected vote shares functions of each candidate are at a *maximum, minimum or a saddle point*. To do so, we use the *Hessian matrix of second order partial derivatives of the vote share functions* evaluated at these critical values as the Hessian determines the *local curvature* of the vote share functions at these critical values. The *sufficient* (*necessary*) second order condition (SOC) for $j$'s expected vote share to be at a maximum at the critical values is that the Hessian be negative (semi-) definite at these critical values which happens only when the eigenvalues of the Hessian of the vote shares are all negative (non-positive) at these critical values.

Denote by $z_j^C$ and $a_j^C$ the *critical* ($C$) policy and advertising campaigns profiles of candidate $j$. The critical policy and campaign advertising profile of all candidates is given by

$$\mathbf{z}^C \equiv (\mathbf{z}_1^C, ..., \mathbf{z}_j^C, ..., \mathbf{z}_c^C) \quad \text{and} \quad \mathbf{a}^C \equiv (\mathbf{a}_1^C, ..., \mathbf{a}_j^C, ..., \mathbf{a}_c^C) \tag{6}$$

The profile $(\mathbf{z}^C, \mathbf{a}^C)$ represent a *prospective*[13] LNE of the election.

## 5 Candidates' Best Response Functions (FONC)

To derive the first order necessary condition (FONC) of candidates' vote shares, we first examine how *voters* respond to changes in candidate $j$'s policies or ad messages then examine how changes in positions and messaging levels affect candidates' *expected vote shares*.

The *marginal* impact of a change in $j$'s policy or message, $z_j$ or $a_j$, on the probability that $i$ votes for $j$ is given by *partial* derivative of $\rho_{ij}$ in (4) with respect to (*wrt*) to each choice variable, ceteris paribus, i.e.,

$$\mathbf{D}\rho_{ij}(\mathbf{z}, \mathbf{a}) \equiv \begin{pmatrix} \frac{\partial \rho_{ij}}{\partial z_j} \\ \frac{\partial \rho_{ij}}{\partial a_j} \end{pmatrix} = 2\rho_{ij}(1 - \rho_{ij}) \begin{pmatrix} \beta(x_i - z_j) \\ e(t_i - a_j) \end{pmatrix}. \tag{7}$$

The effect of a marginal change in one of $j$'s choice variables on $\rho_{ij}$ depends on the *endogenously* determined probability that $i$ votes for $j$, $\rho_{ij}$ in (4), and for any other candidate, $(1 - \rho_{ij})$; on the importance voters give to $j$'s choice variable $\beta$ or $e$; and on how far $i$'s ideal policy or tolerance level, $x_j$ or $t_j$, is from $j$'s corresponding choice $z_j$ or $a_j$.

[12]Called critical as we do not yet know whether at these values the expected vote shares functions are at a maximum.

[13]We say *prospective* because we do not know yet if candidates' critical choices satisfy the SOC for a maximum.

## 5.1 Candidates' Best Response Functions

To find $j$'s best response function take the partial derivative of $j$'s vote share function in (5) with respect to the corresponding variable, $z_j$ or $a_j$ and set it equal to zero, i.e.,

$$\mathbf{D}V_j(\mathbf{z}_j, \mathbf{a}_j) = \frac{1}{n}\sum\nolimits_{i\in\mathcal{N}} \mathbf{D}\rho_{ij} = \frac{1}{n}\sum\nolimits_{i\in\mathcal{N}} 2\rho_{ij}(1-\rho_{ij})\begin{pmatrix}\beta(x_i - z_j)\\ e(t_i - a_j)\end{pmatrix} = 0 \quad (8)$$

where the third term follows from (7) and $\rho_{ij}$ is given by (4). This FONC is satisfied when

$$\sum\nolimits_{i\in\mathcal{N}} \rho_{ij}(1-\rho_{ij})\begin{pmatrix}x_i - z_j\\ t_i - a_j\end{pmatrix} = 0.$$

After some manipulation $j$'s *best response functions* are then given by

$$\begin{pmatrix}z_j^C\\ a_j^C\end{pmatrix} = \sum\nolimits_{i\in\mathcal{N}} \mu_{ij}^C\begin{pmatrix}x_i\\ t_i\end{pmatrix} \quad \text{where} \quad \mu_{ij}^C \equiv \frac{\rho_{ij}^C(1-\rho_{ij}^C)}{\sum_{i\in\mathcal{N}} \rho_{ij}^C(1-\rho_{ij}^C)}. \quad (9)$$

We use superscript $C$ to indicate variables evaluated at $(\mathbf{z}^C, \mathbf{a}^C)$ and in short hand use $\mu_{ij}^C \equiv \mu_{ij}(\mathbf{z}^C, \mathbf{a}^C)$ and $\rho_{ij}^C \equiv \rho_{ij}(\mathbf{z}^C, \mathbf{a}^C)$ in (9).

The result in (9) states that $j$'s best response policy or ad messaging functions, $(z_j^C, a_j^C)$, are a *weighted* average of voters' ideal policy or campaign tolerance levels where the *endogenously* determined *weight* candidate $j$ gives to *voter i* is given by $\mu_{ij}^C$ in (9). The weight $\mu_{ij}^C$ depends on how likely is $i$ to vote for $j$, $\rho_{ij}$ in (4) evaluated at $(\mathbf{z}^C, \mathbf{a}^C)$, $\rho_{ij}^C$, and to vote for any other candidate, $(1-\rho_{ij}^C)$, *relative* to all voters.[14] Since $\rho_{ij}^C$ and $\mu_{ij}^C$ depend on $(\mathbf{z}^C, \mathbf{a}^C)$, so do $j$'s best response functions, that is, $z_j^C$ and $a_j^C$ depend on the policy and advertising choices of all *other* candidates. Note that voters receive the *same* weight in $j$'s policy or advertising campaign.

The weight $\mu_{ij}^C$ in (9) is non-monotonic in $\rho_{ij}^C$. To see this, denote the numerator of $\mu_{ij}^C$ by $W_{ij}^C \equiv \rho_{ij}^C(1-\rho_{ij}^C)$. When $i$ votes for $j$ with probability one ($\rho_{ij}^C = 1$ so that $i$ is a *core* supporter) or when $i$ votes for $j$ with probability zero ($\rho_{ij}^C = 0$), then $W_{ij}^C = 0$ and $j$ gives these two voters zero weight in its critical policy and ad campaign choices, i.e., $\mu_{ij}^C = 0$. When $i$ votes for $j$ with probability $\rho_{ij}^C = \frac{1}{2}$, $i$ is *undecided*, and since $W_{ij}^C = \frac{1}{4}$, $j$ gives $i$ the highest weight in its critical choices. Thus, even though the one-person-one-vote principle applies in this model, candidates weight

[14] Suppose voters are equally likely to vote for $j$ with probability $\rho_{ij} = \upsilon$. The weight $j$ gives each voter is $\mu_{ij} = \dfrac{1}{n}$, as intuitively expected, $j$ weights each voter according to the inverse of the voting population.

voters differently in their electoral campaigns, e.g., $\mu_{ij}^C(\rho_{ij}^C = 0) = \mu_{ij}^C(\rho_{ij}^C = 1) = 0 < \mu_{ij}^C(\rho_{ij}^C = \frac{1}{2})$. In particular, $j$ caters to undecided voters by giving them a higher weight in its best response functions.

Thus, (9) says that when holding $(\mathbf{z}_{-j}^C, \mathbf{a}_{-j}^C)$ constant, $j$'s policy and advertising best response functions, $(z_j^C, a_j^C)$, are weighted averages of voters' ideal policies or campaign tolerance levels where the weight $j$ gives $i$ along the policy and advertising dimensions is given by $\mu_{ij}^C$. Note that when $\mu_{ij}^C$ varies across candidates, candidates adopt different campaigns.

Candidate $j$'s critical ad campaign, $a_j^C$, determines the number of messages $j$ sends to voters. We assume that developing and sending these messages cost money[15]: the more messages $j$'s sends, the more money $j$ spends on advertising.[16]

The weight $j$ gives voters in its critical choices, $\mu_{ij}^C$ in (9), is endogenous since $\mu_{ij}^C$ is the crucial factor determining candidates' policies and ad campaigns *and* depends on the probability voter $i$ chooses candidate $j$, $\rho_{ij}^C$ *which in turn* depends on candidates' positions and advertising campaign levels. These weights change between elections as the parameters $(\beta, e, \alpha_j, \tau_j, \lambda_j)$ for all $j \in C$ in voter $i$'s utility function in (2) as well as the distributions $\Gamma_x$ and $\Gamma_a$ of voters' ideal policies and campaign tolerance levels, $x_i$ and $t_i$ for all $i \in \mathcal{N}$, vary across elections.

## 6 Is There a Common Campaign?

The objective is to find candidates' local Nash equilibrium (LNE) policies and advertising campaigns. We now examine whether candidates *adopt or converge to* a common campaign.

**Definition 3** Let $(z_m, a_m)$ represent electoral mean, i.e., the mean of voters' ideal policies and campaign tolerance levels,

$$\begin{pmatrix} z_m \\ a_m \end{pmatrix} \equiv \frac{1}{n} \sum_{i \in N} \begin{pmatrix} x_i \\ t_i \end{pmatrix} \tag{10}$$

and define $(\mathbf{z}_m, \mathbf{a}_m) = ((z_m, a_m), ..., (z_m, a_m))$ as the joint electoral mean for all candidates.

**Lemma 1 (Common Campaign)** *If candidates adopt a common campaign, they adopt the electoral mean as their electoral campaign strategy.*

*Proof* The proof goes as follows: assume candidates adopt the same campaign. Given identical campaigns, show that since all voters vote with the same probability

[15]E.g., by buying personal data on voters that campaign strategists use to develop the message send to voters.

[16]Recall that $j$'s campaign effectiveness is determined by voters' campaign tolerance and the importance voters give to the ad campaign [see (2)] and not by how much $j$ spends on advertising.

for each candidate they receive the *same* weight in candidates' critical policy and ad campaigns. Finally, show that candidates' best response is to adopt the electoral mean as their campaign strategy.

Suppose candidates adopt the same campaign, i.e., $(z_j, a_j) = (z^O, a^O)$ for $j \in \mathcal{C}$. The probability that voter $i$ chooses candidate $j$, $\rho_{ij}^O$ in (4) at $(z^O, a^O)$ depends on the *difference* between the observable component of $i$'s utility from candidates $h$ and $j$ which using (2) is given by

$$u_{ih}^{*O} - u_{ij}^{*O} \equiv u_{ih}^{*}(z^O, a^O) - u_{ij}^{*}(z^O, a^O) = (\alpha_h - \alpha_j) + (\tau_h - \tau_j) + (\lambda_h - \lambda_j) \quad (11)$$

since under identical campaigns, the policy and ad campaign components cancel out.

Using (11), the probability that voter $i$ chooses candidate $j$ is then independent of candidates' policy and ad campaign since $\rho_{ij}$ in (4) when evaluated at $(\mathbf{z}^O, \mathbf{a}^O)$ reduces to

$$\rho_j^O = \left[ \sum_{h \neq j, h=1}^{c} \exp[u_{ih}^{O*} - u_{ij}^{O*}] \right]^{-1} = \left[ \sum_{h \neq j, h=1}^{c} \exp[(\alpha_h - \alpha_j) + (\tau_h - \tau_j) + (\lambda_h - \lambda_j)] \right]^{-1}, \quad (12)$$

i.e., $\rho_j^O$ depends only on the difference between candidates' mean sociodemographic, traits and competence valences and not on their campaign strategies. In addition, (12) implies that voters choose $j$ with *equal* probability as $\rho_j^O$ does not depend on voters' ideal policies or campaign tolerance levels, i.e., $\rho_{ij}^O = \rho_j^O$ for all $i \in \mathcal{N}$. In this case, $\rho_j^O$ also represents $j$'s expected vote share at $(\mathbf{z}^O, \mathbf{a}^O)$ since using (5), $V_j(\mathbf{z}^O, \mathbf{a}^O)$ is given by

$$V_j^O \equiv V_j(\mathbf{z}^O, \mathbf{a}^O) \equiv \frac{1}{n} \sum\nolimits_{i \in N} \rho_j^O = \rho_j^O. \quad (13)$$

With voters choosing $j$ with the same probability, $\rho_{ij}^O = \rho_j^O$ for all $i \in \mathcal{N}$ in (19), $j$ gives voters the same weight in its policy and ad campaign as $\mu_{ij}^C$ in (9) reduces to

$$\mu^O \equiv \frac{\rho_j^O(1 - \rho_j^O)}{\sum_{i \in N} \rho_j^O(1 - \rho_j^O)} = \frac{1}{n}, \quad (14)$$

i.e., $j$ weights each voter according to the inverse of the voting population.

In this case, $j$'s critical *choices*, $(z_j^C, a_j^C)$ in (9) using (14), are given by

$$\begin{pmatrix} z_j^C \\ a_j^C \end{pmatrix} = \sum\nolimits_{i \in N} \mu^O \begin{pmatrix} x_i \\ t_i \end{pmatrix} = \frac{1}{n} \sum\nolimits_{i \in N} \begin{pmatrix} x_i \\ t_i \end{pmatrix} \equiv \begin{pmatrix} z_m \\ a_m \end{pmatrix} \quad (15)$$

where the last term follows from (10). This says that if *all other* candidates adopt $(z^O, a^O)$ as their campaign strategy, then *j*'s best response is to adopt the electoral mean $(\mathbf{z}_m, \mathbf{a}_m)$ as its campaign strategy. As this is true for all candidates, the Lemma is proven. ■

Using (13), define $V_j^m$ as *j's expected vote share*[17] *at the electoral mean*, $(\mathbf{z}_m, \mathbf{a}_m)$, i.e.,

$$V_j^m \equiv V_j^m(\mathbf{z}_m, \mathbf{a}_m) = \rho_j(\mathbf{z}_m, \mathbf{a}_m) \equiv \rho_j^m \tag{16}$$

where $\rho_j^m$ is $\rho_j^O$ in (12) evaluated at $(\mathbf{z}_m, \mathbf{a}_m)$. $V_j^m = \rho_j^m$ is exogenously given since from (12) $\rho_j^m$ depends only on the exogenously given mean sociodemographic, traits and competence valences. Note that when candidates adopt the electoral mean as their campaign strategy, the campaign profile $(\mathbf{z}_m, \mathbf{a}_m)$ satisfies the FONC for *all* candidates.

Moreover, we can rank candidates' expected vote shares at the electoral mean. From (12) it is clear that to rank $V_j^m$ for all $j \in C$, the mean sociodemographic, traits and competence valences, $(\alpha_j, \tau_j, \lambda_j)$ in (2), of all candidates need to be ranked. To do so, define the *expected composite valence of candidate j*, $\kappa_j$, as

$$\kappa_j = \alpha_j + \tau_j + \lambda_j \quad \text{for all } j \in C. \tag{17}$$

Let candidate 1 (candidate *c*) be the one with the *lowest* (*highest*) expected composite valence so that candidates' expected composite valences satisfy

$$\kappa_1 \leq \ldots \leq \kappa_j \leq \ldots \leq \kappa_c. \tag{18}$$

Using $\kappa_j$ in (17), the probability that voter *i* chooses candidate *j* in (12) when candidates adopt the electoral mean as their campaign strategy is given by

$$\rho_j^m = \left[\sum\nolimits_{h \neq j, h=1}^{c} \exp[(\alpha_h + \tau_h + \lambda_h) - (\alpha_j + \tau_j + \lambda_j)]\right]^{-1} = \left[\sum\nolimits_{h \neq j, h=1}^{c} \exp[\kappa_h - \kappa_j]\right]^{-1}. \tag{19}$$

Given the composite valence ranking in (18), candidates' expected vote shares at the electoral mean $V_j^m$ in (16) satisfy

$$V_1^m = \rho_1^m \leq \ldots \leq V_j^m = \rho_j^m \leq \ldots \leq V_c^m = \rho_c^m. \tag{20}$$

At the electoral mean, $(z_m, a_m)$, candidate 1 with the lowest composite valence, $\kappa_1$, is also the candidate with the *lowest* expected vote share.

Let us now determine the conditions under which candidates adopt the electoral mean as their campaign strategy.

[17]We identify functions evaluated at $(\mathbf{z}_m, \mathbf{a}_m)$ with a subscript or superscript *m*, e.g., the weight given to votes at $(\mathbf{z}_m, \mathbf{a}_m)$ is given by $\mu^m = \frac{1}{n}$.

## 7 Convergence Conditions

We now derive the conditions under which $j$ *adopts the electoral mean as its* critical campaign, $(z_j^C, a_j^C) = (z_m, a_m)$, holding the electoral campaign of all other candidates at the electoral mean.

The Appendix shows that two pivotal vote shares characterize the *necessary* and *sufficient* conditions under which the electoral mean is a local Nash equilibrium (LNE) of the election, i.e., the conditions under which all candidates adopt $(z_m, a_m)$ as their electoral campaign strategy.

**Definition 4** The *necessary (n) and sufficient (s) pivotal (p) vote shares* at $(z_m, a_m)$ are given by

$$\mathcal{V}_{sp}^m \equiv \mathcal{V}_{sp}(\mathbf{z}_m, \mathbf{a}_m) \equiv \frac{1}{2} - \frac{1}{4}\frac{1}{Tr\,[(\mathbf{B})^{-1}\nabla^m]} = \frac{1}{2} - \frac{1}{4}\frac{1}{\frac{1}{\beta}var(x) + \frac{1}{e}var(t)} \tag{21}$$

$$\mathcal{V}_{np}^m \equiv \mathcal{V}_{np}(\mathbf{z}_m, \mathbf{a}_m) \equiv \frac{1}{2} - \frac{1}{4}\frac{Tr(\mathbf{B})}{Tr\,(\nabla^m)} = \frac{1}{2} - \frac{1}{4}\frac{\beta + e}{var(x) + var(t)} \tag{22}$$

where the importance voters give to each dimension are represented by $\mathbf{B}$ defined as

$$\mathbf{B} \equiv \begin{pmatrix} \beta & 0 \\ 0 & e \end{pmatrix} \quad \Rightarrow \quad Tr(\mathbf{B}) = \beta + e \tag{23}$$

so that its trace, $Tr(\mathbf{B})$, measures the aggregate importance voters give to the policy and advertising dimensions.

The matrix $\nabla^m$ in (21) and (22) given by

$$\nabla^m \equiv \frac{1}{n}\sum\nolimits_{i\in\mathcal{N}} \begin{pmatrix} \beta(x_i - z_m)^2\beta & \beta(x_i - z_m)(t_i - a_m)e \\ \beta(x_i - z_m)(t_i - a_m)e & e(t_i - a_m)^2 e \end{pmatrix} \tag{24}$$

measures the weighted variance-covariance matrix of the distribution of voters preferences around the electoral mean $(z_m, a_m)$.

Let $var(x)$ represent the variance of voters' ideal policies, i.e.,

$$var(x) \equiv \frac{1}{n}\sum\nolimits_{i\in\mathcal{N}} \beta(x_i - z_m)^2\beta. \tag{25}$$

and $var(t)$ the variance of voters' campaign tolerance levels, i.e.,

$$var(t) \equiv \frac{1}{n}\sum\nolimits_{i\in\mathcal{N}} e(t_i - a_m)^2 e. \tag{26}$$

Using (24)–(26), the trace of $\boldsymbol{\nabla}^m$, $Tr\,(\boldsymbol{\nabla}^m)$, given by

$$Tr\,(\boldsymbol{\nabla}^m) = \frac{1}{n}\sum\nolimits_{i\in\mathcal{N}} \beta(x_i - z_m)^2\beta + \frac{1}{n}\sum\nolimits_{i\in\mathcal{N}} e(t_i - a_m)^2 e$$
$$= var(x) + var(t), \tag{27}$$

measures the "aggregate" variance of voters' distribution around the electoral mean and is always positive. The trace of $(\mathbf{B})^{-1}\boldsymbol{\nabla}^m$, $Tr\left[(\mathbf{B})^{-1}\boldsymbol{\nabla}^m\right]$ in (21), using (25) and (26),

$$Tr\left[(\mathbf{B})^{-1}\boldsymbol{\nabla}^m\right] = \frac{1}{\beta}\frac{1}{n}\sum\nolimits_{i\in\mathcal{N}} \beta(x_i - z_m)^2\beta + \frac{1}{e}\frac{1}{n}\sum\nolimits_{i\in\mathcal{N}} e(t_i - a_m)^2 e$$
$$= \frac{1}{\beta}var(x) + \frac{1}{e}var(t), \tag{28}$$

scales the weighted variance in each dimension by the importance voters give respectively to the policy, $\beta$, and advertising, $e$, dimensions and is always positive.

**Lemma 2 (Comparing Pivotal Vote Shares)** *It is always the case that the sufficient pivotal vote share is higher than the necessary one, i.e.,* $\mathcal{V}^m_{sp} > \mathcal{V}^m_{np}$ *for all* $j \in \mathcal{C}$.

*Proof* See Appendix A.

Lemma 2 proves that the sufficient condition for convergence to the electoral mean is more stringent than the necessary one, i.e., $\mathcal{V}^m_{sp} > \mathcal{V}^m_{np}$ for all $j \in \mathcal{C}$.

The following proposition gives the sufficient and necessary conditions for the joint electoral mean, $(\mathbf{z}_m, \mathbf{a}_m)$, to be an equilibrium of the election.

**Proposition 1 (Convergence to the Electoral Mean)**

*Assume that candidates adopt the electoral mean* $(z_m, a_m)$ *as their campaign strategy. Let voters' idiosyncratic sociodemographic, traits and competency components of the valences,* $\xi_{ij} \sim E(0, \frac{\pi}{6})$, $\varsigma_{ij} \sim T(0, \frac{\pi}{6})$ *and* $\epsilon_{ij} \sim F(0, \frac{\pi}{6})$ *and their sum* $\varkappa_{ij} = \xi_{ij} + \varsigma_{ij} + \epsilon_{ij} \sim \mathcal{X}(0, \frac{\pi}{6})$ *follow Type-I extreme value distributions.*

- *If* $\mathcal{V}^m_{sp} \leqslant V^m_j$ *for all* $j \in \mathcal{C}$, *the* ***sufficient*** *condition for convergence to* $(z_m, a_m)$ *has been met by all candidates. The joint electoral mean,* $(\mathbf{z}_m, \mathbf{a}_m)$, *constitutes a* ***strick*** *LNE of the election.*
- *If* $\mathcal{V}^m_{np} \leqslant V^m_j < \mathcal{V}^m_{sp}$ *for some* $j \in \mathcal{C}$ *and* $\mathcal{V}^m_{sp} \leqslant V^m_h$ *for* $h \neq j \in \mathcal{C}$, *the* ***necessary*** *but not the sufficient condition for convergence to* $(z_m, a_m)$ *is met by candidate j. The joint electoral mean,* $(\mathbf{z}_m, \mathbf{a}_m)$, *is a* ***weak*** *LNE of the election.*
- *If* $V^m_j < \mathcal{V}^m_{np}$ *for some* $j \in \mathcal{C}$, *the* ***necessary*** *condition for convergence to* $(z_m, a_m)$ *is* ***not*** *met for candidate j. The joint electoral mean,* $(\mathbf{z}_m, \mathbf{a}_m)$, *is* ***not*** *a LNE of the election and at least one candidate adopts a campaign that differs from the electoral mean.*

*Proof* See Appendix.

The Appendix shows that when all candidates' electoral campaigns, except $j$'s, are held constant, the *sufficient* condition for $j$ to adopt the electoral mean, $(z_m, a_m)$, as its electoral campaign, also given in the Appendix in (53), is that

$$\mathcal{V}_{sp}^m \leqslant V_j^m \tag{29}$$

where $\mathcal{V}_{sp}^m$ is given by (21) and $V_j^m$ by (16). Condition (29) says that $j$'s expected vote share at $(\mathbf{z}_m, \mathbf{a}_m)$, $V_j^m$, must be *high* enough for $j$ to adopt $(z_m, a_m)$ as its electoral campaign.

Note that $\mathcal{V}_{sp}^m$ on the left hand side (LHS) of (29), also defined in (21), is independent of $j$'s policy and ad campaign and is the same for all candidates. Thus, (29) gives the *sufficient condition* for all candidates to converge to the electoral mean. Note that if $\mathcal{V}_{sp}^m \leqslant V_1^m$, i.e., if candidate 1 with the lowest composite valence, meets the sufficient condition for convergence, then from (20) so will all the other candidates. When (29) is satisfied for all $j \in \mathcal{C}$, the joint electoral mean $(\mathbf{z}_m, \mathbf{a}_m)$ is a *strick* LNE of the election.

The *necessary* convergence condition for candidate $j$ to adopt $(z_m, a_m)$ as its electoral campaign—also given in the Appendix in (48)—is that

$$\mathcal{V}_{np}^m \leqslant V_j^m \tag{30}$$

where $\mathcal{V}_{np}^m$ is given by (22) and $V_j^m$ by (16). Condition (30) says that the necessary condition for $j$ to adopt $(z_m, a_m)$ as its electoral campaign is that its expected vote share at $(\mathbf{z}_m, \mathbf{a}_m)$, $V_j^m$, is *high* enough, i.e., is greater than $\mathcal{V}_{np}^m$. When (30) is satisfied for some $j \in \mathcal{C}$ while $\mathcal{V}_{sp}^m \leqslant V_h^m$ for $h \neq j \in \mathcal{C}$, the joint electoral mean $(\mathbf{z}_m, \mathbf{a}_m)$ is a *weak* LNE of the election.

If $V_j^m < \mathcal{V}_{np}^m$, $j$'s expected vote share does *not* meet the necessary convergence condition in (30), $j$ moves away from $(z_m, a_m)$ to increase its votes since $j$'s expected vote share is at a minimum or at a saddle-point at $(\mathbf{z}_m, \mathbf{a}_m)$. Once $j$ moves away from $(z_m, a_m)$, other candidates may also find it in their interest move away from the electoral mean. The joint electoral mean, $(\mathbf{z}_m, \mathbf{a}_m)$, is *not* a LNE of the election.

## 8 Comparative Statics

We are interested in finding how the sufficient and necessary pivotal vote shares—given by (21) and (22)—change when the parameters of the model change. The comparative statics carried out are such that if candidates adopt the same electoral campaign, they adopt the electoral mean as their campaign strategy, i.e., maintain the same electoral mean, $(\mathbf{z}_m, \mathbf{a}_m)$, through out the comparative statics exercises. Changes in the importance voters give to each dimension or changes in the variance of voters' policy and ad campaign preferences change the pivotal vote shares leading to changes in the conditions leading to the LNE of the election. We examine changes in the importance given to the policy (ad campaign) dimension then changes to the variance in voters policy and ad campaign preferences. Even though changes in voters' valences affect candidates' expected vote shares but not the pivotal vote shares, they do affect the conditions for the LNE of the election.

**Corollary 1 (Comparative Statics)** *Suppose that the electoral mean is a LNE of the election. The result of the different comparative statics are as follows.*

1. *An increase in the importance voters give to the policy and ad campaign dimensions, $\beta$ and $e$, ceteris paribus,*

- *increases the sufficient pivotal vote share $\mathcal{V}_{sp}^{m}$ thus making it more difficult for the electoral mean to be a strick LNE of the election*
- *has an ambiguous effect on the necessary pivotal vote share $\mathcal{V}_{np}^{m}$. The necessary pivotal vote share increases (decreases) if $\frac{\beta+e}{\beta}var(x) + var(t) > (<) \frac{1}{2}\left[var(x) + var(t)\right]$.*

2. *A mean preserving increase in $var(x)$ or $var(t)$, ceteris paribus,*

- *increases the sufficient and necessary pivotal vote shares.*

3. *An increase in the mean sociodemographic, traits or competence valence has no effect on the pivotal vote shares but changes candidates' expected vote shares and so affects the conditions under which there is a strick (weak) LNE of the election.*

*Proof: See Appendix*

When voters give greater importance to the policy (ad campaign) dimension relative to the other dimension and to the valences, i.e., when $\beta$ increases ($e$ increases), holding every thing else constant, $Tr(\mathbf{B})$ increases. In this case, the *sufficient* pivotal vote share increases (see Appendix). As a consequence, the sufficient condition for convergence to the electoral mean becomes more difficult to satisfy as candidates' expected vote shares must be greater than the now higher sufficient pivotal vote share. Thus, the greater the importance voters give one of the two dimensions the more difficult it becomes for candidates to converge to the electoral mean.

The effect of an increase in $\beta$ or in $e$, holding every thing else constant, may increase or decrease the *necessary* pivotal vote. The effect depends on whether

$$\frac{\beta+e}{\beta}var(x) \gtrless \frac{1}{2}\left[var(x) + var(t)\right]$$

$$\frac{\beta+e}{e}var(t) \gtrless \frac{1}{2}\left[var(x) + var(t)\right].$$

This says that the $\mathcal{V}_{np}^{m}$ rises as $\beta$ or $e$ increase when the variance of voters' ideal policies weighted by $\frac{\beta+e}{\beta}$ is larger than half the aggregate variance of voters' preferences, $Tr(\nabla^{m})$, i.e., greater than half the aggregate variance of voters' preferences, $var(x) + var(t)$. Thus, the more dispersed voters are along either dimension the more difficult it is for candidates' to adopt the electoral mean as their electoral campaign strategy.

These results show that when $\beta$ or $e$ increase, the sufficient condition for convergence to the electoral mean may be harder to meet. In this case, it becomes harder for the electoral mean to be a *strick* LNE of the election. When the necessary pivotal vote share deceases as $\beta$ or $e$ increase, it is more likely that the necessary condition for convergence to the electoral mean is satisfied for a broader range of expected vote shares for all candidates and thus the electoral mean is a weak LNE of the election for a wider range of expected vote share.

When voters' ideal policies or campaign tolerance levels become more dispersed along the corresponding dimension, a higher $var(x)$ or $var(t)$, the pivotal vote shares $\mathcal{V}_{sp}^{m}$ and $\mathcal{V}_{np}^{m}$ rise, thus making it more difficult for candidates to converge to the electoral mean.

Suppose that there is an increase in the mean sociodemographic, traits or competence valence of candidate *j*. The mean valence components do not affect the pivotal vote shares or the electoral mean, however, these mean components increase *j*'s expected vote share while decreasing the expected vote shares of all other candidates and so affect the conditions under which there is convergence to the electoral mean.

## 9 Concluding Comments

We introduce campaign advertising into Schofield's (2007) stochastic election model to examine how candidates choose their advertising campaign strategy by endowing voters with preferences over policies and ad campaigns. Voters evaluate candidates not only by how far the candidate's policy is from their ideal but also by the frequency of candidates' messages relative to their campaign tolerance level. Their choice of candidate is also affected by voters sociodemographic, traits and competence valences.

Lemma 1 shows that if candidates adopt the same campaign they adopt the electoral mean as their campaign strategy. Proposition 1 extends Schofield's (2007) result to include candidates' ad campaigns and voters' campaign tolerance levels. In this extension, we adapt Schofield's results to one where candidates contact voters directly in an effort to give voters' a further impetus to vote for them. The sufficient and necessary conditions for convergence to the electoral mean are presented in terms of pivotal vote shares rather than the convergence coefficient used by Schofield (2007). The pivotal vote shares give intuitive results: candidates' expected vote shares must be *high enough* for candidates to want to rationally adopt the electoral mean, $(\mathbf{z}_m, \mathbf{a}_m)$, as their campaign strategy. The sufficient condition for convergence to the electoral mean requires that all candidates' expected vote shares at the electoral mean be higher than the sufficient pivotal vote share $\mathcal{V}_{sp}^{m}$; the necessary condition that they be higher the necessary pivotal vote share $\mathcal{V}_{np}^{m}$. If the expected vote share of at least one candidate is below $\mathcal{V}_{np}^{m}$, then that candidate moves away from the electoral mean to increase its votes and so the electoral mean cannot be a LNE of the election.

The result on the advertising campaign is new as it shows that the variation in voters' campaign tolerances levels not only determines candidates' advertising campaigns but also affects the electoral outcome. The presence of voters' campaign tolerance levels shows that if a candidate's advertising campaign alienates too many voters by either contacting them too often or not often enough relative to voters' campaign tolerance levels, the utility these voters get from this candidate decreases, thus decreasing the probability that they vote for the candidate. Candidates have then an incentive to find their optimal advertising campaign, one that alienates the fewest number of voters taking into account the effect that their choice of policy has on the expected vote share. In our model, candidates use two instruments to maximize their vote shares: policies and ad campaigns. The implication is that policy and ad campaigns adjustments may be smaller than if tweaking only one of them. That is, the presence of two electoral campaign instruments—policy and advertising—gives candidates more flexibility in their quest to gain votes as well as allowing for smaller adjustments in either instrument. Moreover, ad campaigns allow candidates to use an instrument that only applies during the electoral campaign while allowing them to choose the policies they would like implement after the election. Though not in this model, it is possible that given the longer lasting effect of policies relative to campaigns, that candidates may choose to tweak ad campaigns more than they do policies to achieve their electoral objective.

This model highlights that we must be able to measure, or proxy, voters' campaign tolerance levels. A proxy for campaign tolerance is already available to campaign strategists: as when engaging in door-to-door canvasing, volunteers know whether voters want to talk to them and how much time they spend in each house. As to regards to reading advertising messages this is more difficult. But techniques similar to those used in measuring the effectiveness of TV advertisements, except that they would need to do try to measure messages read on smart phones or on social media. Since Twitter keeps tract of how many times a message has been re-Twitted, this can also be used as a so this could of the importance given to a campaign message.

The comparative statics highlight the conditions under which it becomes more difficult for candidates to converge to the electoral mean. The results show that where voters to give greater importance to one of the two dimensions (an increase in $\beta$ or $e$ holding every thing else constant) the more difficult it is for the expected vote shares to meet the sufficient condition for converge to the electoral mean. If the weighted variance along one dimension is more than half the aggregate variance of voters' preferences around the electoral mean, an increase in $\beta$ or $e$ makes it easier for candidates to meet the necessary condition for convergence to the electoral mean. An increase in the mean sociodemographic, traits and/or competence valences of candidate $j$ increases $j$'s expected vote share, decreases the vote share of all the other candidates but does not affect the pivotal vote shares or the electoral mean, thus affecting the convergence conditions to the electoral mean.

Predictions emanating from these results are that if candidates and their campaign strategists realize that either candidates' positions or ad campaigns are ineffective, in the sense of convincing voters to vote for them, then adjustments to positions and/or ad campaigns are made in an effort to increase the probability that these voters vote for the candidate. This result is important because it shows that when candidates have two campaign instruments—policy and advertising campaigns—at their disposal they may not necessarily adjust their policy position. That is, they may instead adjust their advertising campaign to increase their vote share.

# Appendix

## *Second Order Conditions (SOC)*

Section 5 derived candidates' best response policies and ad campaign functions and the critical values satisfying the first order conditions for a maximum of the vote share functions. Lemma 1 proved that if candidates adopt the same electoral campaign, they adopt the electoral mean, $(\mathbf{z}_m, \mathbf{a}_m)$, as their campaign strategy. To evaluate if candidates' vote share functions are at a maximum at $(\mathbf{z}_m, \mathbf{a}_m)$ we use the *Hessian matrix of second order partial derivatives* of the vote share functions of each candidate evaluated at $(\mathbf{z}_m, \mathbf{a}_m)$.

Before proceeding recall from Sect. 6 that when all candidates adopt the electoral mean as their campaign strategy, the probability that $i$ votes for $j$, $\rho_{ij}$ in (4) when evaluated at $(\mathbf{z}_m, \mathbf{a}_m)$, is independent of voters' ideal policies and campaign tolerance levels as well as independent of candidates' policies and ad campaigns and gives $j$'s expected vote share so that from (16),

$$\rho_{ij}^m = \rho_j^m = V_j^m \text{ for all } i \in \mathcal{N}. \tag{31}$$

### Proof of Proposition 1

We use the *Hessian matrix of second order partial derivatives of* candidate $j$ to determine whether $j$'s vote share function, $V_j$ in (5), is at a maximum at $(z_m, a_m)$ holding the electoral campaigns of all other candidates at $(z_m, a_m)$. To find this Hessian we need the second order partial derivatives of the probability that $i$ votes for $j$, $\mathbf{D}^2\rho_{ij}$ evaluated at $(\mathbf{z}_m, \mathbf{a}_m)$, $\mathbf{D}^2\rho_{ij}^m$, using $\rho_{ij}$ in (4). Then using $\mathbf{D}^2\rho_{ij}^m$ we find the Hessian of $j$'s vote share, $\mathbf{D}^2V_j$ at $(\mathbf{z}_m, \mathbf{a}_m)$.

Suppose all candidates adopt the electoral mean, $(z_m, a_m)$ as their campaign strategy. Using (31), the partial derivatives of (7) wrt $z_j$ and $a_j$ evaluated at $(z_m, a_m)$, i.e., $\mathbf{D}^2\rho_{ij}^m$, is given by

$$\mathbf{D}^2\rho_{ij}^m \equiv \mathbf{D}^2\rho_{ij}(\mathbf{z}_m, \mathbf{a}_m) \equiv \begin{pmatrix} \frac{\partial^2\rho_{ij}}{\partial(z_j)^2} & \frac{\partial^2\rho_{ij}}{\partial z_j \partial a_j} \\ \frac{\partial^2\rho_{ij}}{\partial z_j \partial a_j} & \frac{\partial^2\rho_{ij}}{\partial(a_j)^2} \end{pmatrix}$$
$$= 2\rho_j^m(1-\rho_j^m)\left[2(1-2\rho_j^m)\mathbf{B}\mathbf{E}_{(\mathbf{z}_m,\mathbf{a}_m)}^{ij\mathbf{T}}\mathbf{E}_{(\mathbf{z}_m,\mathbf{a}_m)}^{ij}\mathbf{B} - \mathbf{B}\right]$$
$$= 2\rho_j^m(1-\rho_j^m)\left[2(1-2\rho_j^m)\nabla_i^m - \mathbf{B}\right] \quad (32)$$

$$\text{where} \quad \mathbf{B} \equiv \begin{pmatrix} \beta & 0 \\ 0 & e \end{pmatrix}, \quad \mathbf{E}_{(\mathbf{z}_m,\mathbf{a}_m)}^{ij} \equiv \left[(x_i - z_m)\ (t_i - a_m)\right]. \quad (33)$$

and where a vector with a $\mathbf{T}$ denotes the transpose of that vector. The diagonal of $\mathbf{B}$ shows the importance voters give the policy and advertising dimensions and $\mathbf{E}_{(\mathbf{z}_m,\mathbf{a}_m)}^{ij}$ the differences between $i$'s ideals $(x_i, t_i)$ and $j$'s electoral campaign at the electoral mean $(z_m, a_m)$. The matrix $\nabla_i^m$ in (32) defined by

$$\nabla_i^m \equiv \mathbf{B}\mathbf{E}_{(\mathbf{z}_m,\mathbf{a}_m)}^{ij\mathbf{T}}\mathbf{E}_{(\mathbf{z}_m,\mathbf{a}_m)}^{ij}\mathbf{B} \equiv \begin{pmatrix} \beta(x_i - z_m)^2\beta & \beta(x_i - z_m)(t_i - a_m)e \\ \beta(x_i - z_m)(t_i - a_m)e & e(t_i - a_m)^2 e \end{pmatrix} \quad (34)$$

gives the *weighted variance-covariance matrix* of *i's ideals around the electoral mean* $(z_m, a_m)$.

Using $\mathbf{D}^2\rho_{ij}^m$ in (32), the *Hessian* of second order partial derivatives of $V_j$ in (5) wrt $z_j$ and $a_j$, evaluated at $(\mathbf{z}_m, \mathbf{a}_m)$ is given by

$$\mathbf{D}^2V_j^m \equiv \mathbf{D}^2V_j(\mathbf{z}_m, \mathbf{a}_m) \equiv \frac{1}{n}\sum\nolimits_{i\in\mathcal{N}} \mathbf{D}^2\rho_{ij}^m$$
$$= \frac{1}{n}\sum\nolimits_{i\in\mathcal{N}} 2\rho_j^m(1-\rho_j^m)\left[2(1-2\rho_j^m)\nabla_i^m - \mathbf{B}\right]$$
$$= 2\rho_j^m(1-\rho_j^m)\left\{2(1-2\rho_j^m)\left(\frac{1}{n}\sum\nolimits_{i\in\mathcal{N}}\nabla_i^m\right) - \mathbf{B}\right\}. \quad (35)$$

Note that averaging $\nabla_i^m$ in (34) across all voters, we obtain the matrix

$$\nabla^m \equiv \frac{1}{n}\sum\nolimits_{i\in\mathcal{N}}\nabla_i^m, \quad (36)$$

i.e., *the weighted variance-covariance matrix* of the distribution of voters' ideal policies and campaign tolerance levels around the *electoral mean* where the weights are given by $\mathbf{B}$ in (33).

After substituting $\nabla^m$ from (36) into (35), we get that $j$'s Hessian matrix is given by

$$\mathbf{D}^2V_j^m = 2\rho_j^m(1-\rho_j^m)\left\{2(1-2\rho_j^m)\nabla^m - \mathbf{B}\right\} \quad (37)$$

Candidate $j$ adopts the electoral mean, $(z_m, a_m)$, as its campaign strategy, iff $j$'s vote share is at a maximum at $(z_m, a_m)$. Hence, to show that $(z_m, a_m)$ for all $j \in \mathcal{C}$ is a *strick (weak) local Nash equilibrium* [S(W)LNE] of the election we must find the conditions under which $j$'s Hessian matrix, $\mathbf{D}^2 V_j$ in (35) when evaluated at $(z_m, a_m)$, $\mathbf{D}^2 V_j^m$, is negative definite (semi-definite) for all $j \in \mathcal{C}$ which happens only when its eigenvalues are negative (non-positive) for all $j \in \mathcal{C}$, that is, iff the trace and the determinant of $\mathbf{D}^2 V_j^m$ are respectively negative and positive as in this case the eigenvalues of $\mathbf{D}^2 V_j^m$ are both negative.

Before deriving the trace and the determinant of $\mathbf{D}^2 V_j^m$ and to simplify the calculations, we give the following definitions.

**Definition 5** Let $\mathbf{C}_j^m \equiv \mathbf{C}_j(\mathbf{z}_m, \mathbf{a}_m)$ be the *characteristic matrix* of candidate $j$ when all candidates adopt the electoral mean as their campaign strategy $(\mathbf{z}_m, \mathbf{a}_m)$, i.e.,

$$\mathbf{C}_j^m \equiv \mathbf{C}_j(\mathbf{z}_m, \mathbf{a}_m) \equiv A_j^m \nabla^m - \mathbf{B} \tag{38}$$

$$\text{where} \quad A_j^m \equiv A_j^m(\mathbf{z}_m, \mathbf{a}_m) \equiv 2(1 - 2\rho_j^m) \tag{39}$$

is the *characteristic factor* of $\mathbf{C}_j^m$, $\nabla^m$ is given by (36), $\rho_j^m$ by (31) and $\mathbf{B}$ by (33).

Let $q_j^m$ be the joint probability that voters vote for $j$ and for any other candidate at $(\mathbf{z}_m, \mathbf{a}_m)$,

$$q_j^m \equiv q_j(\mathbf{z}_m, \mathbf{a}_m) \equiv \rho_j^m(1 - \rho_j^m). \tag{40}$$

Using (39) and (40), $\mathbf{D}^2 V_j^m$ in (37) can be expressed as a function of $\mathbf{C}_j^m$ in (38), i.e.,

$$\mathbf{D}^2 V_j^m = 2q_j^m \left\{A_j^m \nabla^m - \mathbf{B}\right\} \equiv 2q_j^m \mathbf{C}_j^m. \tag{41}$$

So that the trace of $\mathbf{D}^2 V_j^m$ in (38) is then given by

$$Tr\left(\mathbf{D}^2 V_j^m\right) = 2q_j^m Tr(\mathbf{C}_j^m). \tag{42}$$

Since $q_j^m$ in (40) is always positive, $Tr\left(\mathbf{D}^2 V_j^m\right) < 0$ iff $Tr(\mathbf{C}_j^m) < 0$.

The trace of $\mathbf{C}_j^m$ in (38) is given by

$$Tr(\mathbf{C}_j^m) \equiv Tr\left[A_j^m \nabla^m - \mathbf{B}\right] = A_j^m Tr\left(\nabla^m\right) - Tr(\mathbf{B}) \tag{43}$$

where $Tr(\nabla^m)$ is given by (27), $Tr(\mathbf{B})$ by (23) and $A_j^m$ by (39). From (43), $Tr(\mathbf{C}_j^m) < 0$ when

$$A_j^m < \frac{Tr(\mathbf{B})}{Tr\left(\nabla^m\right)}. \tag{44}$$

After substituting $A_j^m$ from (39) into (44) and some manipulation, we obtain

$$\frac{1}{2} - \frac{1}{4}\frac{Tr(\mathbf{B})}{Tr\,(\mathbf{\nabla}^m)} < \rho_j^m \tag{45}$$

The left hand side (LHS) of (45) is independent of candidates' policies and ad campaigns and is thus the same for all candidates. Define the LHS of (45) as the *necessary pivotal (np) vote share* at the electoral mean, $(z_m, a_m)$, i.e.,

$$\mathcal{V}_{np}^m \equiv \frac{1}{2} - \frac{1}{4}\frac{Tr(\mathbf{B})}{Tr\,(\mathbf{\nabla}^m)} = \frac{1}{2} - \frac{1}{4}\frac{\beta + e}{\frac{1}{n}\sum_{i\in\mathcal{N}}\beta(x_i - z_m)^2\beta + \frac{1}{n}\sum_{i\in\mathcal{N}}e(t_i - a_m)^2 e} \tag{46}$$

where the numerator in the last term follows from (23) and the denominator from (27).

Since $\rho_j^m$ in (31) is independent of candidates' policies and ad campaigns and is the same for all voters then $\rho_j^m$ also gives *j*'s expected vote share at the electoral mean, i.e., $V_j^m = \rho_j^m$. Therefore, condition (45) can be re-written as

$$\mathcal{V}_{np}^m \equiv \frac{1}{2} - \frac{1}{4}\frac{Tr(\mathbf{B})}{Tr\,(\mathbf{\nabla}^m)} < \rho_j^m = V_j^m \tag{47}$$

This condition says that if *j*'s expected vote share at $(\mathbf{z}_m, \mathbf{a}_m)$, $V_j^m$, is *higher* than the necessary pivotal vote share, $\mathcal{V}_{np}^m$, then $Tr(\mathbf{C}_j^m) < 0$. If for *some* candidate $j \in \mathcal{C}$, $\mathcal{V}_{np}^m > \rho_j^m = V_j^m$, then for this candidate $Tr(\mathbf{C}_j^m) > 0$.

Therefore, the *necessary condition* for *j* to adopt $(z_m, a_m)$ as its electoral campaign, when all other candidates adopt the electoral mean as their electoral campaign, is

$$\mathcal{V}_{np}^m < \rho_j^m = V_j^m. \tag{48}$$

That is, candidate *j* must expect a *high* enough vote share at $(\mathbf{z}_m, \mathbf{a}_m)$ for *j* to adopt the electoral mean as its campaign strategy.

The *sufficient SOC* for *j* to converge to $(z_m, a_m)$ is that the eigenvalues of $\mathbf{D}^2V_j^m$ in (37) be *both* negative implying that the *determinant* of $\mathbf{D}^2V_j^m$ must be positive at $(\mathbf{z}_m, \mathbf{a}_m)$, i.e., $\det\left(\mathbf{D}^2V_j^m\right) > 0$. From (37), the determinant of $\mathbf{D}^2V_j$ when evaluated at $(\mathbf{z}_m, \mathbf{a}_m)$ is given by

$$\det\left(\mathbf{D}^2V_j^m\right) = \det\left[2q_j^m\left\{A_j^m\mathbf{\nabla}^m - \mathbf{B}\right\}\right] = \det\left[2q_j^mA_j^m\mathbf{\nabla}^m - 2q_j^m\mathbf{B}\right]$$

After some manipulation and using (23) and (27), $\det(\mathbf{D}^2V_j^m)$ becomes

$$\det(\mathbf{D}^2V_j^m) = 4\left[q_j^mA_j^m\right]^2\det(\mathbf{\nabla}^m) + 4\left[q_j^m\right]^2\det(\mathbf{B}) \times \left\{1 - A_j^mTr\left((\mathbf{B})^{-1}\mathbf{\nabla}^m\right)\right\} \tag{49}$$

From the triangle inequality, the determinant of the variance-covariance matrix of voters' policies and campaign tolerance levels around the electoral mean, $\det(\nabla^m)$ is always non-negative. The determinant of $\mathbf{D}^2 V_j^m$ is positive iff the last term in (49) is positive. Since $\det(\mathbf{B}) = \beta e > 0$ and since from (40) $q_j^m > 0$, the determinant of $\mathbf{D}^2 V_j^m$ is positive iff the term in curly brackets is positive, i.e., iff

$$A_j^m < \frac{1}{Tr\left[(\mathbf{B})^{-1}\nabla^m\right]}. \tag{50}$$

After substituting $A_j^m$ from (39) into (50) and some manipulation, we obtain

$$\frac{1}{2} - \frac{1}{4}\frac{1}{Tr\left[(\mathbf{B})^{-1}\nabla^m\right]} < \rho_j^m. \tag{51}$$

Note that the LHS of (51) is the same for all candidates. Define the LHS of (51) as the *sufficient pivotal (sp) vote share* at $(\mathbf{z}_m, \mathbf{a}_m)$, i.e.,

$$\mathcal{V}_{sp}^m \equiv \frac{1}{2} - \frac{1}{4}\frac{1}{Tr\left[(\mathbf{B})^{-1}\nabla^m\right]} = \frac{1}{2} - \frac{1}{4}\frac{1}{\frac{1}{\beta}\frac{1}{n}\sum_{i\in\mathcal{N}}\beta(x_i - z_m)^2\beta + \frac{1}{e}\frac{1}{n}\sum_{i\in\mathcal{N}}e(t_i - a_m)^2 e} \tag{52}$$

where the denominator in the last term follows from (28).

As before recall that $j$'s expected vote share at the electoral mean is $V_j^m = \rho_j^m$. Therefore, using (51) we have that the *sufficient condition* for $j$ to adopt $(z_m, a_m)$ as electoral campaign—when all other candidates adopt $(z_m, a_m)$ as their electoral campaign—is that

$$\mathcal{V}_{sp}^m < \rho_j^m = V_j^m, \tag{53}$$

i.e., $V_j^m$ must be *high* enough at $(\mathbf{z}_m, \mathbf{a}_m)$ for $j$ to adopt to the electoral mean as its electoral campaign. ■

For the necessary part of the proof, assume $(\mathbf{z}_m, \mathbf{a}_m)$ is a *weak* local Nash equilibrium of the election. Then for all $j \in \mathcal{C}$, $j$'s Hessian matrix evaluated at $(z_m, a_m)$ must be negative semi-definite. This implies that $Tr(\mathbf{D}^2 V_j^m) \leq 0$ which is true if and only if $Tr(\mathcal{C}_j^m) \leq 0$ for all $j \in \mathcal{C}$. If $\mathcal{V}_{np}^m > V_j^m$ for some $j \in \mathcal{C}$ then it must be the case that $Tr(\mathcal{C}_j^m)$ must be strictly positive violating the weak Nash equilibrium condition. This completes the proof of necessity. ■

## *Proof of Lemma 2*

We only need the proof for candidate $j$. Using (52) and (46), $\mathcal{V}_{sp}^m > \mathcal{V}_{np}^m$ when

$$\frac{1}{2} - \frac{1}{4}\frac{1}{Tr\left[(\mathbf{B})^{-1}\nabla^m\right]} > \frac{1}{2} - \frac{1}{4}\frac{Tr(\mathbf{B})}{Tr\left(\nabla^m\right)}$$

This translates into

$$Tr(\mathbf{B})Tr\left[(\mathbf{B})^{-1}\nabla^{m}\right] > Tr\left(\nabla^{m}\right)$$

which using (23), (28) and (27) and after some manipulation translates into

$$\left(1+\frac{e}{\beta}\right)var(x)+\left(1+\frac{\beta}{e}\right)var(t) > var(x)+var(t).$$

After multiplying this inequality by $[\beta e]^{-1} > 0$ and some manipulation, we get

$$var(x)+var(t) > 0,$$

so that $\mathcal{V}_{sp}^{m} > \mathcal{V}_{np}^{m}$ always as required and this holds for all $j \in \mathcal{C}$. ■

## Comparative Statics

To find the effect of $\beta$ on the *sufficient* pivotal vote share, when all candidates adopt the electoral mean $(\mathbf{z}_m, \mathbf{a}_m)$ as their campaign strategy, $\mathcal{V}_{sp}^{m}$ in (22), take the derivative of $\mathcal{V}_{sp}^{m}$ wrt $\beta$, ceteris paribus, i.e.,

$$\frac{\partial\mathcal{V}_{sp}^{m}}{\partial\beta}=\frac{1}{4}\frac{\frac{1}{\beta^{2}}var(x)}{\left[\frac{1}{\beta}var(x)+\frac{1}{e}var(t)\right]^{2}} > 0,$$

so that $\mathcal{V}_{sp}^{m}$ increases as $\beta$ increases. A similar analysis would show that $\mathcal{V}_{sp}^{m}$ increases when $e$ increases.

The effect of $\beta$ on the *necessary* pivotal vote share, $\mathcal{V}_{sp}^{m}$ in (22), i.e., the derivative of $\mathcal{V}_{np}^{m}$ wrt $\beta$, ceteris paribus, is given by

$$\begin{aligned}\frac{\partial\mathcal{V}_{np}^{m}}{\partial\beta} &= -\frac{1}{4}\frac{var(x)+var(t)-(\beta+e)\frac{1}{n}\sum_{i\in\mathcal{N}}2\beta(x_i-z_m)^2}{[var(x)+var(t)]^2}\\ &= -\frac{1}{4}\frac{var(x)+var(t)-2var(x)-2\frac{1}{n}\sum_{i\in\mathcal{N}}\beta(x_i-z_m)^2 e}{[var(x)+var(t)]^2}\\ &= -\frac{1}{4}\frac{-var(x)+var(t)-2\frac{e}{\beta}var(x)}{[var(x)+var(t)]^2}\\ &= \frac{1}{4}\frac{(1+2\frac{e}{\beta})var(x)-var(t)}{[var(x)+var(t)]^2}\end{aligned}$$

where the fourth line follows after multiplying by $\frac{\beta}{\beta}$. The necessary pivotal vote share increases when $\beta$ increases iff the numerator in the last line of the above equation is positive, i.e., iff

$$(1 + 2\frac{e}{\beta})var(x) > var(t)$$

Adding $var(x)$ on both sides of the above inequality gives

$$2\frac{\beta + e}{\beta} var(x) > var(x) + var(t)$$

The RHS of this inequality equals $Tr\,(\nabla^m)$, given in (27). Thus, when $\beta$ increases the necessary pivotal vote share increases when

$$\frac{\beta + e}{\beta} var(x) > \frac{1}{2} Tr\,(\nabla^m)\,,$$

i.e., when the variance of voters' ideal policies weighted by $\frac{\beta+e}{\beta}$ is larger than half the aggregate variance of voters' preferences about the electoral mean.

A similar analysis shows that $\mathcal{V}^m_{np}$ rises as $e$ increases iff

$$\frac{\beta + e}{e} var(x) > \frac{1}{2} Tr\,(\nabla^m)\,.$$

Suppose voter's preferences become more dispersed along the policy dimension while maintaining the electoral mean $(z_m, a_m)$ constant, so that the distribution of voters' ideal policies undergoes a mean preserving spread. This means that $var(x)$ increases but the electoral mean $(z_m, a_m)$, $var(t)$, $\beta$ and $e$ remain unchanged.

We now examine what happens to the pivotal vote shares as $var(x)$ or $var(t)$ increase. The effect of higher $var(x)$ or $var(t)$ on the sufficient pivotal vote share is given by

$$\frac{\partial \mathcal{V}^m_{sp}}{\partial var(x)} = \frac{1}{4}\frac{1}{\beta}\frac{1}{\left[\frac{1}{\beta}var(x) + \frac{1}{e}var(t)\right]^2}$$

$$\frac{\partial \mathcal{V}^m_{sp}}{\partial var(t)} = \frac{1}{4}\frac{1}{e}\frac{1}{\left[\frac{1}{\beta}var(x) + \frac{1}{e}var(t)\right]^2}$$

which are both always positive. That is, a mean preserving increase in $var(x)$ or $var(t)$ increases the sufficient pivotal vote share. Thus, making it more difficult for the electoral mean to be a strick LNE of the election.

The effect of a mean preserving increase in $var(x)$ or $var(t)$ on the *necessary* pivotal vote in (22) is given by

$$\frac{\partial \mathcal{V}_{np}^m}{\partial var(x)} = \frac{\partial \mathcal{V}_{np}^m}{\partial var(t)} = \frac{1}{4}\frac{(\beta + e)}{[var(x) + var(t)]^2} > 0.$$

The necessary pivotal vote share also increases in the variance of voters' preferences along any dimension. The necessary condition for convergence to the electoral mean is then harder to meet.

## References

Ansolabehere, S., de Figueiredo, J. M., & Snyder, J. (2003). Why is there so little money in U.S. politics? *The Journal of Economic Perspectives, 17*(1), 105–130.

Aragones, E., & Palfrey, T. (2002). Mixed equilibrium in a Downsian model with a favored candidate. *Journal of Economic Theory, 103*, 131–161.

Ashworth, S., & Bueno de Mesquita, E. (2009). Elections with platform and valence competition. *Games Economic Behavior, 67*, 191–216.

Austen-Smith, D., & Banks, J. S. (1999). Cycling of simple rules in the spatial model. *Social Choice and Welfare, 16*(4), 663–672.

Clarke, H. D., Kornberg, A., & Scotto, T. J. (2009). *Making political choices*. Toronto: Toronto University Press.

Condorcet, J. (1785/1984). *Essai sur l'application de l'analyse a la probabilite des decisions rendus a la pluralite des voix*. Paris: Imprimerie Royale.

Downs, A. (1957). *An economic theory of democracy*. New York: Harper and Row.

Gallego, M., & Schofield, N. (2013). The convergence coefficient across political systems. *Scientific World Journal, 2013*, Article ID 653035, 28. http://dx.doi.org/10.1155/2013/653035

Gallego, M., & Schofield, N. (2016). Do parties converge to the electoral mean in all political systems? *Journal of Theoretical Politics, 28*(2), 288–330.

Groseclose, T. (2001). A model of candidate location when one candidate has a valence advantage. *American Journal of Political Science, 45*(4), 862–886.

Hendricks, J. A., & Kaid, L. L. (2011). *Techno politics in presidential campaigning: New voices, new technologies, and new voters*. New York: Routledge.

Herrera, H., Levine, D. K., & Martinelli, C. (2008). Policy platforms, campaign spending and voter participation. *Journal of Public Economics, 92*, 501–513.

Madison, J. (1787/1999). The federalist. In J. Rakove & J. Madison (Eds.) *Writings* (Vol. 10). New York: The Library of America.

Mckelvey, R. D. (1976). Intransitivities in multidimensional voting models and some implications for agenda control. *Journal of Economic Theory, 12*(3), 472–482.

Mckelvey, R. D., & Patty, J. (2006). A theory of voting in large elections. *Games Economic Behavior, 57*, 155–180.

Meirowitz, A. (2008). Electoral contests, incumbency advantages and campaign finance. *Journal of Politics, 70*, 681–699.

Riker, W. H. (1980). Implications from the disequilibrium of majority rule for the study of institutions. *American Political Science Review, 74*, 432–46.

Riker, W. H. (1982). *Liberalism against Populism*. San Francisco: Freeman.

Riker, W. H. (1986). *The art of political manipulation*. New Haven: Yale University Press.

Saari, D. (1997). The generic existence of a core for *q*-rules. *Economic Theory, 9*, 219–260.

Sanders, D., Clarke, H., Stewart, M. C., & Whiteley, P. (2011). Downs, stokes and the dynamics of electoral choice. *British Journal of Political Science, 41*(2), 287–314.
Schofield, N. (1978). Instability of simple dynamic games. *Review of Economic Studies, 45*, 575–594.
Schofield, N. (2002). Evolution of the constitution. *British Journal of Political Science, 32*, 1–20.
Schofield, N. (2006). The spatial stochastic model with party activists. *Review of Economic Design, 10*, 183–203.
Schofield, N. (2007). The mean voter theorem: Necessary and sufficient conditions for convergent equilibrium. *Review of Economic Studies, 74*, 965–980.
Schofield, N., & Gallego, M. (2011). *Leadership or chaos*, Springer: Berlin.
Schofield, N., & Gallego, M., & Jeon, J. S. (2011a). Leaders, voters and activists in elections in the Great Britain 2005 and 2010. *Electoral Studies, 30*(3), 484–496.
Schofield, N., Jeon, J. S., Muskhelishvili, M., Ozdemir, U., & Tavits, M. (2011b). Modeling elections in post-communist regimes: Voter perceptions, political leaders and activists. In N. Schofield & G. Caballero (Eds.), *The political economy of institutions, democracy and voting* (pp. 259–301). Berlin: Springer.
Stokes, D. (1963). Spatial models and party competition. *American Political Science Review, 57*, 368–377.
Stokes, D. (1992). Valence politics. In D. Kavanagh (Ed.), *Electoral politics*. Oxford: Clarendon Press.
Train, K. (2003). *Discrete choice methods for simulation*. Cambridge, Cambridge University Press.
Zakharov, A. V. (2009). A model of candidate location with endogenous valence. *Public Choice, 138*, 347–366.

# Personality Traits and the Gender Gap in Ideology

**Rebecca Morton, Jean-Robert Tyran, and Erik Wengström**

## 1 Introduction

Considerable evidence shows that women are more likely to express left-leaning political preferences than men.[1] Furthermore, although variations across countries exist, these differences cannot generally be explained by gender differences in religiosity, class, age, or labor force participation.[2]

However, these standard demographic variables are not the only differences between men and women that might explain gender differences in ideological

[1]See Giger (2009) for a study of 12 European countries, Inglehart and Norris (2000) for a study of over 60 countries, and Norrander and Wilcox (2008) for evidence from the United States.

[2]See Inglehart and Norris (2000).

R. Morton (✉)
Department of Politics, NYU NYC and NYU Abu Dhabi, 19 West 4th Street, 2nd Floor, New York, NY 10012, USA
e-mail: rebecca.morton@nyu.edu

J.-R. Tyran
Department of Economics, Faculty of Business, Economics and Statistics, University of Vienna, Oskar-Morgenstern-Platz 1, 1090 Vienna

Department of Economics, University of Copenhagen, Øster Farimagsgade 5, 1353 Copenhagen, Denmark
e-mail: jean-robert.tyran@univie.ac.at

E. Wengström
Department of Economics, Lund University, P.O. Box 7082, S-220 07 Lund, Sweden

Department of Economics, University of Copenhagen, Øster Farimagsgade 5, 1353 Copenhagen, Denmark
e-mail: erik.wengstrom@nek.lu.se



M. Gallego, N. Schofield (eds.), *The Political Economy of Social Choices*, Studies in Political Economy, DOI 10.1007/978-3-319-40118-8_7

preferences. Substantial data suggests that men and women differ in their personality traits. In particular, cross-country research shows that women are less likely to be emotionally stable and more likely to be extraverts, agreeable, and conscientious.[3] Moreover, recent research has highlighted the importance of personality traits in understanding ideological preferences.[4] If women's personality trait differences affect their ideological preferences, then it may be the case that the ideological gender gap is a consequence of these personality differences.

Given that personality trait differences are argued to be generally formed in childhood with a large genetic component, examining whether the gender gap in ideology can be explained by such differences speaks to the existing debate over whether the ideological gender gap is a decision by women to choose more leftist positions in response to structural institutional changes in marriage that affect adults such as the increase in divorce and unmarried parenthood or a consequence of gender socialization and evolutionary biological differences that are manifested in childhood which affect men and women.[5] In addition, evidence suggests that personality trait differences are larger in countries with higher levels of human development, equal access to knowledge, education, and economic wealth, which suggests that these differences across countries could be an explanation for why the gender gap may have increased in the latter half of the twentieth century in a number of countries as development has increased.[6]

In this paper we examine the possibility that personality differences explain the ideological gender gap. We model personality traits as mediating variables between the effect of gender and ideological preferences. Thus, we are able to estimate the indirect effect of gender on ideological preferences through personality trait differences between the sexes. We find that the effects via personality trait differences are generally much more sizeable than those independent of such trait differences, comprising an overwhelming majority of the effects of gender on ideological differences. Thus, our results suggest that personality trait differences can explain a substantial portion of the ideological gender gap.

Our study is distinctive for the data we use to examine these relationships:

First, we use data from Denmark, a westernized developed country where there is a significant gender gap in political attitudes and a high divorce rate.[7] Hence,

---

[3]See Del Giudice (2009), Feingold (1994), Lippa (2010), and Schmitt et al. (2008).

[4]For work on the so-called Big Five personality traits which we explain in the next section see Alford and Hibbing (2007), Barbaranelli et al. (2007), Caprara et al. (1999, 2006), Carney et al. (2008), Gerber et al. (2010), Gosling et al. (2003), Jost et al. (2003), McCrae (1996), Mehrabian (1996), Mondak and Halperin (2008), Rentfrow et al. (2009), Riemann et al. (1993), Schoen and Schumann (2007), Stenner (2005), and Van Hiel et al. (2000).

[5]We review these literatures in the next section.

[6]See Schmitt et al. (2008, 2009) for cross-country comparisons of personality trait differences and Inglehart and Norris (2000) for changes in the gender gap over time.

[7]We present evidence of the gender gap in our empirical analysis. Although calculating cross-country divorce rates is complicated, Divorce Magazine, http://www.divorcemag.com/statistics/statsWorld.shtml estimates that 44.5 % of Danish new marriages end in divorce at a divorce rate of

our data come from a nation in which nonpersonality trait explanations of the gender gap have been argued to be especially relevant and our data provide a strong test of the personality trait explanation for the gender gap.[8] Furthermore, in Denmark political issues are relatively centralized without significant variations at the regional level in educational systems, issues of concern to voters, social groupings, and other factors which can also affect ideological preferences, which allows us to focus more exclusively on the effects of gender and personality traits on ideology and the mediating effect of personality traits. Yet our results can be viewed as generalizable given that the nature of the educational system, economic conditions, and ideological divides are plainly characteristic of other western developed countries.

Second, our personality measures are drawn from a 60 question set unlike the more limited questions used in most other studies of the relationship between personality and ideology.[9] We therefore have a more precise measure of personality trait differences.

Third, our other data is also more accurate in other ways than in previous studies. Our income data is verified by taxation authorities in contrast to standard self-reported responses to broad income categories. Thus, we do not assign subjects midpoint estimates of income or some other arbitrary income level. We also have verified data from governmental authorities on residency, educational attainment, church membership, age, marital status, and gender of our subjects.

In the next section we review the theoretical and empirical evidence on the gender gap and ideology and the role that personality trait differences may play in explaining that gap. In Sect. 3 we present our empirical analysis and in Sect. 4 we discuss the implications of our study for the existing debate over whether the ideological gender gap is due to structural changes in the institution of marriage or a consequence of genetic and socialization differences between men and women.

## 2 How Gender Differences in Personality May Affect Ideological Preferences

Much research suggests that individuals' personality traits can be classified into five basic dispositional traits which are typically called the Big Five or OCEAN: Openness, Conscientiousness, Extraversion, Agreeableness, and Emotional Stability (sometimes denoted as the opposite, Neuroticism) [see Borghans et al. (2008)

2.7 per 1000 marriages per year. According to either measure, Denmark is in the set of countries they study with relatively high divorce rates.

[8]Iversen and Rosenbluth (2006) discuss how the gender gap arose earlier in Scandinavian countries than other western countries and trace the relationship between structural and economic changes in these countries and the gender gap.

[9]Gerber et al. (2010), for example, use only ten questions.

and Gerber et al. (2010)]. Openness represents Openness to experience and curiosity about the desire for new experiences and ideas. Conscientiousness indicates the extent that an individual is organized, persistent, and goal motivated. Extraversion is a measure of an individual's sociability, warmth, assertiveness, and activity. Agreeableness refers to an individuals' interpersonal orientation and can range from being good-natured, trusting, and gullible to cynical, rude, suspicious, and manipulative. Finally, Emotional Stability is the degree to which an individual is low in anxiety and has a high tolerance for stress. Considerable evidence from psychology suggests that these personality traits have a genetic component and are relatively stable throughout adulthood (although there are some changes in Conscientiousness into late adulthood) [see Borghans et al. (2008, p. 976)].[10]

Gender differences in personality traits have been the subject of considerable empirical and theoretical research in the psychology literature since the 1930s as reviewed by Feingold (1994). Women are lower on Emotional Stability, higher on Agreeableness, Extraversion and Conscientiousness than men in most countries.[11] Two main theories have been proffered as to the sources of these personality trait differences: social role explanations and evolutionary explanations.[12]

According to the social role story, personality trait differences arise because women and men are socialized into different roles and appropriate ways of thinking. In early life experiences women are encouraged to not compete or be aggressive, but to be more social and "caring." Women are likely to be encouraged to carry more of the burden in terms of household chores and domestic duties involving taking care of children, the sick, and the elderly. In contrast, men are allowed more freedom to pursue individualistic goals outside the household.[13]

The evolutionary biological story suggests that the differences in how women and men value factors such as sensitivity to others, social status, and risk taking, evolved as women and men faced different pressures and adapted to these pressures in distinctive ways. Women are argued to be more sensitive to others and be more empathetic because their core evolutionary pressure was childbirth and childcare, while men are believed to be more competitive and risk takers because of the pressure to compete with others to attract fertile women. These evolutionary

[10]See Else-Quest et al. (2006), Feingold (1994), McCrae and Costa (1984), and Wilgenbusch and Merrell (1999).

[11]See Brody and Hall (2000), Byrnes et al. (1999), Feingold (1994), Kring and Gordon (1998), Lippa (2010), Lynn and Martin (1997), Maccoby and Jacklin (1974), and Schmitt et al. (2008, 2009).

[12]See Feingold (1994) and Schmitt et al. (2008) for reviews of the theoretical literature. A third theory is that sex differences in personality traits are largely due to measurement errors. As Schmitt et al. (2008) review, the empirical evidence provides only limited support for this explanation. See also Lippa (2010).

[13]See Eagly (1987) and Ruble and Martin (1998).

forces are argued to have led men and women to develop different psychological mechanisms.[14]

Both explanations are compelling, however, they make different predictions about how personality trait differences may be manifested across countries. That is, the social role explanation suggests that as a country becomes more egalitarian and more prosperous, such that women and men occupy more similar roles, the gender differences in personality traits should become less significant. Thus, these differences should be least evident in industrialized, wealthy countries with high levels of human development. On the other hand, some theorists working from the evolutionary biology explanation suggest that these differences may be more pronounced in such countries because these societies are psychologically closer to hunter-gatherer cultures than are less-developed agricultural or pastoral cultures.[15] Schmitt et al. (2008, p. 170) explain the reasoning for this view: "Agricultural and pastoral cultures, with extremely large disparities in resource distribution, familial isolation, and relative gender inequality, may represent the largest psychological deviations from our hunter-gatherer past . . . "

In support of the evolutionary theory, Schmitt et al. (2008, 2009) find in a large cross-cultural study of 55 countries which varied significantly in terms of development that gender differences in personality traits were *larger* with higher levels of human development.[16] They contend that the personality traits of men and women diverge in more developed countries because when social and economic conditions are fortunate, natural divergence due to genetic differences are more likely to be manifested, but that in less developed countries these differences are constrained by environmental conditions. They find that the cross-country differences are mainly explained by differences in personality traits of men rather than women. That is, in less developed countries, men's personality traits are more like women's traits, while women's personality traits vary less across countries. Apparently, in less developed countries the personality traits of men are more constrained than those of women.[17]

[14]See Buss (1997), Gangestad and Simpson (2000), Gangestad and Thornhill (1997), and Tooby and Cosmides (1990).

[15]See Korotayev and Kazankov (2003), Lamb and Hewlett (2005), Lee and Daly (1999), and Schmitt (2005).

[16]In a similar cross-country comparison, Lippa (2010) also concludes that his results conformed more strongly to the biological model than to social, structural, attributional, or social comparison models.

[17]Schmitt et al. (2008) suggest that these sex differences in personality traits (p. 179) "reflect a more general biological trend toward greater dimorphism in resource rich environments and reduced dimorphism in constrained or high stress environmental conditions. . . . It is mainly men, not women, who became less neurotic but also less agreeable and conscientious in their self-descriptions. In less fortunate conditions, the innate personality differences between men and women are attenuated. Although, speculative, another illustration of this principle may be seen regarding sex differences in competitiveness: Even when opportunities and incentives for achieving in sport grow in a way to become more equitable, sex differences in the proportion of men to women who run relatively fast increase with greater opportunity. . . . " .

In this paper we are agnostic as to whether the social role or evolutionary theories are better explanations of gender differences in personality traits. Our starting point is the evidence that such personality trait differences exist, and that they appear to be larger in more developed, westernized cultures, countries in which there is also more likely to exist an ideological gender gap. We now turn to how personality trait differences relate to ideology.

## 2.1 Personality and Ideology

As Knight (2006) documents, political scientists have largely converged on three essential components of a core definition of ideology—coherence, stability, and contrast. Specifically, she finds that political scientists think of ideology as (p. 619) "the way a system–a single individual or even a whole society–rationalizes itself. Ideologies may be idiosyncratic ..., impractical, or even delusional, but they still share the characteristics of coherence and temporal stability." Furthermore, Knight finds that formal theorists and empirical researchers (p. 625, italics in original) "have converged on a *spatial conceptualization of ideology* as a left-right or liberal-conservative continuum." In our analysis we investigate revealed ideological preferences which we consider as observed responses that individuals provide when asked to either place themselves on such a continuum, designate which political party they support, or take a position on an issue that has a known left/right dimension.

Since Adorno et al.'s (1950) work on the "authoritarian personality," a number of researchers have contended that there is a link between personality traits and ideological preferences.[18] Recently researchers have begun to investigate the relationship between the Big Five core personality traits and such preferences. Two of the Big Five apparently have strong effects on ideology: Conscientiousness and Openness, in opposite directions. Evidence suggests that individuals who are more Conscientious are more likely to express an overall rightwing ideology, vote for rightwing parties, and express rightwing views on economic policies.[19] As Gerber et al. (2010) point out, these results are consistent with the view that those who are Conscientious are more likely to adhere to traditional norms and rules, and thus should not surprisingly support them. On the other hand, individuals who are high on Openness have been shown to be significantly more likely to be leftist, vote for leftist parties, and support leftist economic policies.[20] The rationale for this relationship is

[18]See for example Altemeyer (1996), Block and Block (2005), Jost et al. (2003), Jost (2006), Lakoff (2004), McClosky (1958), and Pratto et al. (1994).

[19]See Barbaranelli et al. (2007), Caprara et al. (1999, 2006), Carney et al. (2008), Gerber et al. (2010), Gosling et al. (2003), Mondak and Halperin (2008), and Stenner (2005). In contrast, Alford and Hibbing (2007) and Mehrabian (1996) do not find a significant relationship.

[20]See Barbaranelli et al. (2007), Caprara et al. (1999, 2006), Carney et al. (2008), Gerber et al. (2010), Gosling et al. (2003), Jost et al. (2003), McCrae (1996), Mehrabian (1996), Mondak and

that individuals who are high on Openness are more likely to be accepting of new ideas and proposals that overturn traditional and existing ones.

The other three personality traits have less consistent relationships with ideological preferences. Emotional stability has been sometimes shown to be positively related with conservatism and voting for rightist parties.[21] Here, the reasoning is that individuals who are more neurotic and worried will prefer more leftist policies, particularly economic ones. Most research finds little relationship between Agreeableness and Extraversion and ideological preferences with some notable exceptions. In particular, Morton et al. (2011) find a positive association between Agreeableness and leftist positions using a more extensive personality test than that used in other samples and Gerber et al. (2010) found that Extraverts are more rightwing.[22]

Morton et al. (2011) also show that income serves as a mediating variable between some personality traits and ideological preferences. Specifically, they find that individuals who have higher levels of Emotional Stability are more likely to make higher incomes which indirectly leads these individuals to express more rightwing preferences. They discover weaker evidence that Agreeableness has an indirect effect through income on ideological preferences; as expected, individuals who are more agreeable make lower incomes and this indirectly leads them to express more leftist preferences. Similarly, they find that Conscientiousness has an indirect effect through income on economic policy measures of ideology; individuals who are more conscientious make higher incomes which leads them to express more rightwing views on economic policies. However, these effects are small relative to the direct effects of these variables on ideological preferences and in the same direction.

## 2.2 *The Ideological Gender Gap and Personality Traits*

Given the widespread interpretation of ideology as a system by which individuals place themselves on a left/right continuum, the ideological gender gap is a difference in how individuals' place themselves on this continuum that appears to be largely explained by gender. That is, when asked questions that attempt to measure how

Halperin (2008), Rentfrow et al. (2009), Riemann et al. (1993), Schoen and Schumann (2007), Stenner (2005), and Van Hiel et al. (2000).

[21]See Barbaranelli et al. (2007), Carney et al. (2008), Gerber et al. (2010), Gosling et al. (2003), Mondak and Halperin (2008), and Morton et al. (2011).

[22]Carney et al. (2008) found in some samples that Agreeableness and Extraversion were associated with conservatism. Gerber et al. (2010) find that although Agreeableness has no significant effect on overall ideology in their more limited personality trait questionnaire, they find that it is likely to make one more liberal on economic attitudes, but conservative on social attitudes. Thus, they contend that the previous inconsistent and insignificant findings may be because of these countervailing effects on overall ideology.

leftist or rightist individuals are by asking their party preferences, their reported preferences, or views on issues that vary in a well established way along the standard left-right division, researchers have found significant differences between the average responses of men and those of women. As noted in the Introduction, these ideological gender gaps exist across a number of countries and cannot be entirely explained by other demographic control variables (Giger 2009; Inglehart and Norris 2000; Norrander and Wilcox 2008).

A number of explanations have been proffered for the so-called gender gap in politics. On the one hand, researchers have highlighted structural institutional changes such as the rise in divorce and unmarried parenthood.[23] These factors are argued to reduce private transfers women receive from men and result in an increase in female support for redistribution by governments and more leftist policies. Evidence which suggests that the ideological gender gap is a relatively recent phenomenon (until the mid 1960s in the United States and in the late 1970s in Europe women were more rightist than men) supports the structural explanation since the institutional changes in marriage occurred concurrently. Edlund and Pande (2002) find that indeed there is a greater gender gap in voting that is explained by variations in divorce risk in a cross-country comparison. Iversen and Rosenbluth (2006) present a more nuanced view, pointing out that the preferences of married women should depend both on their divorce risks and their outside options. Furthermore, they point out that considering outside options might explain gender ideological differences in Scandinavian countries where the size of government is a major issue (p. 13). In particular, in these countries women are more likely to be employed in the public sector than men. If both spouses are considering their outside options more than family income, then "... since pay in the public sector is financed by taxing the private sector, the policies affecting relative pay are a perfect example of an area where gender conflict is likely to be intense." They find in a similar cross-country comparison that unmarried women and married women in the labor force are more likely to be leftist, supporting their argument that differences in outside options for men and women coupled with divorce risk explains much of the ideological gender gap.[24]

On the other hand, scholars have suggested that women are more likely to express leftist preferences because of gender socialization or biological differences between the genders, much as researchers have also explained gender differences in personality traits.[25] In support of this argument, Fridkin and Kenney (2007) discover a significant political gender gap in eighth grade students in Arizona and contend

[23]See Edlund and Pande (2002), Edlund et al. (2005), and Iversen and Rosenbluth (2006).

[24]Norrander and Wilcox (2008) find significant differences in the ideological preferences of married and single women in the United States, with single women more leftist, and argue that in the United States the gap is as much a divide between different types of women as between women and men.

[25]See Chaney et al. (1998), Chodorow (1978), Conover (1988), Conover and Sapiro (1993), Diekman and Schneider (2010), Jelen et al. (1994), Kaltenthaler et al. (2008), and Studlar et al. (1998) for studies of the ideological gender gap which focus on such explanations.

that such evidence is most reasonably explained by gender socialization and not by economic factors. However, the differences observed in children might also reflect evolutionary biological factors. Hatemi et al. (2009), in a study of twins, find support for a genetic explanation of the gender gap for some political preferences. Both the socialization and evolutionary biological explanations suggest then that personality trait differences play an important role in our understanding of the ideological gender gap. That is, these two types of theories suggest that personality traits may act as mediating variables through which gender affects ideological preferences. Furthermore, the research of Schmitt et al. (2008, 2009), discussed above, which shows that personality trait differences between the sexes are more pronounced in more developed countries also provides an explanation for why the gender gap in ideological preferences has become more pronounced in recent years.

In our empirical analysis below we demonstrate that personality traits can explain a substantial portion of the ideological gender gap. Later we address the extent that our results speak to the debate over whether the ideological gender gap is a consequence of structural changes in marriage or due to gender socialization and/or biological differences between men and women.

# 3 Empirical Analysis

## 3.1 Data

Our data are from a paid survey conducted through iLEE (internet Laboratory for Experimental Economics) at the Center for Experimental Economics at the University of Copenhagen in spring 2008.[26] A random sample of 22,207 subjects were drawn based on socioeconomic characteristics from the general voting-age population of Denmark, with the assistance of Statistics Denmark (SD) a Danish governmental agency. Hard-copy letters were sent by SD through regular mail, inviting participants to log into iLEE's webpage using a random ID number and to complete the survey within a week. Once subjects logged in, they were informed that the study would take approximately 50 min and that they would only be paid if they completed the entire study. The survey was conducted in Danish. After completion

[26]The internet platform is supported by the Carlsberg Foundation. The survey was part of a larger survey/internet experiment (for more details on the "virtual lab approach" used here, see Thöni et al. (2012) or Andersson et al. (2016)). Some parts were explicitly paid (and payments depended on choices), while others, like the survey reported in this paper, were not explicitly paid. However, overall earnings were only paid out if all parts were completed. Thus, participants had an incentive to complete the entire study, but they did not have an incentive to answer the questions used in this paper in a particular way. Average earnings were approximately 276 Danish Kroner (which is equivalent to about 37 Euro). 114 of the subjects participated in an experiment in which they were not paid. If we omit these subjects from the analysis the qualitative results are unchanged and the quantitative estimated coefficients are extremely similar.

of the survey, the survey data were sent to SD where they were matched by SD with detailed socioeconomic data (income, education, etc.). The data were then made available to the researchers in a completely anonymous format. Of the subjects contacted, 3584 logged into the survey, and 1823 completed all the questions.

The Appendix summarizes the data used in the analysis and compares the sample to the Danish population 18 years old and older. There is a high level of internet penetration in Denmark. Therefore, in general the match between the Danish population and our subjects is good, despite the high drop off rate, with a somewhat over-representation of middle aged individuals.

We use the four questions on ideological preferences of the subjects as our principal dependent variables, which we label our Ideology Measures. All of the questions provided subjects with the choice "Prefer not to answer." We asked two types of questions–general ideological questions and two questions related to economic policy. The first question, which was one of our general ideological measures, concerned voting behavior: "What party would you vote for if there was an election tomorrow?" The subjects could choose one of the nine parties who were eligible for parliament in the 2007 elections, along with the options "Would not vote" and "Would vote blank." The nine parties were ranked on a scale from right to left, with 1 denoting the most extreme rightist party (Danish People's Party) and 9 representing the most extreme leftist party (Unity List–Red–Green Alliance).[27] This variable is labeled *Party* and 1893 subjects completed this question.

The remaining three questions were taken from the European Values Study (1999). The second question, which was also a general ideological measure, was a classic ideology question in which the subject had to place herself on a left-right scale: "In political matters, people talk of 'the left' and 'the right.' How would you place your views on this scale, generally speaking?" A scale from 1 to 10 was presented with 1 representing extreme left and 10 representing extreme right. For our empirical analysis the scale was reversed such that 10 represented an extreme left position. The re-scaled variable is labeled *Leftist* and 2028 subjects completed this question.

Our general ideological measures, *Party* and *Leftist*, then, are used in our empirical analysis to determine if an overall or general ideological gender gap exists and can be explained by personality trait differences.

The third and fourth questions related to economic policy of the government and were our economic policy measures. The third question concerned government responsibility for individuals' wellbeing. Subjects were presented with two opposing views placed on either end of a 1–10 scale and asked to place themselves on the scale in accordance with their own views. The two statements were: "People should

[27]The full ordering of the parties is as follows: 1 = Danish People's Party, 2 = Conservatives, 3 = Liberals, 4 = New Alliance, 5 = Christian Democrats, 6 = Social Liberals, 7 = Social Democrats, 8 = Socialist Peoples Party, and 9 = Unity List–Red–Green Alliance. The ranking was validated by Jørgen Goul Andersen, an expert on Danish political parties. The ranking also corresponds to a ranking of the same Danish parties in Benoit and Laver (2006); Benoit and Laver exclude one new party which we include and include two parties which were no longer electoral contenders when we conducted our analysis.

take more responsibility to provide for themselves" and "The government should take more responsibility to ensure that everyone is provided for." These statements were placed at 1 and 10 respectively. We used answers to this question to create the variable labeled *Econ1* and 2107 subjects completed this question.

The fourth question concerned preferences for competition and had the same structure as the third question which we used to create the variable labeled *Econ2*. The two opposing statements were: "Competition is good. It stimulates people to work hard and develop new ideas" and "Competition is harmful. It brings out the worst in people." Again, the first statement was placed at 1 and the second statement was placed at 10. 2106 subjects completed *Econ2*.

We use our economic policy measures, *Econ1* and *Econ2* to determine if an ideological gender gap in economic policy exists and whether it can be explained by personality trait differences. *Econ1* is a question that is especially relevant to structural explanations of the ideological gender gap. That is, the structural explanations argue that when divorce rates and unmarried parenthood are high, women prefer a more interventionist government and a stronger "safety net." *Econ2* picks up differences in views about the role of competition. Other research has found that women appear to have a distaste for competition; specifically, Niederle and Vesterlund (2007) find that controlling for ability of men and women, women were half as likely to participate in a competition based compensation scheme compared to a noncompetitive one.

Once these questions were completed, subjects were given a Big Five personality test. We used the Big Five personality test NEO PI-R Short Version and the IQ test I-S_T 200R which are copyrighted and administered by the Danish Psychological Publishing Company (Dansk Psykologisk Forlag). The Danish NEO-PI-R Short Version consists of five 12-item scales measuring each domain. The 12 items for each domain are chosen from the original 48 items (of the full NEO-PI-R test) as follows: for each facet the two items (out of eight) with the highest correlation with the total factor score are chosen (this is different from the American 60-item version of NEO-PI-R, called NEO-FFI, where the 12 items with the highest correlation with the total factor score is picked, without regard to which facets the single items belong to). In the Danish short version, all facets are therefore represented equally within each domain. Furthermore, the short version has been shown to be highly correlated with the long one such that Costa and McCrae (2004) conclude that if one only wants to examine the Big Five factors, the short test can be used as easily as the full version.[28]

Subjects were presented with 60 statements and asked to state whether they agree or disagree with each on a 0–4 scale which contained the options "Strongly disagree," "Disagree," " Neutral," "Agree," and "Strongly agree." Using these answers we constructed the five variables that measured personality traits, *BigFiveO* for

[28]The correlations between the short and long tests are as follows: 0.93 for Emotional Stability, 0.90 for Extraversion, 0.93 for Openness, 0.89 for Agreeableness, and 0.91 for Conscientiousness. These estimates are based on a sample of 600 observations (Costa and McCrae 2004).

Openness, *BigFiveC* for Conscientiousness, *BigFiveE* for Extraversion, *BigFiveA* for Agreeableness, and*BigFiveN* for Emotional Stability. In our analysis the first four of these variables, *BigFiveO, BigFiveC, BigFiveE,* and *BigFiveA* are measured with higher values representing higher levels of the respective trait, but *BigFiveN* is measured with higher values representing lower levels of Emotional Stability (higher levels of Neuroticism). After completing the survey and tests, subjects were then asked to provide their bank account numbers and were told that their payment would be transferred to them after the study was concluded.

The data from the survey was combined with demographic data from the Danish government and the results of an IQ test.[29] Our income data is from 2006 and refers to the logged value of total gross income of the individual (salary, pension, capital income, etc.) in thousand Danish Kroner, which is labeled as the continuous variable *Income*. As noted above, our income data is unique for studies of the effects of income on ideology. Furthermore, our income data is individualistic rather than family income and thus allows us to consider explicitly whether women's ideological preferences are affected by their own income, even if married. We omitted individuals with negative values of income due to large capital losses. For robustness, we also considered truncating the data at the lower and upper percentiles, replacing the continuous variable with dummies and including a squared income variable. The qualitative results were robust to these variations.[30]

We also include age, age squared, and age cubed, in order to capture nonlinear relationships between age and our dependent variables. As with income, we also tested the robustness of our results by replacing the continuous age variable with dummies.

Education levels are divided into six categories.[31] Other demographic variables included in the analysis whose names are self-explanatory include: *Female*, *Urban,*

[29]In the IQ test, subjects were presented with 20 questions or puzzles and asked to solve as many as possible within a 10 min period. The IQ test we used is part of a more extensive test called "IST 2000 R". It contains several modules and we chose to use a part that is a variation of Raven's Progressive Matrices. The advantage of using Raven's Progressive Matrices Test is that it does not depend heavily on verbal skills or other kinds of knowledge taught during formal education. A subject's number of correct answers was used to create our variable measuring performance which we labeled *IQ*. We omitted subjects who scored less than 5 on the IQ test.

[30]Since our income variable includes capital income it arguably measures wealth as well. We also have an independent measure of wealth that includes the value of many assests such as stocks and bonds, housing, and cash holdings at bank accounts. Including the wealth variable does not change our qualitative results.

[31]*EduBasic* for individuals who did not complete a high school degree, *EduHS* for individuals who completed an academic high school degree, but no advanced education; *EduVoc* for individuals who completed vocational education instead of attending an academic high school, and three variables for various levels for advanced university training: *EduAdv1* for individuals who have completed some academic post-secondary education less than a Bachelor's degree, *EducAdv2* for individuals who have completed a Bachelor's degree, and *EducAdv3* for individuals who have completed a Masters or Ph.D. degree. *EduVoc* was omitted in the empirical analysis and served as our baseline case since previous research has suggested that individuals with vocational education have the most rightist ideological preferences. *Student* designated individuals who are currently

*Parttime, Retired, Unemployed, Married,* and *Divorced.* We also include *Couple* for unmarried couples and *Church* for individuals who are members of the Danish Lutheran Church.

## 3.2 Empirical Models

Our main goal is to examine the extent that personality traits serve as mediating variables for *Female* in determining observed ideological preferences. In order to estimate the mediation effects, we used the following procedure. We first estimated four sets of seemingly unrelated regressions. In each set we estimated simultaneously seven equations. In the first five equations the five personality traits serve as the dependent variables. In the sixth equation, *Income* serves as the dependent variable, since given earlier research by Morton et al. (2011), we expect that *Income* serves as a mediating variable for personality traits on ideological preferences. We also expect that Income may serve as a mediating variable between gender and ideological preferences independent of personality traits. In the seventh equation one of the four ideological measures serves as the dependent variable, *Party, Leftist, Econ1,* or *Econ2.* For example, to determine the mediation effect of the Big Five personality traits for the ideological measure *Party* we estimated the simultaneously the following seven seemingly unrelated regressions (other covariates included our demographic variables and measures of education and IQ above):

$$BigFiveO = \alpha_O + \beta_{OF} Female + OtherCovariates + \mu_O \quad (1)$$

$$BigFiveC = \alpha_C + \beta_{CF} Female + OtherCovariates + \mu_C \quad (2)$$

$$BigFiveE = \alpha_E + \beta_{EF} Female + OtherCovariates + \mu_E \quad (3)$$

$$BigFiveA = \alpha_A + \beta_{AF} Female + OtherCovariates + \mu_A \quad (4)$$

$$BigFiveN = \alpha_N + \beta_{NF} Female + OtherCovariates + \mu_N \quad (5)$$

$$\begin{aligned} Income = {} & \alpha_I + \beta_{IA} BigFiveA + \beta_{IC} BigFiveC + \beta_{IO} BigFiveO \\ & + \beta_{IE} BigFiveE + \beta_{IN} BigFiveN + \beta_{IF} Female + OtherCovariates + \mu_I \end{aligned} \quad (6)$$

$$\begin{aligned} Party = {} & \alpha_p + \beta_{pA} BigFiveA + \beta_{PC} BigFiveC + \beta_{pO} BigFiveO + \beta_{pE} BigFiveE \\ & + \beta_{pN} BigFiveN + \beta_{pI} Income + \beta_{pF} Female + OtherCovariates + \mu_P \end{aligned} \quad (7)$$

studying and spanned all the education categories, depending on the extent that they had completed their education.

The coefficient $\beta_{pF}$ in Eq. (7) is an estimate then of the direct effect on *Party* of *Female*, controlling for *Income* and personality traits. To determine the indirect effect or mediated effect of *Female* on *Party* through Agreeableness that is not mediated through income, for example, we multiply the coefficient on *Female* in Eq. (4), $\beta_{AF}$ times the coefficient on *BigFiveA* in the Eq. (7), $\beta_{pA}$. The indirect effect or mediated effect of *Female* on *Party* through Agreeableness that is also mediated through income is given by $\beta_{AF} \times \beta_{IA} \times \beta_{pI}$. The total indirect effect or mediated effect of *Female* on *Party* through Agreeableness is the sum of these two terms. If a mediation effect has occurred, we expect the indirect effect to be nonzero and have the same sign as the direct effect. If a suppression effect has occurred, we expect the indirect effect to be nonzero and to have the opposite sign from the direct effect.[32] The estimates of these indirect effects via personality trait differences are our estimates of the degree to which personality trait differences can explain the ideological gender gap. The indirect effects of *Female* on *Party* via the other personality traits is similarly defined.

In order to compare the personality trait explanation with the structural explanation of the gender gap, we also estimate the indirect effects of *Female* through *Female*'s direct effect on *Income*, independent of personality trait differences. This effect is estimated by multiplying the coefficient on *Female* in Eq. (6), $\beta_{IF}$, times the coefficient on *Income* in Eq. (7), $\beta_{PI}$. Some may argue that this is not an appropriate test of the structural explanation and that the structural explanation attempts to explain why a gender gap exists controlling for income differences between men and women. Our estimate of the coefficient on *Female* in Eq. (7), $\beta_{PF}$, tells us if there is an effect on ideology independent of such differences and may be seen as a more appropriate comparison. Since we do not have variation within our data over those structural variables such as divorce risk, child care provision, etc., which are said to be relevant in cross-country considerations of the structural explanation, we are not able to directly test the structural explanation. However, as discussed in the Introduction, the divorce rate in Denmark is relatively high which implies that our data provide a particularly strong test for nonstructural explanations of the ideological gender gap.

The direct and indirect effects of *Female* on the other ideological measures are similarly defined. Although the estimation of Eqs. (1)–(7) provides estimates of the standard errors of the components of the indirect effects, it does not provide an estimate for the standard error of the products of these components. Moreover, the products are likely to be nonnormal and skewed when either a mediation or suppression effect exists. As Shrout and Bolger (2002) discuss, ignoring the skew reduces the power to detect mediation or suppression effects. We therefore used bootstrap measures to determine the standard errors and confidence intervals for the indirect effects as developed by Efron and Tibshirani (1993) and described in Appendix A of Shrout and Bolger.[33]

[32]See MacKinnon et al. (1995, 2000).

[33]See also Preacher and Hayes (2004).

In our analysis we compute the indirect effects through each of the personality variables, treating the other traits as covariates. We therefore assume that there are no interaction effects among the Big Five variables in determining ideological preferences. Although we use ordinary least squares for this analysis, we also estimated ordered logit and tobit models of these equations independently with little difference in the qualitative results.[34]

We summarize the relationships we expect to find between *Female,* personality traits, *Income*, and *Party* in our estimations in Table 1 below. We expect the other ideological measures to have similar relationships. Recall that our ideology measures are coded such that higher values represent more leftist preferences. The top half of the table reports our expectations with regard to the effects of personality traits. The second column in the top half reports our expectations about the effects of *Female* on each of the five personality traits. That is, we expect to find that women are higher on Agreeableness, Conscientiousness, and Extraversion, and lower on Emotional Stability (recall that our measure of Emotional Stability, *BigFiveN*, decreases with higher levels of Stability). The third and fourth columns in the top half report our expectations concerning the effects of each personality trait on *Income* and *Party*, respectively; we expect that higher levels of Conscientiousness and Emotional Stability and lower levels of Agreeableness lead to higher incomes and that higher levels of Agreeableness and Openness will lead to more leftwing preferences, while higher levels of Emotional Stability and Conscientiousness will result in more rightwing preferences.

**Table 1** Predicted relationships in estimation for *Party* (other ideological measures are expected to be similar)

| | Effects of | | | Indirect effects of | |
|---|---|---|---|---|---|
| | *Female* on | Traits on | | *Female* on *Party* through traits | |
| | Traits | *Income* | *Party* | Indep. of *Income* | Through *Income* |
| Coefficients | $\beta_{xF}$ | $\beta_{Ix}$ | $\beta_{Px}$ | $\beta_{xF}\beta_{Px}$ | $\beta_{xF}\beta_{Ix}\beta_{PI}$ |
| *BigFiveO* | 0 | 0 | + | 0 | 0 |
| *BigFiveC* | + | + | – | – | – |
| *BigFiveE* | + | 0 | 0 | – | 0 |
| *BigFiveA* | + | – | + | + | + |
| *BigFiveN* | + | – | + | + | + |
| Residual effects of *Female* | | $\beta_{IF}$ | $\beta_{PI}$ | | |
| (Indep. of traits) | | – | ? | | |
| Indirect effects of *Female* | | | | | $\beta_{IF}\beta_{PI}$ |
| (Indep. of traits) | | | | | + |

[34]An alternative estimation strategy would be to use an instrumental variable approach as in two-stage least squares. But doing so would require us to make exclusion restrictions—arbitrary assumptions about the variables which would be contrary to our theoretical expectations.

The fifth column in the top half summarizes our predictions about the indirect effects of *Female* via each personality trait on our ideology measures that is not mediated by income. We expect that women's increased tendency to be more Agreeable and less Emotionally Stable will have the indirect effect of increasing their tendency to be leftist, while women's increased tendency to be more Conscientious and more Extraverted will have the indirect effect of decreasing their tendency to be leftist. The sixth column in the top half summarizes the predictions about the indirect effects of *Female* via each personality trait's indirect effect on income. We expect that women's tendencies to be more Agreeable and less Emotionally Stable will lead to lower incomes and thus more leftist preferences, while we expect that women's tendencies to be more Conscientious will lead to higher incomes and thus less leftist preferences.

Finally, the last four rows provide predictions of the effect of *Female* independent of personality trait differences. That is, we expect that women will earn lower incomes than men independent of personality trait differences. We are agnostic as to whether *Female* will have a direct effect on our ideological measures after controlling for the indirect effects via personality traits or income or both. However, we expect that the tendency of women to earn less income, independent of personality traits, will have the indirect effect of making women more leftist than men as well given the structural explanation discussed above. To the extent that we find that there are indirect effects of *Female* on ideological preferences via personality traits (either independent of *Income* or via *Income*) then we find support for the socialization and biological explanations for the gender gap. But to the extent that we find that there are either direct effects of *Female* on ideological preferences independent of personality traits (either independent of *Income* or via *Income*), then we find support for the structural explanations for the gender gap.

## 3.3 Results

### 3.3.1 Does an Ideological Gender Gap Exist?

Before reporting our estimations of Eqs. (1)–(7), we first establish that a gender gap in ideology exists in our data in Tables 2 and 3 below. Table 2 summarizes the raw differences in our ideological measures between men and women and Table 3 reports on four simple regressions with each ideological measure as a dependent variable and *Female* as an independent variable. We also report on regressions which include our demographic variables as controls (but exclude our personality trait measures). We find indeed an ideological gender gap across all four of our measures and that it is significant. Women are generally about 2/3 of a point more leftist on our ideological scales (recall that *Party* is measured on a 1–9 point scale, while the other measures are on a 10 point scale). When we control for demographic differences, the gap still remains and women are approximately 1/2 of a point more

**Table 2** Ideology by gender

| | | Ideological measures | | | |
|---|---|---|---|---|---|
| | | *Party* | *Leftist* | *Econ1* | *Econ2* |
| Women | Mean | 5.42 | 5.93 | 5.45 | 4.41 |
| | Obs. | 909 | 965 | 1029 | 1027 |
| Men | Mean | 4.73 | 5.29 | 4.91 | 3.75 |
| | Obs. | 984 | 1063 | 1078 | 1079 |

**Table 3** Effect of *Female* on ideology (excluding personality trait measures)

| | Ideological measures | | | | | | | |
|---|---|---|---|---|---|---|---|---|
| | *Party* | | *Leftist* | | *Econ1* | | *Econ2* | |
| Controls Incl. | No | Yes | No | Yes | No | Yes | No | Yes |
| Coeff. | 0.69*** | 0.54*** | 0.65*** | 0.52*** | 0.54*** | 0.48*** | 0.66*** | 0.64*** |
| Std. Err. | 0.12 | 0.12 | 0.10 | 0.10 | 0.10 | 0.10 | 0.08 | 0.09 |
| $R$ squared | 0.02 | 0.10 | 0.02 | 0.09 | 0.01 | 0.04 | 0.03 | 0.08 |
| Observations | 1893 | | 2028 | | 2107 | | 2106 | |

***Significant at 1 % level

leftist than men across all ideological measures, holding demographics constant.[35] We now turn to our investigation of whether personality trait differences between men and women can explain the gender gap in ideology.[36]

### 3.3.2 Gender Differences in Personality Traits

We estimated versions of Eqs. (1)–(7) for each of our four measures of ideology. First we report the results of the estimations of Eqs. (1)–(5) in Table 4. Since we used four different measures of ideology and estimated separate seemingly unrelated regressions for each measure, we report only the estimated coefficients for *Female* in each of the four different estimations of Eqs. (1)–(5), depending on the ideology measure considered in the seemingly unrelated regression estimation, as identified by the column headings in the table. That is, column 3 presents the results of the estimation of Eqs. (1)–(5) when the dependent variable in Eq. (7) is *Party*, column 4 presents the results when the dependent variable in Eq. (7) is *Leftist*, etc. Although our sample size varies as not all subjects responded to all of the ideological measures we find little variation in results across ideological measures. We included the

[35]We also estimated the tobit and ordered logit regressions with the same qualitative results.

[36]Furthermore, we estimated separate effects for married and unmarried women in order to determine if the direct effects of *Female* on our ideological measures depend on marital status, as found in previous studies (Iversen and Rosenbluth 2006; Norrander and Wilcox 2008). We found that unmarried women were more leftist on the ideological measure *Party*, controlling for other direct and indirect effects, but little evidence of a difference on the three other ideological measures.

**Table 4** Effect of female on personality ($\beta_{xF}$ in Eqs. (1)–(5)) (controls included)

| | | Ideological measure in Eq. (7) | | | |
|---|---|---|---|---|---|
| Dependent variable in equation | | *Party* | *Leftist* | *Econ1* | *Econ2* |
| Equation (1): *BigFiveO* | $\beta_{OF}$ | 0.64** | 0.62** | 0.51** | 0.52** |
| | Std. Err. | 0.27 | 0.26 | 0.26 | 0.26 |
| | R-Sq. | 0.11 | 0.11 | 0.11 | 0.11 |
| | Effect Size[a] | 0.10 | 0.10 | 0.08 | 0.09 |
| Equation (2): *Big FiveC* | $\beta_{CF}$ | 0.24 | 0.25 | 0.26 | 0.26 |
| | Std. Err. | 0.26 | 0.25 | 0.24 | 0.24 |
| | R-Sq. | 0.04 | 0.04 | 0.04 | 0.04 |
| | Effect size[a] | 0.04 | 0.04 | 0.05 | 0.05 |
| Equation (3): *BigFiveE* | $\beta_{EF}$ | −0.46 | −0.31 | −0.49* | −0.48* |
| | Std. Err. | 0.29 | 0.28 | 0.28 | 0.28 |
| | R-Sq. | 0.05 | 0.05 | 0.05 | 0.05 |
| | Effect size[a] | −0.07 | −0.05 | −0.08 | −0.08 |
| Equation (4): *BigFiveA* | $\beta_{AF}$ | 2.61*** | 2.64*** | 2.68*** | 2.66*** |
| | Std. Err. | 0.25 | 0.24 | 0.24 | 0.24 |
| | R-Sq. | 0.09 | 0.09 | 0.10 | 0.09 |
| | Effect size[a] | 0.47 | 0.48 | 0.49 | 0.48 |
| Equation (5): *BigFiveN* | $\beta_{NF}$ | 3.53*** | 3.37*** | 3.39*** | 3.42*** |
| | Std. Err. | 0.31 | 0.30 | 0.30 | 11.46 |
| | R-Sq. | 0.10 | 0.10 | 0.10 | 0.10 |
| | Effect size[a] | 0.50 | 0.48 | 0.48 | 0.49 |
| Observations | | 1893 | 2028 | 2107 | 2106 |

[a]Std. deviation difference in trait between women and men
***Sig. at 1 % level, **Sig. at 5 % level, *Sig. at 10 % level

demographic controls discussed above in the estimation and the full results are available from the authors.

We find that women are significantly more likely to be Agreeable and to have lower levels of Emotional Stability which shows support for some of the previous work on sex differences in personality traits. However, we also find that the women in our sample are significantly more likely to be Open to Experience, in contrast to previous work which typically finds no relationship. Furthermore, unlike previous research which finds that women are more Conscientious and Extraverted, we do not find any evidence of significant differences between men and women in these traits.

In order to determine whether the estimated relationships are meaningful, we also estimated the sizes of the effects of *Female* in terms of the standard deviation difference in each of our personality traits which are also reported in Table 4. We find that the largest effects are by far for the traits of Agreeableness and Emotional Stability. In general, women are approximately half a standard deviation higher in Agreeableness and half a standard deviation lower in Emotional Stability than men. Thus, we find that there are significant and sizeable personality trait differences between women and men in our sample and that the principal differences appear to be on traits that also have been shown to matter in determining ideological preferences.

### 3.3.3 Direct Effects on Ideology

We now turn to the results in estimating ideology in the four versions of Eq. (7) which are summarized in Table 5 below. The coefficients in this table report the direct effects on each ideology measure of each of the variables listed. Recall that our ideological measures are constructed such that higher values represent more leftist ideological preferences. Somewhat surprising, we find no significant direct or residual effect of *Female* on three of our ideology measures, but a significant positive effect for *Econ2*. The lack of a significant residual effect for three of the ideological measures suggests possible full mediation through the effects of *Female* on our personality trait variables and *Income.*

We report the estimated sizes of the effects of a standard deviation change in personality traits on our ideological measures in Table 6 below. Since our variable *Income* is logged, the effect size is nonlinear for changes in non-logged income. We report the effect sizes for a one standard deviation change for income at the 25th percentile, 50th percentile, and 75th percentile. The effect size diminishes with increases in relative income, which is reasonable.

**Table 5** Determinants of ideology equation (7) (direct effects) (by ideological measure in SUR Est., controls included)

| | | *Party* | *Leftist* | *Econ1* | *Econ2* |
|---|---|---|---|---|---|
| *Female* | $\beta_{pF}$ | 0.09 | 0.11 | 0.09 | 0.29*** |
| | Std. Err. | 0.12 | 0.10 | 0.10 | 0.09 |
| *Income* | $\beta_{PI}$ | −0.52*** | −0.51*** | −0.24** | −0.27*** |
| | Std. Err. | 0.11 | 0.09 | 0.10 | 0.09 |
| *BigFiveO* | $\beta_{PO}$ | 0.06*** | 0.06*** | 0.04*** | 0.02*** |
| | Std. Err. | 0.01 | 0.01 | 0.01 | 0.01 |
| *Big FiveC* | $\beta_{PC}$ | −0.04*** | −0.05*** | −0.02** | −0.02*** |
| | Std. Err. | 0.01 | 0.01 | 0.01 | 0.01 |
| *BigFiveE* | $\beta_{PE}$ | −0.01 | −0.005 | 0.01 | −0.02*** |
| | Std. Err. | 0.01 | 0.01 | 0.01 | 0.01 |
| *BigFiveA* | $\beta_{PA}$ | 0.09*** | 0.09*** | 0.09*** | 0.06*** |
| | Std. Err. | 0.01 | 0.01 | 0.01 | 0.01 |
| *BigFiveN* | $\beta_{NA}$ | 0.03*** | 0.02*** | 0.03*** | 0.03*** |
| | Std. Err. | 0.01 | 0.01 | 0.01 | 0.01 |
| Observations | | 1893 | 2028 | 2107 | 2106 |
| R-Squared | | 0.19 | 0.21 | 0.12 | 0.14 |

***Sig. at 1 % level, **Sig. at 5 % level, *Sig. at 1 % level

**Table 6** Estimated direct effect on leftwing preferences (recall *Party* measured on a 1–9 scale, others on a 1–10 scale)

| | Ideological measure in Eq. (7) | | | |
|---|---|---|---|---|
| One standard deviation change in: | Party | Leftist | Econ1 | Econ2 |
| Non-logged income at 25th percentile | −0.43 | −0.41 | −0.19 | −0.22 |
| Non-logged income at 50th percentile | −0.32 | −0.31 | −0.14 | −0.16 |
| Non-logged income at 75th percentile | −0.26 | −0.25 | −0.12 | −0.13 |
| BigFiveO | 0.36 | 0.35 | 0.23 | 0.12 |
| BigFiveC | −0.23 | −0.28 | −0.11 | −0.12 |
| BigFiveE | −0.04 | −0.03 | 0.04 | −0.15 |
| BigFiveA | 0.51 | 0.50 | 0.52 | 0.32 |
| BigFiveN | 0.18 | 0.14 | 0.20 | 0.23 |

In keeping with previous research, we find that *Income* has a significant and negative relationship with all four of our ideological measures, although the effect is twice as high for our overall ideological measures than for our economic policy measures. We also find, as in previous research, that increases in Openness (higher values of *BigFiveO*) and Agreeableness (higher values of *BigFiveA*), decreases in Conscientiousness (lower values of *BigFiveC*) and Emotional Stability (higher values of *BigFiveN*), have the direct effect of leading to higher levels of leftwing preferences across our ideological measures. In contrast to previous research, we find that Extraversion (*BigFiveE*) has no apparent direct effect on our ideology measures with the exception of *Econ2*, where we find the expected negative relationship.[37] Notably, the size of the effect on ideology of a standard deviation change is greatest for Agreeableness and also relatively large for Emotional Stability, the two personality traits on which we find the greatest differences between women and men in the previous section.

### 3.3.4 Indirect Effects of *Female* on Ideology Independent of Income

Our main contribution is our investigation of the indirect effects of *Female* on ideology through personality traits, which are provided in Tables 7 and 8 below. Table 7 presents the results for our general ideological measures *Party* and *Leftist* and Table 8 presents the results for economic policy measures *Econ1* and *Econ2*. As noted in the previous section, we use bootstrap standard errors (with 5000 bootstrap-samples for each estimation) and corrected for bias in computing the reported 95 % confidence intervals of the indirect effects.[38] As Efron and Tibshirani (1993) explain, standard percentile confidence intervals tend to be too narrow. Thus

[37]Gerber et al. (2010) similarly find that Extraversion has no significant effect on their ideology measures.

[38]Given the nonnormal distribution of the test statistic, we do not report other levels of significance.

**Table 7** Indirect effects of *Female* on general ideology independent of *Income* (controls included, bias corrected confidence intervals)

| Indirect effects of *Female* via personality traits on *Party* independent of *Income* | | | | | |
|---|---|---|---|---|---|
| Trait | $\beta_{xF}\beta_{Px}$ | Bias | Bootstrap Std. Err. | 95 % Conf. interval | |
| *BigFiveO* | 0.0372** | 0.0031 | 0.0173 | 0.0057 | 0.0718 |
| *BigFiveC* | −0.0098 | −0.0002 | 0.0108 | −0.0349 | 0.0088 |
| *BigFiveE* | 0.0031 | −0.0008 | 0.0054 | −0.0025 | 0.0244 |
| *BigFiveA* | 0.2409** | −0.0032 | 0.0340 | 0.1823 | 0.3159 |
| *BigFiveN* | 0.0912** | 0.0030 | 0.0361 | 0.0193 | 0.1577 |
| Total effect | 0.3626** | 0.0028 | 0.0535 | 0.2568 | 0.4669 |
| Indirect effects of *Female* via personality traits on *Leftist* independent of *Income* | | | | | |
| *BigFiveO* | 0.0353** | 0.0011 | 0.0160 | 0.0070 | 0.0694 |
| *BigFiveC* | −0.0125 | −0.0004 | 0.0127 | −0.0389 | 0.0116 |
| *BigFiveE* | 0.0015 | 2.88*e*−08 | 0.0040 | −0.0027 | 0.0156 |
| *BigFiveA* | 0.2370** | −0.0013 | 0.0306 | 0.1821 | 0.3025 |
| *BigFiveN* | 0.0688** | 0.0013 | 0.0284 | 0.0148 | 0.1269 |
| Total Effect | 0.3301** | 0.0010 | 0.0465 | 0.2408 | 0.4236 |

**Sig. at 5 % level established through confidence interval estimation

**Table 8** Indirect effects of *Female* on economic ideology independent of *Income* (controls included, bias corrected confidence intervals, *EduVoc* omitted)

| Indirect effects of *Female* via personality traits on *Econ1* independent of *Income* | | | | | |
|---|---|---|---|---|---|
| Traits | $\beta_{xF}\beta_{Px}$ | Bias | Bootstrap Std. Err. | 95 % Conf. Interval | |
| *BigFiveO* | 0.0192 | 0.0023 | 0.0116 | −0.0006 | 0.0429 |
| *BigFiveC* | −0.0051 | −0.0008 | 0.0062 | −0.0208 | 0.0040 |
| *BigFiveE* | −0.0032 | −0.0001 | 0.0052 | −0.0190 | 0.0037 |
| *BigFiveA* | 0.2504** | 0.0002 | 0.0315 | 0.1931 | 0.3155 |
| *BigFiveN* | 0.0949** | −0.0041 | 0.0304 | 0.0430 | 0.1651 |
| Total effect | 0.3562** | −0.0023 | 0.0463 | 0.2697 | 0.4523 |
| Indirect effects of *Female* via personality traits on *Econ2* independent of *Income* | | | | | |
| *BigFiveO* | 0.0099** | −0.00001 | 0.0067 | 0.0008 | 0.0288 |
| *BigFiveC* | −0.0057 | −7.90*e*−06 | 0.0059 | −0.0210 | 0.0032 |
| *BigFiveE* | 0.0005 | 4.30*e*−06 | 0.0007 | −0.0003 | 0.0030 |
| *BigFiveA* | 0.1546** | −0.0004 | 0.0240 | 0.1090 | 0.2035 |
| *BigFiveN* | 0.1142** | 0.0002 | 0.0272 | 0.0649 | 0.1712 |
| Total effect | 0.2845** | 0.00004 | 0.0400 | 0.2075 | 0.3663 |

**Sig. at 5 % level established through confidence interval estimation

we use the bias corrected bounds which take into account the asymmetries in the distributions of the bootstrap estimates.

As expected, we find significant positive indirect effects through increased Agreeableness and decreased Emotional Stability of *Female* on all four of our ideological measures, independent of income. That is, we find robust evidence that women's increased tendency to be Agreeable and decreased tendency to be

Emotionally Stable, leads to women being more leftist than men. The indirect effect through Agreeableness is the largest, leading to a 0.15–0.24 point difference in ideological preferences that is explained by gender, while the indirect effect through Emotional Stability leads to a 0.07–0.11 point difference in ideological preferences.

Other results we find are unexpected. We find that *Female* has a positive significant indirect effect independent of income through increased Openness in three of our four ideological measures (*Econ1* is the exception). We also find no significant indirect effects independent of income through increased Conscientiousness or Extraversion. As these indirect effects were expected to be negative, lead women to be less leftist than men, then it is not surprising that we find that the total indirect effect independent of *Income* for all personality traits (as reported in the rows marked Total Effect) is significantly positive. Hence, we can conclude that personality trait differences between men and women indirectly lead women to be significantly more leftist by men independent of income changes. The total indirect effect point difference independent of income ranges between 0.28 and 0.36. We now turn to measuring the indirect effects of female via personality trait differences through income changes in the next subsection.

### 3.3.5 Indirect Effects of *Female* on Ideology Through Income

We now turn to our estimates of the indirect effects of *Female* through personality trait differences' effects on *Income* and directly through *Income*. We first report our estimations of each version of Eq. (6) for each ideological measure in Table 9 below. Columns 3 presents the results of the estimation of Eq. (6) when the dependent variable in Eq. (7) is *Party*, column 4 presents the results when the dependent variable in Eq. (7) is *Leftist*, etc. First, we find significant evidence of direct effects of *Female* on *Income* independent of effects through personality trait differences.[39] Second, we find strong evidence that individuals who are more Agreeable and less Emotionally Stable earn significantly less income, as predicted. We also find significant evidence that individuals who are more Conscientious earn more income in estimations with three of our four ideological measures (*BigFiveC* is significant at the 10 % level in the estimation combined with *Party*), which is also expected. As in the previous analysis, we estimate the sizes of these effects which are also reported in Table 9. We find that a standard deviation change in each personality trait has a small effect on income, either increasing or decreasing income by only a few percentage points. In comparison, independent of personality trait differences and controlling for other demographic variables, women in our sample earn on average 20 % less than men.

[39]In unreported regressions we found that a difference between the earnings of married and unmarried women. Married women generally earned 29 % less than married and unmarried men, while unmarried women generally earned 8 % less than men.

**Table 9** Determinants of Income (Eq. (6))
(by ideological measure in SUR estimation, controls included)

| Indep. Variables | | Ideological measure in Eq. (7) | | | |
|---|---|---|---|---|---|
| | | *Party* | *Leftist* | *Econ1* | *Econ2* |
| *BigFiveO* | $\beta_{IO}$ | −0.003 | −0.003 | −0.003 | −0.003 |
| | Std. Err. | 0.002 | 0.002 | 0.002 | 0.002 |
| | Effect size[a] | −0.02 | −0.02 | −0.02 | −0.02 |
| *Big FiveC* | $\beta_{IC}$ | 0.004* | 0.004** | 0.005** | 0.005** |
| | Std. Err. | 0.002 | 0.002 | 0.002 | 0.002 |
| | Effect size[a] | 0.02 | 0.02 | 0.03 | 0.03 |
| *BigFiveE* | $\beta_{IE}$ | 0.002 | 0.002 | 0.002 | 0.002 |
| | Std. Err. | 0.002 | 0.002 | 0.002 | 0.002 |
| | Effect size[a] | 0.01 | 0.01 | 0.01 | 0.02 |
| *BigFiveA* | $\beta_{IA}$ | −0.005** | −0.005*** | −0.005** | −0.005*** |
| | Std. Err. | 0.002 | 0.002 | 0.002 | 0.002 |
| | Effect size[a] | −0.03 | −0.03 | −0.03 | −0.03 |
| *BigFiveN* | $\beta_{IN}$ | −0.006*** | −0.006*** | −0.006*** | −0.006*** |
| | Std. Err. | 0.002 | 0.002 | 0.002 | 0.002 |
| | Effect size[a] | −0.04 | −0.04 | −0.04 | −0.04 |
| *Female* | $\beta_{IF}$ | −0.209*** | −0.202*** | −0.199*** | −0.199*** |
| | Std. Err. | 0.024 | 0.023 | 0.022 | 0.023 |
| | Effect size[a] | −0.20 | −0.19 | −0.19 | −0.19 |
| Controls included | | Yes | Yes | Yes | Yes |
| Observations | | 1893 | 2028 | 2107 | 2106 |
| R-Squared | | 0.61 | 0.61 | 0.61 | 0.61 |

[a]Percentage change in *Income* from a standard deviation change in personality trait or from being *Female*
***Sig. at 1 % level, **Sig. at 5 % level, *Sig. at 1 % level

From our estimations in Tables 4, 6, and 9, we calculate the indirect effects of *Female* on ideology through *Income* both through the indirect effect of *Female* via *Income* independent of personality trait differences and indirectly through personality trait differences' effects on *Income*, which are reported in Tables 10 and 11 below. The organization of these tables is similar to Tables 7 and 8, with Table 10 reporting the indirect effects through *Income* on our general measures of ideology, *Party* and *Leftist* and Table 11 reporting the indirect effects through *Income* on our economic policy measures, *Econ1* and *Econ2*.

Our results of the indirect effects of *Female* through *Income* on our ideology measures show support for both the personality trait and structural explanations of the gender gap. First, we find a significant positive indirect effect of *Female* on ideology via reduced *Income* independent of personality trait differences across all four of our ideology measures. Thus, we find support for the structural explanation of the gender gap via income differences; women make less income and thus they

**Table 10** Indirect effects of *Female* on general ideology through *Income* (controls included, bias corrected confidence intervals)

| Indirect effects of *Female* on *Party* through *Income* | | | | | | |
|---|---|---|---|---|---|---|
| | | *Obs. Coeff.* | Bias | Bt. Std. Err. | 95 % Conf. Interval | |
| Indep. of traits ($\beta_{IF}\beta_{IP}$) | | 0.1081** | 0.0023 | 0.0284 | 0.0560 | 0.1676 |
| Via traits | *BigFiveO* | 0.0009 | −7.60*e*−06 | 0.0009 | −0.0003 | 0.0034 |
| ($\beta_{xF}\beta_{IX}\beta_{IP}$) | *BigFiveC* | −0.0005 | −6.86*e*−06 | 0.0007 | −0.0031 | 0.0003 |
| | *BigFiveE* | 0.0005 | 4.30*e*−06 | 0.0007 | −0.0003 | 0.0030 |
| | *BigFiveA* | 0.0066 | 0.0005 | 0.0042 | −0.0010 | 0.0156 |
| | *BigFiveN* | 0.0107** | 0.0001 | 0.0048 | 0.0031 | 0.0224 |
| | Total | 0.0182** | 0.0007 | 0.0071 | 0.0058 | 0.0333 |
| Indirect effects of *Female* on *Leftist* through *Income* | | | | | | |
| Indep. of traits ($\beta_{IF}\beta_{IP}$) | | 0.1034** | 0.0005 | 0.0236 | 0.0608 | 0.1540 |
| Via traits | *BigFiveO* | 0.0009 | −0.0001 | 0.0008 | −0.00004 | 0.0033 |
| ($\beta_{xF}\beta_{IX}\beta_{IP}$) | *BigFiveC* | −0.0006 | −2.26*e*−06 | 0.0007 | −0.0030 | 0.0003 |
| | *BigFiveE* | 0.0003 | 0.00001 | 0.0006 | −0.0002 | 0.0023 |
| | *BigFiveA* | 0.0069** | 0.0001 | 0.0038 | 0.0003 | 0.0155 |
| | *BigFiveN* | 0.0104** | −0.0002 | 0.0041 | 0.0039 | 0.0207 |
| | Total | 0.0180** | −0.0001 | 0.0061 | 0.0074 | 0.0318 |

**Sig. at 5 % level established through confidence interval estimation

**Table 11** Indirect effects of *Female* on economic ideology through *Income* (controls included, bias corrected confidence intervals)

| Indirect effects of *Female* on *Econ1* through *Income* | | | | | | |
|---|---|---|---|---|---|---|
| | | *Obs.Coeff.* | Bias | Bt. Std. Err. | 95 % Conf. Interval | |
| Indep. of traits ($\beta_{IF}\beta_{IP}$) | | 0.0474** | 0.0018 | 0.0227 | 0.0033 | 0.0922 |
| Via traits | *BigFiveO* | 0.0003 | −8.95*e*−06 | 0.0003 | −0.00002 | 0.0016 |
| ($\beta_{xF}\beta_{IX}\beta_{IP}$) | *BigFiveC* | −0.0003 | −0.00002 | 0.0004 | −0.0018 | 0.0001 |
| | *BigFiveE* | 0.0003 | −7.24*e*−06 | 0.0003 | −0.00004 | 0.0019 |
| | *BigFiveA* | 0.0031 | 0.0001 | 0.0022 | −4.23*e*−06 | 0.0090 |
| | *BigFiveN* | 0.0049** | 0.0001 | 0.0030 | 0.0008 | 0.1333 |
| | Total | 0.0083** | 0.0002 | 0.0045 | 0.0012 | 0.0190 |
| Indirect effects of *Female* on *Econ2* through *Income* | | | | | | |
| Indep. of traits ($\beta_{IF}\beta_{IP}$) | | 0.0543** | 0.0004 | 0.0190 | 0.0181 | 0.0927 |
| Via traits | *BigFiveO* | 0.0004 | −7.18*e*−06 | 0.0004 | −1.69*e*−07 | 0.0017 |
| ($\beta_{xF}\beta_{IX}\beta_{IP}$) | *BigFiveC* | −0.0003 | −9.85*e*−06 | 0.0004 | −0.0019 | 0.0001 |
| | *BigFiveE* | 0.0003 | −0.00001 | 0.0004 | −0.0001 | 0.0017 |
| | *BigFiveA* | 0.0038** | −0.0001 | 0.0022 | 0.0005 | 0.0097 |
| | *BigFiveN* | 0.0056** | 0.0001 | 0.0032 | 0.0013 | 0.0144 |
| | Total | 0.0098** | 0.0001 | 0.0046 | 0.0025 | 0.0209 |

**Sig. at 5 % level established through confidence interval estimation

are more likely to express leftist preferences. However, the size of the effect is small, women's reduced income results in an approximately a 0.1 point difference in the general ideological measures of *Party* and *Leftist* and an approximately 0.05 point difference in the economic ideological measures of *Econ1* and *Econ2*.[40]

Second, as expected, we find the indirect effect of *Female* on ideology via the effect of decreased Emotional Stability on *Income* is also significantly positive across all four measures of ideology. Thus, the effect of differences in Emotional Stability between men and women on income differences between the sexes leads to women having more leftist preferences than men. We also find for two of our measures of ideology (*Leftist* and *Econ2*) a significant positive indirect effect via the effect of increased Agreeableness, providing weak evidence that the effect of differences in Agreeableness between men and women on income differences between the sexes also leads to women having more leftist preferences. Yet, these indirect effects through *Income* changes are even smaller than the indirect effects independent of personality traits, ranging from an approximately 0.02 point difference in the general ideological measures and an approximately 0.01 point difference in the economic ideological measures. Furthermore, contrary to predictions, we find no significant indirect effects via *Income* for Conscientiousness, which is not surprising given that we find no significant differences between men and women in Conscientiousness in our sample.

### 3.3.6 Structural Changes Versus Personality Differences

Our analysis has found significant support for both the personality trait and structural explanations of the gender gap in ideology. That is, we find support for indirect effects of *Female* on ideology through income differences independent of personality trait differences which supports the structural explanation and we find support for indirect effects of *Female* on ideology via personality traits directly and via personality traits' effects on income, which supports the personality trait explanation. Which of these effects (and explanations) are larger in explaining the gender gap?

Table 12 and Fig. 1 decompose the ideological gender gap according to whether it is explained by personality traits or is unexplained by personality traits by ideological measure.[41] As noted above, only for our ideological measure *Econ2* do we find a significant direct effect of *Female* on ideology independent of either personality trait differences or income differences. Furthermore, as the figure shows,

[40]We also estimated whether there is a difference in the effect of gender on ideology via *Income* between women based on marital status. In unreported analyses available from the authors, we find that the indirect effect of gender on ideology via *Income* was positive and significant for married women, but insignificant for unmarried women. This is largely explained by the fact that married women earned significantly less than unmarried women in our data.

[41]Since the direct effects of *Female* on *Party*, *Leftist*, and *Econ1* were insignificant, they are given a 0 value.

**Table 12** Decomposition of ideological gender gap (estimated significant effects)

| | | Ideological measure | | | |
|---|---|---|---|---|---|
| | | *Party* | *Leftist* | *Econ1* | *Econ2* |
| Explained by traits | Indep. of income | 0.36 | 0.33 | 0.36 | 0.28 |
| | Via income | 0.02 | 0.02 | 0.01 | 0.01 |
| | Total | 0.38 | 0.35 | 0.37 | 0.29 |
| Unexplained by traits | Indep. of income | 0 | 0 | 0 | 0.29 |
| | Via income | 0.11 | 0.10 | 0.05 | 0.05 |
| | Total | 0.11 | 0.10 | 0.05 | 0.34 |
| Total ideological gender gap | | 0.49 | 0.45 | 0.42 | 0.63 |
| Percent explained by traits | | 0.78 | 0.78 | 0.88 | 0.46 |

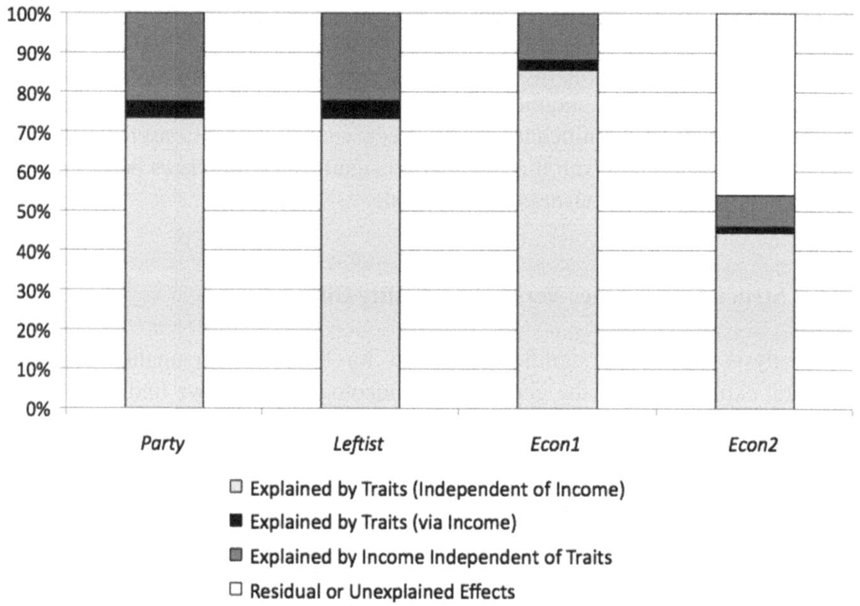


**Fig. 1** Decomposition of the ideological gender gap

the indirect effects via personality trait differences (adding together the effects via *Income* and independent of *Income*) are much larger and a more sizeable explanation of the gender gap than the indirect effects via *Income* independent of personality trait differences. For our general measures of ideology, *Party* and *Leftist*, the indirect effects via personality trait differences comprise approximately 78 % of the gender gap. Hence, we find strong evidence that personality trait differences are a larger factor in explaining the general gender gap in ideology than structural or others influences independent of income differences.

We find somewhat mixed results for our economic policy measures. For *Econ1*, the gap that can be explained by personality trait differences comprise over 88 % of the ideological gender gap. Recall that this measure of preferences is particularly relevant to structural explanations of the gap in that it asks individuals their views of government as a "safety net." That is, the structural explanations argue that due to divorce and unmarried parenthood, women are more likely to prefer interventionist policies. While we find indeed that women are more leftist on such policies, we find that the vast majority of the difference between the sexes can be explained by differences in personality traits.

In contrast, for *Econ2* personality trait differences explain only about 46 % of the gender gap for *Econ2*. Our slightly different results from the ideological measure *Econ2* are intriguing. Recall that this measure asks subjects their opinions on competition and whether it is "good" or not. As noted earlier, other research has found that women appear to have a distaste for competition, see Niederle and Vesterlund (2007). The residual effect of *Female* on *Econ2* may reflect differences in tastes for competition that are not picked up in our personality trait or other measures.

In summary, we find strong evidence across ideological measures that personality trait differences are a large factor in explaining the gender gap. And with the exception of *Econ2*, we find strong evidence across ideological measures that personality trait differences are a larger factor in explaining the gender gap in ideology than structural or others influences independent of income differences.

## 4 Discussion: Are Male Preference Changes the Source of the Gender Gap?

We have found that differences in personality traits explain a large portion of the ideological gender gap. Women in our sample are more open to experience, more agreeable, and less emotionally stable than men. Individuals with these trait differences tend to be more leftist, largely through a direct effect on ideology but also indirectly through the negative effects these traits have on income. We found indeed evidence which suggests that women are more leftist because of these trait differences.

We also discovered that independent of personality trait differences, women earn less income which makes them more leftist as well. Thus gender has an indirect effect on ideology via income differences holding personality traits constant. This evidence provides some support for alternative structural explanations of the ideological gender gap. Nevertheless, the effects through personality trait differences are in most of our analysis larger than the effects independent of personality trait differences, outweighing non-personality trait effects.

As discussed above, evidence suggests that personality trait differences between men and women are larger in more developed societies. Recall that in particular men's personality traits are different, while personality traits of women are largely similar across degrees of development. Schmitt et al. (2008, 2009) found that among other things, men were less agreeable and more emotionally stable in developed societies. Hence the finding that such differences might explain the gender gap in ideology is not inconsistent with the fact that the gender gap has newly emerged and has become more pronounced in westernized societies over time.

Moreover, given these variations across countries which are related to institutional and economic development, then the personality trait differences explanation of the gender gap is also structural since personality trait differences appear to be affected by economic institutional changes taking place in westernized societies. Yet, the mechanism by which institutional structure affects ideological preferences is distinctive from the standard structural explanations in which women are more leftist directly because of increased divorce risk and the possible increased negative consequences for women with less viable outside options. That is, changes in institutions affect personality trait development of men, which results in men having more rightist preferences than women, rather than women choosing more leftist positions in response to the changes in institutions governing marriage.

Our results therefore suggest that the focus on how institutional changes have affected women and, as a result, their ideological preferences, should be reconsidered. Attention should also be paid to the sources of the changes in male personality traits that have occurred with institutional change and economic development and how these changes in traits have led to greater gender differences in ideological preferences in westernized societies.

**Acknowledgements** Support for this research from the Carlsberg Foundation is gratefully acknowledged. We thank Isabella Mares for helpful suggestions.

## Appendix: Data Summary and Comparison to Danish Population

In Table 13 we present a summary of the variables used in the analysis. Note that since all subjects did not answer all of the ideology questions, we have different numbers of observations for each ideology question estimation. Thus, we present the means and standard deviations for the variables used in each estimation separately, by ideological question. As the data shows, there is little variation across estimations in these variables.

Table 14 below compares our subjects to the Danish population 18 years old and older. Again, we present separate comparisons for the set of subjects who answered each of the four ideological questions. Note that the Danish census numbers are measured for those who have completed a given educational level and thus we report from our survey the comparison numbers, which are slightly different from

**Table 13** Variables used in analysis

| | *Party* question | | *Leftist* question | | *Econ1* question | | *Econ2* question | |
|---|---|---|---|---|---|---|---|---|
| Variable | Mean | Std. Dev. | Mean | Std. Dev. | Mean | Std. Dev. | Mean | Std. Dev. |
| *Age* | 47.04 | 13.98 | 47.25 | 13.91 | 47.05 | 13.89 | 47.07 | 13.89 |
| *Female* | 0.48 | 0.50 | 0.48 | 0.50 | 0.49 | 0.50 | 0.49 | 0.50 |
| *Urban* | 0.13 | 0.33 | 0.13 | 0.33 | 0.13 | 0.33 | 0.13 | 0.33 |
| *Church* | 0.86 | 0.35 | 0.86 | 0.35 | 0.86 | 0.35 | 0.86 | 0.35 |
| *Married* | 0.63 | 0.48 | 0.63 | 0.48 | 0.63 | 0.48 | 0.63 | 0.48 |
| *Couple* | 0.11 | 0.32 | 0.12 | 0.32 | 0.12 | 0.32 | 0.12 | 0.32 |
| *Divorced* | 0.08 | 0.28 | 0.08 | 0.27 | 0.08 | 0.28 | 0.08 | 0.28 |
| *EduBasic* | 0.12 | 0.32 | 0.11 | 0.32 | 0.11 | 0.32 | 0.11 | 0.32 |
| *EduHS* | 0.06 | 0.23 | 0.06 | 0.23 | 0.06 | 0.23 | 0.05 | 0.23 |
| *EduAdv1* | 0.07 | 0.25 | 0.07 | 0.25 | 0.07 | 0.25 | 0.07 | 0.25 |
| *EduAdv2* | 0.27 | 0.44 | 0.27 | 0.45 | 0.27 | 0.44 | 0.27 | 0.44 |
| *EduAdv3* | 0.15 | 0.35 | 0.15 | 0.35 | 0.14 | 0.35 | 0.14 | 0.35 |
| *Student* | 0.13 | 0.34 | 0.13 | 0.33 | 0.13 | 0.34 | 0.13 | 0.34 |
| *Parttime* | 0.12 | 0.32 | 0.12 | 0.32 | 0.12 | 0.32 | 0.12 | 0.32 |
| *Retired* | 0.13 | 0.33 | 0.13 | 0.33 | 0.13 | 0.33 | 0.13 | 0.33 |
| *Unemployed* | 0.09 | 0.28 | 0.08 | 0.28 | 0.08 | 0.28 | 0.08 | 0.28 |
| *Income* | 12.47 | 0.76 | 12.48 | 0.75 | 12.47 | 0.75 | 12.47 | 0.75 |
| *BigFiveA* | 32.40 | 5.56 | 32.41 | 5.54 | 32.43 | 5.53 | 32.43 | 5.52 |
| *BigFiveC* | 33.00 | 5.63 | 33.01 | 5.57 | 33.00 | 5.56 | 33.00 | 5.56 |
| *BigFiveE* | 30.61 | 6.36 | 30.53 | 6.35 | 30.52 | 6.34 | 30.51 | 6.35 |
| *BigFiveN* | 19.01 | 7.04 | 19.00 | 7.01 | 19.07 | 7.02 | 19.07 | 7.03 |
| *BigFiveO* | 27.24 | 6.16 | 27.28 | 6.12 | 27.17 | 6.14 | 27.16 | 6.14 |
| *IQ* | 8.85 | 2.88 | 8.86 | 2.88 | 8.86 | 2.89 | 8.86 | 2.89 |
| *Party* | 5.06 | 2.57 | | | | | | |
| *Leftist* | | | 5.59 | 2.21 | | | | |
| *Econ1* | | | | | 5.17 | 2.26 | | |
| *Econ2* | | | | | | | 4.08 | 1.98 |
| Observations | 1893 | | 2028 | | 2107 | | 2106 | |

those used in the analysis which classifies individuals in an educational category if they have completing or are in the process of completing a particular level. Unfortunately we do not have comparison numbers from the Danish population for our personality trait measures and *IQ*. In fact, the Danish Psychological Publishing Company agreed partly to allow us to use their measures in order to provide them with better population estimates of these traits.

**Table 14** Comparison to Danish population

| Characteristic | Danish Population (%) | Ideological measure *Party* (%) | *Leftist* (%) | *Econ1* (%) | *Econ2* (%) |
|---|---|---|---|---|---|
| *Gender*[a] | | | | | |
| Women | 50.2 | 48.0 | 47.6 | 48.8 | 48.8 |
| *Age*[a] | | | | | |
| 18–30 | 20.2 | 14.3 | 13.5 | 13.3 | 13.9 |
| 31–40 | 19.0 | 17.3 | 16.8 | 16.8 | 17.5 |
| 41–50 | 19.6 | 26.9 | 25.5 | 25.5 | 27.5 |
| 51–60 | 17.8 | 23.2 | 23.6 | 23.6 | 23.2 |
| 61–70 | 15.0 | 14.1 | 15.0 | 15.1 | 13.8 |
| 71–80 | 8.4 | 4.1 | 5.7 | 5.7 | 4.1 |
| *Education*[b] | | | | | |
| Basic (up to 10 years) | 26.3 | 11.8 | 11.4 | 11.3 | 11.4 |
| High school (up to 12 years) | 6.4 | 5.7 | 5.7 | 5.5 | 5.5 |
| Vocational (up to 12 years) | 39.1 | 34.2 | 34.0 | 34.9 | 34.9 |
| Short tertiary (less than 3 years) | 5.4 | 6.7 | 7.0 | 6.8 | 7.0 |
| Medium tertiary (between 3 & 4 years) | 15.7 | 27.0 | 27.3 | 27.0 | 26.8 |
| Long tertiary (more than 4 years) | 7.2 | 14.5 | 14.7 | 14.5 | 14.4 |
| *Income (DKK per year)* | | | | | |
| Less than 100,000 | 13.7 | 8.8 | 8.7 | 8.7 | 8.8 |
| 100,000–199,999 | 26.2 | 16.7 | 16.4 | 16.8 | 16.7 |
| 200,000–299,999 | 23.2 | 24.9 | 25.1 | 25.5 | 25.4 |
| 300,000–399,999 | 20.0 | 26.2 | 26.3 | 26.3 | 26.4 |
| more than 400,000 | 16.9 | 23.3 | 23.5 | 22.7 | 22.7 |
| *Church membership*[c] | | | | | |
| Member of the state church | 82.7 | 85.8 | 85.7 | 86.0 | 86.0 |

[a]Gender and age is based on individuals aged 18–80
[b]Education is based on individuals aged 18–69. The education variables for the survey include ongoing education, but these figures are for completed education only
[c]Church members is based on individuals aged 20–79

# References


Adorno, T. W., Frenkel-Brunswick, E., Levinson, D. J., & Sanford, N. S. (1950). *The authoritarian personality*. New York: Harper.

Alford, J. R., & Hibbing, J. R. (2007). Personal, interpersonal, and political temperaments. *The Annals of the American Academy of Political and Social Science, 614*, 196–212.

Altemeyer, R. A. (1996). *The authoritarian specter*. Cambridge, MA: Harvard University Press.

Andersson, O., Holm, H. J., Tyran, J-R., & Wengström, E. (2016). Deciding for others reduces loss aversion. *Management Science, 62*(1), 29–36.

Barbaranelli, C., Caprara, G. V., Vecchione, M., & Fraley, C. R. (2007). Voters' personality traits in presidential elections. *Personality and Individual Differences, 42*, 1199–1208.

Benoit, K., & Laver, M. (2006). *Party policy in modern democracies*. London: Routledge.

Block, J., & Block, J. H. (2005). Nursery school personality and political orientation two decades later. *Journal of Research in Personality, 39*, 395–422.

Borghans, L., Duckworth, A. L., Heckman, J. J., & Weel, B. (2008). The economics and psychology of personality traits. *Journal of Human Resources, 43*, 972–1059.

Brody, L. R., & Hall, J. A. (2000). Gender, emotion, and expression. In M. Lewis & J. M. Haviland-Jones (Eds.), *Handbook of emotions: Part IV: Social/personality issues* (2nd ed., pp. 325–414). New York: Guilford Press.

Buss, D. M. (1997). Evolutionary foundations of personality. In R. Hogan (Ed.), *Handbook of personality psychology* (pp. 317–344). London: Academic Press.

Byrnes, J. P., Miller, D. C., & Schafer, W. D. (1999). Gender differences in risk-taking: A meta-analysis. *Psychological Bulletin, 125*, 367–383.

Caprara, G. V., Barbaranelli, C., & Zimbardo, P. G. (1999). Personality profiles and political parties. *Political Psychology, 20*, 175–197.

Caprara, G. V., Schwartz, S., Capanna, C., Vecchione, M., & Barbaranelli, C. (2006). Personality and politics: Values, traits, and political choice. *Political Psychology, 27*, 1–28.

Carney, D. R., Jost, J. T., Gosling, S. D., & Potter, J. (2008). The secret lives of liberals and conservatives: Personality profiles, interaction styles, and the things they leave behind. *Political Psychology, 29*, 807–840.

Chaney, C. K., Alvarez, M. R., & Nagler, J. (1998). Explaining the gender gap in U.S. presidential elections, 1980–1992. *Political Research Quarterly, 51*(2), 229–311.

Chodorow, N. (1978). *The reproduction of mothering: Psychoanalysis and the sociology of gender.* Berkeley, CA: University of California Press.

Conover, P. J. (1988). Feminists and the gender gap. *Journal of Politics, 50*(4), 985–1010.

Conover, P. J., & Sapiro, V. (1993). Gender, feminist consciousness, and war. *American Journal of Political Science, 37*, 1079–1099.

Costa, P. T., & McCrae, R. R. (2004). *NEO PI-R manual*. Copenhagen: Dansk Psykologisk Forlag.

Del Giudice, M. (2009). On the real magnitude of psychological sex differences. *Evolutionary Psychology, 7*(2), 264–279.

Diekman, A. B., & Schneider, M. C. (2010). A social role theory perspective on gender gaps in political attitudes. *Psychology of Women Quarterly, 34*, 486–497.

Eagly, A. H. (1987). *Sex differences in social behavior: A social-role interpretation*. Hillsdale, NJ: Erlbaum.

Edlund, L., Haider, L., & Pande, R. (2005). Unmarried parenthood and redistributive politics. *Journal of the European Economic Association, 3*(1, March), 95–119.

Edlund, L., & Pande, R. (2002), Why have women become left-wing: The political gender gap and the decline in marriage. *Quarterly Journal of Economics, 111*, 917–961.

Efron, B., & Tibshirani, R. (1993). *An Introduction to the Bootstrap*. New York: Chapman & Hall/CRC.

Else-Quest, N. M., Hyde, J. S., Goldsmith, H. H., & Van Hulle, C. A. (2006). Gender differences in temperament: A meta-analysis. *Psychological Bulletin, 132*, 33–72.

European Values Study (1999) European Values Study and GESIS Data Archive for the Social Sciences. Variable Report: Bilingual Documentation Denmark (English-Danish). GESIS-Variable Report 2012/166. Tilburg University. https://dbk.gesis.org/dbksearch/file.asp?file=ZA3785_cdb_dk-dn.pdf

Feingold, A. (1994). Gender differences in personality: A meta analysis. *Psychological Bulletin, 116*(3), 429–456.

Fridkin, K. L., & Kenney, P. J. (2007). Examining the gender gap in children's attitudes towards politics. *Sex Roles, 56*, 133–140.

Gangestad, S. W., & Simpson, J. A. (2000). The evolution of human mating: Trade-offs and strategic pluralism. *Behavioral and Brain Sciences, 23*, 573–587.

Gangestad, S. W., & Thornhill, R. (1997). The evolutionary psychology of extrapair sex: The role of fluctuating asymmetry. *Evolution and Human Behavior, 18*(2), 69–88.

Gerber, A. S., Huber, G. A., Doherty, D., Dowling, C. M., & Ha, S. E. (2010). Personality and political attitudes: Relationships across issue domains and political contexts. *American Political Science Review, 104*(1), 111–133.

Giger, N. (2009). Towards a modern gender gap in Europe? A comparative analysis of voting behavior in 12 countries. *The Social Science Journal, 46*, 474–492.

Gosling, S. D., Rentfrow, P. J., & Swann, W. B. Jr. (2003). A very brief measure of the big-five personality domains. *Journal of Research in Personality, 37*, 504–528.

Hatemi, P. K., Medland, S. E., & Eaves, L. J. (2009). Do genes contribute to the 'gender gap'? *Journal of Politics, 71*(1, January), 262–276.

Inglehart, R., & Norris, P. (2000). The developmental theory of the gender gap: Women's and men's voting behavior in global perspective. *International Political Science Review/Revue Internationale de Science Pol, 21*(4), 441–463.

Iversen, T., & Rosenbluth, F. (2006). The political economy of gender: Explaining cross-national variation in the gender division of labor and the gender voting gap. *American Journal of Political Science, 50*(1), 1–19.

Jelen, T. G., Thomas, S., & Wilcox, C. (1994). The gender gap in comparative perspective. Gender differences in abstract ideology and concrete issues in western Europe. *European Journal of Political Research, 25*(2), 171–186.

Jost, J. T. (2006). The end of the end of ideology. *American Psychologist, 61*, 651–670.

Jost, J. T., Glaser, J., Kruglanski, A. W., & Sulloway, F. J. (2003). Political conservatism as motivated social cognition. *Psychological Bulletin, 129*, 339–375.

Kaltenthaler, K., Ceccoli, S., & Geleny, R. (2008). Attitudes toward eliminating income inequality in Europe. *European Union Politics, 9*(2), 217–241.

Knight, K. (2006). Transformations of the concept of ideology in the twentieth century. *American Political Science Review, 100*(4, November), 619–626.

Korotayev, A. V., & Kazankov, A. A. (2003). Factors of sexual freedom among foragers in cross-cultural perspective. *Cross-Cultural Research, 37*, 29–61.

Kring, A. M., & Gordon, A. H. (1998). Sex differences in emotion: Expression, experience, and physiology. *Journal of Personality and Social Psychology, 74*, 686–703.

Lakoff, G. (2004). *Don't think of an elephant: Know your values and frame the debate*. White River Junction, VI: Chelsea Green.

Lamb, M. E., & Hewlett, B. S. (Eds.). (2005). *Hunter-gatherer childhoods: Evolutionary, developmental, and cultural perspectives*. New York: Transaction Publishers.

Lee, R. B., & Daly, R. (Eds.). (1999). *The Cambridge encyclopedia of hunters and gatherers*. Cambridge: Cambridge University Press.

Lippa, R. A. (2010). Sex differences in personality traits and gender-related occupational preferences across 53 nations: Testing evolutionary and social-environmental theories. *Archives of Sexual Behavior, 39*, 619–636.

Lynn, R., & Martin, T. (1997). Gender differences in extraversion, neuroticism, and psychoticism in 37 countries. *Journal of Social Psychology, 137*, 369–373.

Maccoby, E. E., & Jacklin, C. N. (1974). *The Psychology of Sex Differences*. Stanford, CA: Stanford University Press.

MacKinnon, D. P., Warsi, G., & Dwyer, J. H. (1995). A simulation study of mediated effect measures. *Multivariate Behavioral Research, 30*, 41–62.

MacKinnon, D. P., Krull, J. L., & Lockwood, C. M. (2000). Equivalence of the mediation, confounding and suppression effects. *Prevention Science, 1*, 173–181.

McClosky, H. (1958). Conservatism and personality. *American Political Science Review, 52*, 27–45.

McCrae, R. R. (1996). Social consequences of experiential openness. *Psychological Bulletin, 120*, 323–337.

McCrae, R. R., & Costa, P. T. (1984). *Emerging lives, enduring dispositions*. Glenview, IL: Scott Foresman.

Mehrabian, A. (1996). Relations among political attitudes, personality, and psychopathology assessed with new measures of libertarianism and conservatism. *Basic and Applied Social Psychology, 18*, 469–491.

Mondak, J. J., & Halperin, K. D. (2008). A framework for the study of personality and political behaviour. *British Journal of Political Science, 38*, 335–362.

Morton, R., Tyran, J.-R., & Wengström, E. (2011). Income and ideology: How personality traits, cognitive abilities, and education shape political attitudes. Unpublished manuscript.

Niederle, M., & Vesterlund, L. (2007). Do women shy away from competition? Do men compete too much? *Quarterly Journal of Economics, 122*(3), 1067–1101.

Norrander, B., & Wilcox, C. (2008). The gender gap in ideology. *Political Behavior, 30*, 503–523.

Pratto, F., Sidanius, J., Stallworth, L. M., & Malle, B. F. (1994). Social dominance orientation: A personality variable predicting social and political attitudes. *Journal of Personality and Social Psychology, 67*, 741–763.

Preacher, K. J., & Hayes, A. F. (2004). SPSS and SAS procedures for estimating indirect effects in simple mediation models. *Behavioral Research Methods, Instruments, & Computers, 36*(4), 717–731.

Rentfrow, P. J., Jost, J. T., Gosling, S. D., & Potter, J. (2009). Statewide differences in personality predict voting patterns in 1996–2004 U.S. presidential elections. In J. T. Jost, A. C. Kay, & H. Thorisdottir (Eds.), *Social and psychological bases of ideology and system justification* (pp. 314–347). New York: Oxford University Press.

Riemann, R., Grubich, C., Hempel, S., Mergl, S., & Richter, M. (1993). Personality and attitudes towards current political topics. *Personality and Individual Differences, 15*, 313–321.

Ruble, D. N., & Martin, C. L. (1998). Gender development. In W. Damon & N. Eisenberg (Eds.), *Handbook of child psychology* (5th ed., pp. 933–1016). New York: Wiley.

Schoen, H., & Schumann, S. (2007). Personality traits, partisan attitudes, and voting behavior. Evidence from Germany. *Political Psychology, 28*, 471–498.

Schmitt, D. P. (2005). Measuring sociosexuality across people and nations: Revisiting the strengths and weaknesses of cross-cultural sex research [response]. *Behavioral and Brain Sciences, 28*, 297–311.

Schmitt, D. P., Realo, A., Voracek, M., & Allik, J. (2008). Why can't a man be more like a woman? Sex differences in big five personality traits across 55 countries. *Journal of Personality and Social Psychology, 94*(1), 168–182.

Schmitt, D. P., Realo, A., Voracek, M., & Allik, J. (2009). Correction to 'Why Can't a Man Be More Like a Woman? Corrected online January 7, 2009.

Shrout, P. E., & Bolger, N. (2002). Mediation in experimental and nonexperimental studies: New procedures and recommendations. *Psychological Methods, 7*(4), 422–445.

Stenner, K. (2005). *The authoritarian dynamic*. Cambridge: Cambridge University Press.

Studlar, D., McAllister, I., & Hayes, B. (1998). Explaining the gender gap in voting: A cross-national analysis. *Social Science Quarterly, 79*(4), 779–798.

Thöni, C., Tyran, J.-R., & Wengström, E. (2012). Microfoundations of Social Capital. *Journal of Public Economics, 96*(7), 635–643.

Tooby, J., & Cosmides, L. (1990). On the universality of human nature and the uniqueness of the individual: The role of genetics and adaptation. *Journal of Personality, 58*(1), 17–67.

Van Hiel, Alan, Kossowska, Malgorzata, & Mervielde, Ivan. (2000). The relationship between openness to experience and political ideology. *Personality and Individual Differences, 28*, 741–751.

Wilgenbusch, T., & Merrell, K. W. (1999). Gender differences in selfconcept among children and adolescents: A meta-analysis of multidimensional studies. *School Psychology Quarterly, 14*, 101–120.

# Statistical Utilitarianism

**Marcus Pivato**

## 1 Introduction

Utilitarianism may be philosophically attractive, but as a practical method for making collective decisions, it faces at least three major problems.

(Pr1) Interpersonal comparisons of utility are problematic. Even if we accept that such interpersonal comparisons are meaningful in principle, it is not clear how precise interpersonal comparisons could be made in practice.

(Pr2) It is difficult for the social planner to obtain accurate information about the voters' utility functions. (It is not generally feasible to obtain a precise utility assessment from every voter for every possible social alternative.)

(Pr3) Due to epistemic failures, a voter may have incorrect beliefs about the long-term consequences of various policy alternatives. Furthermore, people fail to correctly predict their own future utility level, even in apparently straightforward decision problems (Loewenstein and Schkade 1999). Indeed, there is ample empirical evidence that people's beliefs about their own past, present, and future happiness are surprisingly unreliable, and subject to systematic biases, errors, and illusions (Kahneman et al. 1999). For instance, Sen (1999, Chap. 1, p. 15) argues that the poor may become "habituated" to misery, and hence overstate their level of well-being. In short: voters do not even correctly perceive their *own* utility functions.

M. Pivato (✉)
THEMA/UFR d'Economie et Gestion, Université de Cergy-Pontoise, 33 Boulevard du Port, 95011 Cergy-Pontoise Cedex, France
e-mail: marcuspivato@gmail.com



M. Gallego, N. Schofield (eds.), *The Political Economy of Social Choices*, Studies in Political Economy, DOI 10.1007/978-3-319-40118-8_8

However, we will show that these three problems almost disappear in large populations of voters satisfying certain mild statistical assumptions.[1] In Sect. 2, we show that averaging utility data (even noisy or miscalibrated data) from a large population of voters will yield a good approximation of utilitarianism with high probability, despite problems (Pr1)–(Pr3). In Sect. 3, we enhance the speed and quality of the convergence results from Sect. 2. These results are not merely a straightforward application of the Law of Large Numbers, for two reasons. First, we deal with multiplicative noise as well as additive noise (to account for errors in interpersonal utility comparisons). Second, in Sect. 4, we show that the same conclusion holds even when the measurement errors associated with different voters are highly correlated. In all cases, we not only show that the probability of a socially suboptimal decision decays to zero, but *how quickly* it decays to zero as the population becomes large. We do not assume the measurement errors follow a particular probability distribution; we only assume that they have zero mean and bounded variance.

Surprisingly, there is very little previous literature which deploys statistical methods to obviate the technical difficulties of utilitarianism. Lerner (1944, pp. 29–32) was perhaps the first to suggest such an approach. Under the plausible assumption that all people have diminishing marginal utility for wealth, Lerner argued that, even in a state of *total ignorance* about the precise structure of people's cardinal utility functions, an egalitarian distribution of wealth would maximize the *expected* aggregate utility for society, because the expected utility gains of the poor under such a wealth redistribution would more than cancel the expected utility losses of the rich. Lerner's original argument was obscure and generated much confusion; it was later clarified by Breit and Culbertson Jr. (1970).[2] For Lerner's argument to work, his vague hypothesis of "total ignorance" about individual utility functions must be formalized in terms of quite specific assumptions about the probability distribution of the utility functions. The results of this paper can be seen as a far-reaching extension of Lerner's approach.

## 2 Basic Model

First we will fix some notation that will be maintained throughout the paper. Let $\mathbb{R}$ denote the set of real numbers. Let $\mathcal{A}$ denote a finite set of social alternatives, and let $A := |\mathcal{A}|$. Let $\mathcal{I}$ be a set of voters, and let $I := |\mathcal{I}|$. (We will typically suppose that $I$ is very large; indeed, we will be mainly interested in asymptotic behaviour as $I \to \infty$.) For every $i$ in $\mathcal{I}$, let $u_i : \mathcal{A} \longrightarrow \mathbb{R}$ be a cardinal utility function for voter $i$, and let $c_i > 0$ be a "calibration constant", which we will use to make

[1] A fourth problem is strategic voting. We discuss some solutions in the conclusion of the paper.

[2] See also Lerner (1970), Breit and Culbertson Jr. (1972), McCain (1972), and McManus et al. (1972).

cardinal interpersonal utility comparisons. We suppose that the functions $c_i\,u_i$ and $c_j\,u_j$ are interpersonally comparable for all voters $i$ and $j$ in $\mathcal{I}$. In other words, for any alternatives $a$, $b$, $c$, and $d$ in $\mathcal{A}$, if $c_i\,u_i(b) - c_i\,u_i(a) = c_j\,u_j(d) - c_j\,u_j(c)$, then the welfare that voter $i$ gains in moving from alternative $a$ to alternative $b$ is exactly the same as the welfare that voter $j$ gains in moving from $c$ to $d$. We would therefore like to maximize the utilitarian social welfare function $U_{\mathcal{I}} : \mathcal{A} \longrightarrow \mathbb{R}$ defined by

$$U_{\mathcal{I}}(a) \quad := \quad \frac{1}{I}\sum_{i\in\mathcal{I}} c_i\,u_i(a), \qquad \text{for every alternative } a \text{ in } \mathcal{A}. \tag{1}$$

Let $\operatorname{argmax}_{\mathcal{A}}(U_{\mathcal{I}})$ denote the set of alternatives in $\mathcal{A}$ which maximize $U_{\mathcal{I}}$—we will refer to these as *utilitarian optima*.[3] A utilitarian social planner wants to find a utilitarian optimum, but she may not have enough information to do this, because of the aforementioned problems (Pr1)–(Pr3). We will formalize her informational problems with two assumptions:

(U1) The interpersonal calibration constants $\{c_i\}_{i\in\mathcal{I}}$ are unknown. The social planner regards $\{c_i\}_{i\in\mathcal{I}}$ as independent (but not necessarily identically distributed) real-valued random variables.[4] There is a constant $\sigma_c^2 > 0$ (independent of $I$) such that, for every voter $i$ in $\mathcal{I}$, the random variable $c_i$ has a variance less than $\sigma_c^2$, and an expected value of 1.

(U2) The utility functions $\{u_i\}_{i\in\mathcal{I}}$ are not precisely observable. Instead, for each voter $i$ in $\mathcal{I}$, the planner can only observe the function $v_i := u_i + \epsilon_i$, where $\epsilon_i : \mathcal{A} \longrightarrow \mathbb{R}$ is a random "error" term. For each alternative $a$ in $\mathcal{A}$, the random errors $\{\epsilon_i(a)\}_{i\in\mathcal{I}}$ are independent.[5] They are not necessarily identically distributed, but they all have an expected value of 0, and variance less than or equal some constant $\sigma_\epsilon^2 > 0$ (independent of $I$).
Finally, the random variables $\{c_i\}_{i\in\mathcal{I}}$ are independent of the random functions $\{\epsilon_i\}_{i\in\mathcal{I}}$.

Assumption (U1) encodes problem (Pr1), while (U2) encodes both (Pr2) and (Pr3). Note that, while we assume that $\{v_i\}_{i\in\mathcal{I}}$ and $\{c_i\}_{i\in\mathcal{I}}$ are random variables, we make no assumptions about the mechanism generating the underlying profile of utility functions $\{u_i\}_{i\in\mathcal{I}}$. These utility functions might be fixed in advance, or they might themselves be generated by some other random process.[6] If they *are* randomly generated, then we do *not* need to assume that $\{u_i\}_{i\in\mathcal{I}}$ are identically distributed, or assume that the random variables $u_i(a)$ and $u_i(b)$ are independent for a given

[3]Of course, $U_{\mathcal{I}}$ and its maximizer(s) depend on the specific profile of utility functions $\{u_i\}_{i\in\mathcal{I}}$. For simplicity, we do not indicate this explicitly in our notation.

[4]Presumably, $c_i > 0$ for all $i \in \mathcal{I}$. But we do not need this hypothesis to prove our results.

[5]Note that we do *not* assume that, for a fixed voter $i$ in $\mathcal{I}$, the random errors $\epsilon_i(a)$ and $\epsilon_i(b)$ are independent for different alternatives $a$ and $b$ in $\mathcal{A}$.

[6]In this case, the results in this paper should be interpreted as statements which hold for any specific realization of these random utility functions.

voter $i$ in $\mathcal{I}$ and distinct alternatives $a$ and $b$ in $\mathcal{A}$. However, we will assume the utility profile $\{u_i\}_{i\in\mathcal{I}}$ satisfies two boundedness conditions:

(U3) There is a constant $M > 0$ (independent of $I$) such that, for every alternative $a$ in $\mathcal{A}$, we have $\frac{1}{I}\sum_{i\in\mathcal{I}} u_i(a)^2 < M^2$.

(U4) There is a constant $\Delta > 0$ (independent of $I$) such that $\max_{\mathcal{A}}(U_{\mathcal{I}})-U_{\mathcal{I}}(a) > \Delta$ for every alternative $a \notin \operatorname{argmax}_{\mathcal{A}}(U_{\mathcal{I}})$.

The constant $\Delta$ in condition (U4) is the minimum "social welfare cost" of failing to choose a utilitarian optimum. Condition (U3) is clearly satisfied if $|u_i(a)| < M$ for every alternative $a$ in $\mathcal{A}$ and every voter $i$ in $\mathcal{I}$. Alternately, suppose that, for each $a$ in $\mathcal{A}$, the utility values $\{u_i(a)\}_{i\in\mathcal{I}}$ are independent, identically distributed random variables drawn from a distribution with finite variance. If $I$ is large, then the Law of Large Numbers says that (U3) will hold with very high probability.

Given the data $\{v_i\}_{i\in\mathcal{I}}$, the social planner can compute the "observed" social welfare function $V_{\mathcal{I}}$, which we define as follows:

$$V_{\mathcal{I}}(a) \quad := \quad \frac{1}{I}\sum_{i\in\mathcal{I}} v_i(a), \qquad \text{for every alternative } a \text{ in } \mathcal{A}. \tag{2}$$

Note that $V_{\mathcal{I}}$ is based only on observable data (the functions $\{v_i\}_{i\in\mathcal{I}}$), and does not require the planner to know the true values of the utility functions $\{u_i\}_{i\in\mathcal{I}}$ or the interpersonal calibration constants $\{c_i\}_{i\in\mathcal{I}}$. Therefore, the social planner can compute $V_{\mathcal{I}}$, and identify the alternatives in $\mathcal{A}$ which maximize $V_{\mathcal{I}}$. Our first result says that, if the population is sufficiently large, then $V_{\mathcal{I}}$ is a good approximation of $U_{\mathcal{I}}$, so that any social alternative which maximizes $V_{\mathcal{I}}$ will also maximize the value of $U_{\mathcal{I}}$, with very high probability.

**Theorem 1** *For every voter $i$ in $\mathcal{I}$, let $u_i : \mathcal{A}\longrightarrow\mathbb{R}$ be a utility function. Suppose that the profile $\{u_i\}_{i\in\mathcal{I}}$ satisfies conditions* (U3) *and* (U4), *and suppose $\{c_i\}_{i\in\mathcal{I}}$, $\{\epsilon_i\}_{i\in\mathcal{I}}$ and $\{v_i\}_{i\in\mathcal{I}}$ are randomly generated according to rules* (U1) *and* (U2). *Define $U_{\mathcal{I}}$ and $V_{\mathcal{I}}$ as in Eqs.* (1) *and* (2). *Then* $\lim_{I\to\infty} \operatorname{Prob}\left[\operatorname*{argmax}_{\mathcal{A}}(V_{\mathcal{I}}) \subseteq \operatorname*{argmax}_{\mathcal{A}}(U_{\mathcal{I}})\right] = 1.$

We can refine this result in three ways. First, we can drop condition (U4). Second, and relatedly, instead of demanding that a maximizer of $V_{\mathcal{I}}$ *exactly* maximizes $U_{\mathcal{I}}$, we could entertain the possibility that it only *almost* maximizes $U_{\mathcal{I}}$—something which would be almost as good, for practical purposes. Third, we can estimate how *large* the population needs to be in order to achieve such "almost-maximization" with a certain probability. To achieve these refinements, we need some more notation. For any utility profile $\{u_i\}_{i\in\mathcal{I}}$, if $U_{\mathcal{I}}$ is as in Eq. (1), then let

$$U^*_{\mathcal{I}} \quad := \quad \max\{U_{\mathcal{I}}(a)\ ;\ a \in \mathcal{A}\} \tag{3}$$

be the optimum utilitarian social welfare. Let $\delta > 0$ represent a "social suboptimality tolerance", and let $p > 0$ represent the probability that this tolerance will be exceeded (the planner wants both of these to be small). For any values of $\delta$ and $p$, we define

$$\overline{I}(\delta, p) \quad := \quad 4A\,\frac{M^2\,\sigma_c^2 + \sigma_\epsilon^2}{p\,\delta^2}. \tag{4}$$

Our next result says that, for any population larger than $\overline{I}(\delta, p)$, any $V_{\mathcal{I}}$-maximizing social alternative will produce a social welfare within $\delta$ of the theoretical optimum, with probability at least $1 - p$.

**Theorem 2** *Suppose* $\{u_i\}_{i\in\mathcal{I}}$, $\{c_i\}_{i\in\mathcal{I}}$, $\{\epsilon_i\}_{i\in\mathcal{I}}$ *and* $\{v_i\}_{i\in\mathcal{I}}$ *satisfy* (U1), (U2) *and* (U3). *Define* $U_{\mathcal{I}}$, $V_{\mathcal{I}}$, *and* $U_{\mathcal{I}}^*$ *as in Eqs.* (1)–(3). *For any* $\delta > 0$ *and any* $p \in (0, 1)$, *if* $I \geq \overline{I}(\delta, p)$, *then for any* $a_V^*$ *in* $\text{argmax}_{\mathcal{A}}(V_{\mathcal{I}})$, *we have* $\text{Prob}\left[U_{\mathcal{I}}(a_V^*) < U_{\mathcal{I}}^* - \delta\right] < p$.

To illustrate the meaning of Theorem 2 and formula (4), we will now consider a simple and somewhat stylized example, involving six social alternatives.

*Example 3* Suppose $|\mathcal{A}| = 6$. For every alternative $a$ in $\mathcal{A}$, suppose that the utilities $\{u_i(a);\ i \in \mathcal{I}\}$ are independent, uniformly distributed random variables ranging over some interval of length at most 10, contained within the interval $[-9, 9]$ (with perhaps different subintervals of $[-9, 9]$ for different alternatives in $\mathcal{A}$). Let $M := 5$; then for a large population of voters, condition (U3) will be satisfied with very high probability.[7] Suppose $\sigma_c^2 = 1$ and $\sigma_\epsilon^2 = 5$, and let $\delta := 0.2$ (i.e. 1.1 % of the total utility range) and $p := 0.01$. Then $\overline{I}(\delta, p) = 1{,}800{,}000$. Thus, for a polity of two million voters, Theorem 1 says that, with 99 % probability, the $V_{\mathcal{I}}$-maximal alternative will yield a $U_{\mathcal{I}}$-value within 1.1 % of the theoretical optimum $U_{\mathcal{I}}^*$.

**A Remark on the Informational Assumptions** There are at least four methods by which the planner could acquire the data $\{v_i\}_{i\in\mathcal{I}}$ posited in assumption (U2). The first and most obvious method is through verbal surveys, interviews, or polls (e.g. with questions of the form, "On a scale of 1 to 100, how happy would you be with each of the following outcomes?"). To fix a quantitative scale, we might formulate such questions in terms of "willingness to pay" (e.g. "How much money would you pay to obtain each of the following outcomes?") or some other "willingness to sacrifice" (e.g. "What percent of your income would you give up?", "How many hours would you work?", "How many days would you go without food?"). Since utility is presumably a nonlinear function of these variables, we could get more accurate estimates by eliciting preferences over *lotteries* over these variables (assuming von Neumann-Morgenstern preferences). Of course, all of these methods are vulnerable to strategic misrepresentation (which is only partially mitigated if the sacrifices in question are actual and not merely hypothetical); we will discuss some solutions to this problem in the conclusion of this paper.

[7] *Proof.* The variance of any such uniform distribution is at most $8\frac{1}{3} < 9$, and the square of its mean is at most $4^2 = 16$. Thus, its second moment will be at most $9 + 16 = 25 = 5^2$.

In addition to deliberate misrepresentation, surveys and other verbal methods may suffer from involuntary misreporting, due to poor communication, cognitive biases, and/or problem (Pr3). So a second approach is to estimate an individual's utility by examining physiological data such as her blood pressure, muscle tension, pupil dilation, EEG, EKG and/or MRI patterns, and/or blood concentrations of dopamine, endorphin, epinephrine, and cortisol. Such data could be used to estimate people's *actual* utility levels (e.g. to detect correlations between utility and various socioeconomic, health, and/or lifestyle factors), but it could also be used to estimate an individual's utility for various *hypothetical* scenarios (e.g. by measuring her physiological responses to visual portrayals or verbal vignettes of these scenarios). This approach has been extensively investigated in the hedonic psychology literature (Kahneman et al. 1999, Part V), but it is still speculative, and unlikely to be feasible for a large population.

A third approach is to predict a population's utility response to hypothetical social alternatives with a mathematical model, perhaps informed by data collected using the two previous approaches. For example, suppose we had empirical data about the utility levels of a large population of individuals (e.g. estimated using verbal surveys and/or physiological measurements), along with putative causal variables such as their income, consumption of various private and public goods, health status, occupation, lifestyle, leisure activities, social situation, and place of residence, as well as fixed demographic variables such age, gender, family status, and ethnic background. For each value of the fixed demographic variable, we could construct the utility function (in terms of the putative causal variables) which best fits the empirical data.[8] We could then use this ensemble of utility functions to predict the utility response this population to a hypothetical perturbation of the status quo (e.g. a policy which shifts the equilibrium of the economy by some specified vector.) Such predictive utility models would be indispensable when the decision has important long-term consequences, such as in environmental policy.

Finally, we can obtain utility information from individuals by asking them to vote. Although seemingly crude, this method can sometimes be surprisingly effective at estimating the utilitarian social welfare function (Pivato 2015, 2016a).

Clearly, all of these methods are prone to considerable error, and provide little insight into interpersonal comparisons. This is the reason for the random variables $\{c_i\}_{i\in\mathcal{I}}$ and $\{\epsilon_i\}_{i\in\mathcal{I}}$ in assumptions (U1) and (U2). As should be clear from this informal discussion, we consider $\{c_i\}_{i\in\mathcal{I}}$ and $\{\epsilon_i\}_{i\in\mathcal{I}}$ as random variables *from the perspective of the social planner*, not from the perspective of some third party. We suppose that the planner has already obtained as much information as is feasible about the utility functions of the voters (e.g. through surveys, mathematical models, etc.), and that $\{c_i\}_{i\in\mathcal{I}}$ and $\{\epsilon_i\}_{i\in\mathcal{I}}$ represent the residual uncertainty in her utility estimates, given all this information. Thus, Theorems 1 and 2 (and the results later in this paper) describe the *planner's* probability estimates, based on her information.

[8]For this curve-fitting problem to be well defined, we would need to make some assumptions about the functional form of the utility function—e.g. that it is Cobb-Douglas, CES, logarithmic, etc.

It is for this reason that we can assume without loss of generality that the random variables $\{\epsilon_i(a);\ i \in \mathcal{I} \text{ and } a \in \mathcal{A}\}$ all have zero expected value; if some $\epsilon_i(a)$ had nonzero expected value $\overline{\epsilon}$ (from the *planner's* perspective), then she could correct her estimate of $v_i(a)$ by adding $\overline{\epsilon}$ to it, leaving a residual error with zero expected value. Likewise, we can assume that $\{c_i\}_{i\in\mathcal{I}}$ all have expected value 1; if there was some $i$ in $\mathcal{I}$ such that $\mathbb{E}[c_i] = \overline{c}_i \neq 1$, then the planner could replace $v_i$ with $\tilde{v}_i := \overline{c}_i\, v_i$, which we could regard as an estimate of the underlying utility function $\tilde{u}_i := \overline{c}_i\, u_i$, with random error $\tilde{\epsilon}_i := \overline{c}_i\, \epsilon_i$ (so that $\tilde{v}_i = \tilde{u}_i + \tilde{\epsilon}_i$). If we replace $c_i$ with $\tilde{c}_i := c_i/\overline{c}_i$, then we would have $\tilde{c}_i\, \tilde{u}_i = c_i\, u_i$ (so the sum (1) would be unchanged), but $\mathbb{E}[\tilde{c}_i] = 1$, as desired.

Of course, the social planner may be systematically *wrong* in her estimates. For example, due to some hidden bias or error in her methodology, she might systematically underestimate the preference intensities (as reflected in $\{c_i\}_{i\in\mathcal{I}}$) of some demographic group (e.g. low-income women). In this case, her estimates of the utilitarian social welfare function would be systematically incorrect. While this is possible, it is outside the scope of our model. We have assumed that the planner has already incorporated all information which is available to her. So if there is some remaining bias in her estimates, then this bias must be detectable only through information which is somehow *not* available to the planner—even in principle—yet which *is* available to some neutral third party. To be credible, a critique based on such "hidden bias" would need to explain how such information could remain hidden from the planner, yet be available to another observer.

## 3 Enhanced Convergence

In this brief section, we discuss how the speed and/or quality of the convergence in Theorems 1 and 2 can be enhanced, by making stronger hypotheses about the distribution of the random measurement errors.

**Faster Convergence** Theorem 2 shows that the population size $I$ that is required to ensure a welfare loss less than $\delta$ with probability $1-p$ has growth rate $\mathcal{O}(A/p\delta^2)$ as $p, \delta \to 0$ and $A \to \infty$. This estimate arises from Chebyshev's inequality. If we strengthen hypotheses (U1) and (U2), then we can obtain a bound on $I$ that grows much more slowly. For example, suppose there are constants $C > 0$ and $E > 0$ such that $|c_i| \leq C$ and $|\epsilon_i(a)| < E$ for all $i \in \mathcal{I}$ and $a \in \mathcal{A}$. Let $M$ be as in hypothesis (U3), and let $B := M\,C + E$. Then using Hoeffding's inequality in place of Chebyshev's, we get a variant of Theorem 2 where

$$\overline{I}(\delta, p) \quad = \quad \frac{8\,B^2}{\delta^2} \log\left(\frac{2A}{p}\right). \tag{5}$$

For example, let $A := 6$, $M = 5$ and $\delta = 0.2$ as in Example 3, and suppose $C = 1$ and $E = 5$. Then $\overline{I}(\delta, p) = 20\,000 \log(2A/p)$. Thus, for $p = 0.01$, we have

$\bar{I}(\delta, p) \approx 142\,000$. So 1150 voters are enough to ensure a 99 % probability that the $V_{\mathcal{I}}$-maximizing alternative has a $U_{\mathcal{I}}$-value within 1.1 % of the theoretical optimum. Note that the population bound in formula (5) grows much more slowly than the one in formula (4) as $A \to \infty$ and $p \to 0$. We would also get a formula very similar to (5) if $\{c_i\}_{i \in \mathcal{I}}$ and $\{\epsilon_i(a)\}_{a \in \mathcal{A}}$ were independent normal random variables.

**Almost-Sure Convergence** Theorems 1 and 2 follow from the fact that, for any $\delta > 0$,

$$\lim_{I \to \infty} \text{Prob}\,[\|U_{\mathcal{I}} - V_{\mathcal{I}}\|_{\infty} > \delta] = 0, \text{ where } \|U_{\mathcal{I}} - V_{\mathcal{I}}\|_{\infty} := \max_{a \in \mathcal{A}} |U_{\mathcal{I}}(a) - V_{\mathcal{I}}(a)|\,.$$

This is a statement about convergence *in probability*. With stronger hypotheses, it is possible to establish a similar result for convergence *almost surely*. To do this, let $v_i := u_i + \epsilon_i$ as in assumption (U2), but now instead suppose:

(U2*) For each alternative $a$ in $\mathcal{A}$, the random errors $\{\epsilon_i(a)\}_{i \in \mathcal{I}}$ are independent and *identically distributed*, with expected value zero (but possibly infinite variance).

An application of Kolmogorov's Strong Law of Large Numbers yields the following result.

**Theorem 4** *Suppose that* $c_i = 1$ *for all* $i \in \mathcal{I}$, *while the random errors* $\{\epsilon_i\}_{i \in \mathcal{I}}$ *satisfy condition* (U2*), *and define* $U_{\mathcal{I}}$ *and* $V_{\mathcal{I}}$ *as in Eqs.* (1) *and* (2). *Then*

$$\lim_{I \to \infty} \|U_{\mathcal{I}} - V_{\mathcal{I}}\|_{\infty} = 0, \quad \textit{with probability 1.} \tag{6}$$

*Let* $U_{\mathcal{I}}^*$ *be as in Eq.* (3), *and define* $U_{\mathcal{I}}^{\dagger} := \min\{U_{\mathcal{I}}(a); a \in \operatorname{argmax}_{\mathcal{A}}(V_{\mathcal{I}})\}$. *Then*

$$\lim_{I \to \infty} \left(U_{\mathcal{I}}^* - U_{\mathcal{I}}^{\dagger}\right) = 0, \quad \textit{with probability 1.} \tag{7}$$

In Eq. (7), the expression $U_{\mathcal{I}}^* - U_{\mathcal{I}}^{\dagger}$ is the maximum "social welfare loss" that the social planner risks if she chooses a social alternative which maximizes the observed social welfare function $V_{\mathcal{I}}$. The limit (7) says that, as the population size becomes very large, this maximum social welfare loss is almost certain to vanish.

However, Theorem 4 delivers too much and not enough at the same time. On the one hand, almost-sure convergence is actually *stronger* than the convergence in probability that we need in order to be confident of obtaining close-to-optimal policies. On the other hand, the "identically distributed" part of hypothesis (U2*) may be somewhat implausible, under some circumstances. Also, Theorem 4 assumes away problem (Pr1).[9] A third shortcoming of Theorem 4 is that it says

[9]To extend Theorem 4 to cover (Pr1), we must replace (U1) with the assumption (U1*): "The random variables $\{c_i\}_{i \in \mathcal{I}}$ are independent and *identically distributed*, and furthermore, their

nothing about the *speed* of convergence. If the population is "big enough", then $V_{\mathcal{I}}$ and $U_{\mathcal{I}}$ are essentially the same. But how big is "big enough"? Theorem 2 gives a precise answer to this question.[10]

## 4 Correlated Voters

One obvious weakness of the model in Sect. 2 is the assumption that $\{c_i\}_{i\in\mathcal{I}}$ and $\{\epsilon_i\}_{i\in\mathcal{I}}$ are *independent* random variables. This may not be entirely realistic, because people belonging to the same social group might have similar characteristics, leading to correlation amongst their preference intensities and measurement errors. For example, suppose we estimated people's utility functions using physiological data. It is quite possible that members of different demographic groups (e.g. different age groups, different genders, etc.) might exhibit, on average, different physiological responses for the same utility function. Alternately, suppose we tried to elicit utility data using a verbal questionnaire. Since members of different subcultures have slightly different understandings of the language, they may respond to the same survey questions in slightly different ways. Of course, a well-designed utility-measurement technology would recognize and correct for such group-specific biases. But it is still possible that some residual bias will remain, and this would lead to a correlation of measurement errors within each social group.

To obtain results comparable to Theorems 1 and 2, we need some constraints on the covariance of the random errors. For any $I \in \mathbb{N}$ and any $I\times I$ matrix $\mathbf{S} = [s_{i,j}]_{i,j=1}^{I}$, let

$$\|\mathbf{S}\|_1 \;:=\; \sum_{i,j=1}^{I} |s_{i,j}|. \tag{8}$$

For all $I \in \mathbb{N}$, let $\mathcal{S}_I$ be a set of real-valued $I \times I$ matrices. (The covariance matrices for our random variables will be drawn from these sets.) We will assume that

$$\lim_{I\to\infty} \overline{\sigma}_I \;=\; 0, \tag{9}$$

where, for all $I \in \mathbb{N}$, we define $\overline{\sigma}_I := \sup\{\|\mathbf{S}\|_1/I^2;\ \mathbf{S} \in \mathcal{S}_I\}$.

---

distribution is *symmetric* about 1." We must also replace (U3) with the assumption (U3′) from Sect. 4. We can then extend Theorem 4 to obtain the limits (6) and (7) under hypothesis (U1*), (U2*), and (U3′).

[10] We do have an *asymptotic* estimate on the rate of convergence in Theorem 4. The Law of the Iterated Logarithm says that, with probability 1, there is some $I_0$ such that $\|U_{\mathcal{I}} - V_{\mathcal{I}}\|_\infty < \sqrt{2\log[\log|\mathcal{I}|]/|\mathcal{I}|}$ for all $\mathcal{I}$ with $|\mathcal{I}| \geq I_0$. However, $I_0$ itself is a random variable (it depends on the particular realization of the random error functions $\{\epsilon_i\}_{i\in\mathcal{I}}$), so it cannot be known in advance. Hence, for practical purposes, this asymptotic convergence rate is not very useful.

As in Sect. 2, we will suppose that, for each $i$ in $\mathcal{I}$, the planner can only observe a function $v_i := u_i + \epsilon_i$, where $\epsilon_i : \mathcal{A} \longrightarrow \mathbb{R}$ is a random "error" term. For every alternative $a$ in $\mathcal{A}$, let $\boldsymbol{\epsilon}(a) := [\epsilon_i(a)]_{i \in \mathcal{I}}$, an $I$-dimensional vector of these random errors. (The entries are not necessarily identically distributed.) Likewise, if $\{c_i\}_{i \in \mathcal{I}}$ are the interpersonal comparison coefficients from Sect. 2, then the planner regards $\mathbf{c} := [c_i]_{i \in \mathcal{I}}$ as another $I$-dimensional random vector. (Again, the entries are not necessarily identically distributed.) We now replace assumptions (U1)–(U3) with the following:

(U1′) The covariance matrix of the random vector $\mathbf{c}$ is an element of $\mathcal{S}_I$.[11]

(U2′) For every alternative $a$ in $\mathcal{A}$, the covariance matrix of the random vector $\boldsymbol{\epsilon}(a)$ is an element of $\mathcal{S}_I$.

(U3′) There is some constant $M > 0$ (independent of $I$) such that $|u_i(a)| \leq M$ for every $i$ in $\mathcal{I}$ and $a$ in $\mathcal{A}$.

We must also supplement these with:

(U0′) For every $a$ in $\mathcal{A}$, the covariance matrix between the random vectors $\boldsymbol{\epsilon}(a)$ and $\mathbf{c}$ is an element of $\mathcal{S}_I$.[12]

For example, if $\mathbf{c}$ and $\boldsymbol{\epsilon}(a)$ are independent of one another (as we assumed in Sect. 2), then assumption (U0′) is satisfied as long as $\mathcal{S}_I$ contains the zero matrix. Likewise, assumptions (U1) and (U2) from Sect. 2 are just special cases of (U0′)–(U2′). (To see this, let $\mathcal{S}_I$ be the set of all non-negative diagonal matrices with entries bounded by $\max\{\sigma_c^2, \sigma_\epsilon^2\}$.) Note that, like (U2), assumption (U2′) does *not* require the random error vectors $\boldsymbol{\epsilon}(a)$ and $\boldsymbol{\epsilon}(b)$ to be independent or identically distributed for $a \neq b$. The dependency structure between $\boldsymbol{\epsilon}(a)$ and $\boldsymbol{\epsilon}(b)$ is arbitrary. The covariance matrix of $\boldsymbol{\epsilon}(a)$ describes the pattern of correlations we expect to see in the society $\mathcal{I}$ with respect to utility-measurement errors concerning alternative $a$. We do *not* assume that all societies will exhibit the same pattern of correlations for all social decision problems. The same society might exhibit different patterns of correlations for different social decisions. Meanwhile, two different societies of the same size might exhibit different patterns of correlations for the *same* social decision (e.g. because they have a different demographic structure, or a different patchwork of overlapping subcultures). The set $\mathcal{S}_I$ represents the set of *possible* patterns of correlations we could expect to see for any possible social decision, in any possible society of size $I$. For any $\mathbf{S} \in \mathcal{S}_I$, the ratio $\|\mathbf{S}\|_1/I^2$ measures, roughly, the *average amount of correlation* between citizens in a society described by matrix $\mathbf{S}$.[13] Thus,

---

[11]Recall: If $X$ and $Y$ are two real-valued random variables with expected values $\overline{X}$ and $\overline{Y}$, then their *covariance* is defined $\mathrm{cov}(X, Y) := \mathbb{E}\left[(X - \overline{X})(Y - \overline{Y})\right]$. If $\mathbf{X} = (X_1, \ldots, X_I)$ is an $I$-dimensional random vector, then its *covariance matrix* is the symmetric, positive definite, $I \times I$ matrix $\mathbf{S}$ where $s_{i,j} = \mathrm{cov}(X_i, X_j)$.

[12]If $\mathbf{X} = (X_1, \ldots, X_I)$ and $\mathbf{Y} = (Y_1, \ldots, Y_I)$ are $I$-dimensional random vectors, then $\mathrm{cov}(\mathbf{X}, \mathbf{Y})$ is the the $I \times I$ matrix $\mathbf{S}$ where $s_{i,j} = \mathrm{cov}(X_i, Y_j)$.

[13]This is not exactly true, because the diagonal entries of the matrices in (U1′) and (U2′) are variances. But if $I$ is large, then the sum of diagonals is dominated by the sum of the off-diagonal covariance terms.

the limit (9) says that the average correlation between citizens becomes smaller as their societies become larger. For example, suppose there are constants $J \in \mathbb{N}$ and $\varsigma > 0$ such that, for every $I \in \mathbb{N}$ and every $\mathbf{S} \in \mathcal{S}_I$, each row of $\mathbf{S}$ has at most $J$ nonzero entries, and each of these has magnitude at most $\varsigma$ (i.e. any person in any society can be correlated with at most $J$ other people, and each correlation can have strength at most $\varsigma$). Then the limit (9) is satisfied. Or more generally, suppose there is some constant $K > 0$ such that the absolute sum of each row in $\mathbf{S}$ is at most $K$, for every $I \in \mathbb{N}$ and every $\mathbf{S} \in \mathcal{S}_I$ (i.e. any person can be correlated with as many other people as she wants, but she has a fixed "covariance budget" which she must distribute amongst them). Then again the limit (9) is satisfied. The next result says that the conclusion of Theorem 1 still holds in this setting.

**Theorem 5** *For every voter $i$ in $\mathcal{I}$, let $u_i : \mathcal{A} \longrightarrow \mathbb{R}$ be a utility function. Suppose that the profile $\{u_i\}_{i\in\mathcal{I}}$ satisfies conditions* (U3′) *and* (U4), *and suppose $\{c_i\}_{i\in\mathcal{I}}$, $\{\boldsymbol{\epsilon}(a)\}_{a\in\mathcal{A}}$ and $\{v_i\}_{i\in\mathcal{I}}$ are randomly generated according to rules* (U0′)–(U2′). *Define $U_\mathcal{I}$ and $V_\mathcal{I}$ as in Eqs.* (1) *and* (2). *Then* $\lim_{I\to\infty} \text{Prob}\left[\underset{\mathcal{A}}{\text{argmax}}(V_\mathcal{I}) \subseteq \underset{\mathcal{A}}{\text{argmax}}(U_\mathcal{I})\right] = 1.$

As in Sect. 2, we can refine this result by dropping condition (U4), and by computing the size of the population needed to ensure that a $V_\mathcal{I}$-maximizer *almost* maximizes $U_\mathcal{I}$ with some specified probability. Let $M$ be as in assumption (U3′). For any $\delta > 0$ and any probability $p \in (0, 1)$, the limit Eq. (9) implies that, if $I$ is large enough, then we will have

$$\overline{\sigma}_I \quad \leq \quad \frac{p\,\delta^2}{4A\,(M+1)^2}. \tag{10}$$

Let $\widetilde{I}(\delta, p)$ be such that $\overline{\sigma}_I$ satisfies the inequality (10) for all $I \geq \widetilde{I}(\delta, p)$.[14] The next result plays a role similar to Theorem 2.

**Theorem 6** *Suppose $\{u_i\}_{i\in\mathcal{I}}$, $\{c_i\}_{i\in\mathcal{I}}$, $\{\boldsymbol{\epsilon}(a)\}_{a\in\mathcal{A}}$ and $\{v_i\}_{i\in\mathcal{I}}$ satisfy* (U0′)–(U3′). *Define $U_\mathcal{I}$, $V_\mathcal{I}$, and $U^*_\mathcal{I}$, as in Eqs.* (1)–(3). *For any $\delta > 0$ and any $p \in (0, 1)$, if $I \geq \widetilde{I}(\delta, p)$, then for any $a^*_V$ in* $\text{argmax}_\mathcal{A}(V_\mathcal{I})$, *we have* $\text{Prob}\left[U_\mathcal{I}(a^*_V) < U^*_\mathcal{I} - \delta\right] < p.$

As we have noted, assumptions (U1) and (U2) are special cases of (U0′)-(U2′). Thus, if we replace (U3) with (U3′) in Theorems 1 and 2, then they become corollaries of Theorems 5 and 6. One possible shortcoming of Theorems 5 and

[14]To provide a precise formula for $\widetilde{I}(\delta, p)$, analogous to formula (4), we would need more specific assumptions about the families $\mathcal{S}_I$ of admissible covariance matrices posited in assumptions (U0′)-(U2′). In the "covariance budget" example, we have $\overline{\sigma}_I = K/I$; thus we obtain $\widetilde{I}(\delta, p) = 4\,KA\,(M+1)^2/\delta^2\,p$.

If $(\mathbf{c}, \boldsymbol{\epsilon})$ was a multivariate normal random variable, then (10) could be replaced with the inequality $\overline{\sigma}_I \leq \delta^2/8\,(M+1)^2 \ln(2A/p)$. Then we would have $\widetilde{I}(\delta, p) = 8\,K\,(M+1)^2 \ln(2A/p)/\delta^2$ in the covariance budget example. So for normal random variables, $\widetilde{I}(\delta, p)$ grows very slowly as $A \to \infty$ and $p \to 0$.

6 is that the boundedness assumption (U3′) is more restrictive than (U3). But if we strengthen (U1′) to (U1), and strengthen (U0′) to the assumption that $\mathbf{c}$ and $\boldsymbol{\epsilon}(a)$ are independent (as in Sect. 2), then we can weaken (U3′) to (U3) and still have the same conclusions.[15]

One might also worry that assumptions *(U0′)–(U2′)*, while certainly much weaker than independence, still restrict the amount of correlation between the variables. An alternative approach, which avoids this concern, is to identify the elements of $\mathcal{I}$ with points on a grid (representing, e.g. geographic locations) and suppose that the random variables $\{\epsilon_i(a)\}_{i\in\mathcal{I}}$ form a stationary random field on this grid, for each $a \in \mathcal{A}$.[16] We must also assume that $\{c_i\}_{i\in\mathcal{I}}$ and $\{u_i\}_{i\in\mathcal{I}}$ are *independent* stationary random fields on this grid.[17] Then variants of Theorems 1 and 4 can be obtained, using the Birkhoff Ergodic Theorem in place of the Law of Large Numbers (Krengel 1985, Chap. 6). However, it is more difficult to obtain a convergence speed result like Theorem 2 in this "grid" model. Furthermore, a grid is perhaps not the best way to model interpersonal correlations.

## 5 Conclusion

The model in this paper assumes that the social planner can observe the cardinal utility functions of the voters (at least, up to some measurement error), and it abstracts from the problem of strategic misrepresentation. But in a series of companion papers, we show that the main conclusions of this paper continue to hold even if the social planner only has access to *ordinal* data, and also when voters can be strategically dishonest. For example, Pivato (2015) shows that, if the statistical distribution of the voters' utility functions satisfies plausible hypotheses, then with a very high probability, the profile of the voters' ordinal preferences will admit a Condorcet winner, and this winner will maximize the utilitarian social welfare function. This is important, because the Condorcet winner is determined by purely ordinal data, and is quite resistant to strategic voting. (For example, it is the subgame-perfect Nash equilibrium outcome in any binary voting agenda.) Meanwhile, Pivato (2016a) deals with scoring rules (e.g. the Borda rule, approval voting, and evaluative voting); by applying Theorem 1 from the present paper, it shows that, under certain reasonable statistical assumptions, these rules have a very high probability of selecting the utilitarian-optimal outcome in a large population. Pivato (2016a) does not deal with strategic voting, but in a companion paper (Núñez

---

[15]Indeed, in this case, we can redefine $\|\mathbf{S}\|_1$ as $\sum_{i,j=1}^{I} s_{i,j}$ in Eq. (8), which makes the limit condition (9) much easier to achieve.

[16]To be precise, the "grid" here is the group $\mathbb{Z}^2$. In fact, we could use any amenable group.

[17]Alternately, if $c_i = 1$ for all $i \in \mathcal{I}$, then $\{u_i\}_{i\in\mathcal{I}}$ could be arbitrary.

and Pivato 2016), we show that almost any voting rule $F$ (including those discussed in Pivato 2016a) can be *asymptotically approximated* by a stochastic voting rule $\widetilde{F}$ which is *asymptotically truth-revealing*. This means that, if the population is sufficiently large, then $\widetilde{F}$ will select the same outcome as $F$, with very high probability; however, in the rule $\widetilde{F}$, each voter will find it optimal to vote honestly (with very high probability). Thus, the problem of strategic voting essentially vanishes, in large populations. Finally, Pivato (2016b) applies the results of the present paper to develop a dominant-strategy implementation of utilitarianism, using a variant of the Pivotal Mechanism first proposed by Groves (1973) and Clarke (1971).

**Acknowledgements** I am grateful to Gustaf Arrhenius, Miguel Ballester, Marc Fleurbaey, Ori Heffetz, Christophe Muller, and Clemens Puppe for helpful discussions and comments on earlier versions of this paper. I would also like to thank the participants of the 2015 Workshop on the Political Economy of Social Choices in Oaxaca, Mexico, and especially the workshop organizers, Maria Gallego and Norman Schofield. None of these people are responsible for any errors. Most of this research was done when I was at the Department of Mathematics at Trent University, in Canada. This research was supported by NSERC grant #262620-2008, and by Labex MME-DII (ANR11-LBX-0023-01).

## Appendix: Proofs

Let $\|U_{\mathcal{I}} - V_{\mathcal{I}}\|_\infty := \max_{a \in \mathcal{A}} |U_{\mathcal{I}}(a) - V_{\mathcal{I}}(a)|$. Theorem 1 will follow from Theorem 2, which, in turn, will follow from the next result.

**Proposition A1** *Let $U_{\mathcal{I}}$ and $V_{\mathcal{I}}$ be as in Theorem 2, and let $\delta > 0$ and $p \in (0,1)$. If $I \geq \overline{I}(\delta, p)$, then* $\text{Prob}\left[\|U_{\mathcal{I}} - V_{\mathcal{I}}\|_\infty > \frac{\delta}{2}\right] < p$.

*Proof* For all $a \in \mathcal{A}$, the quantity $U_{\mathcal{I}}(a) - V_{\mathcal{I}}(a)$ is a random variable. We will first compute its expected value and the variance of its distribution.

**Claim 1** *For all $a \in \mathcal{A}$, $\mathbb{E}[U_{\mathcal{I}}(a) - V_{\mathcal{I}}(a)] = 0$.*

*Proof* For all $a \in \mathcal{A}$,

$$\begin{aligned} U_{\mathcal{I}}(a) - V_{\mathcal{I}}(a) &\underset{(\diamond)}{=} \frac{1}{I}\sum_{i \in \mathcal{I}} c_i\, u_i(a) - \frac{1}{I}\sum_{i \in \mathcal{I}} v_i(a) \\ &\underset{(\ddagger)}{=} \frac{1}{I}\sum_{i \in \mathcal{I}} c_i\, u_i(a) - \frac{1}{I}\sum_{i \in \mathcal{I}} \Big(u_i(a) + \epsilon_i(a)\Big) \\ &= \frac{1}{I}\sum_{i \in \mathcal{I}} \Big((c_i - 1)\, u_i(a) - \epsilon_i(a)\Big), \end{aligned} \tag{11}$$

where ($\diamond$) is by defining Eqs. (1) and (2), and ($\ddagger$) is by assumption (U2). Thus,

$$\mathbb{E}[U_{\mathcal{I}}(a) - V_{\mathcal{I}}(a)] = \frac{1}{I}\sum_{i\in\mathcal{I}} \mathbb{E}\Big[(c_i - 1)\, u_i(a) - \epsilon_i(a)\Big]$$

$$= \frac{1}{I}\sum_{i\in\mathcal{I}} \Big(\mathbb{E}[(c_i - 1)\, u_i(a)] - \mathbb{E}[\epsilon_i(a)]\Big)$$

$$\underset{(*)}{=} \frac{1}{I}\sum_{i\in\mathcal{I}} \Big(u_i(a)\cdot \mathbb{E}[c_i - 1] + 0\Big) \underset{(\dagger)}{=} \frac{1}{I}\sum_{i\in\mathcal{I}} u_i(a)\cdot 0 = 0.$$

as desired. Here, ($*$) is because $u_i(a)$ is a constant and $\mathbb{E}[\epsilon_i(a)] = 0$ for all $i \in \mathcal{I}$, while ($\dagger$) is because $\mathbb{E}[c_i] = 1$ for all $i \in \mathcal{I}$, by assumption (U1). $\diamond$ Claim 1

**Claim 2** *For all* $a \in \mathcal{A}$, $\mathrm{var}[U_{\mathcal{I}}(a) - V_{\mathcal{I}}(a)] \leq \dfrac{M^2\cdot\sigma_c^2 + \sigma_\epsilon^2}{I}$.

*Proof* For all $a \in \mathcal{A}$,

$$\mathrm{var}[U_{\mathcal{I}}(a) - V_{\mathcal{I}}(a)] \underset{(a)}{=} \mathrm{var}\left[\frac{1}{I}\sum_{i\in\mathcal{I}}\Big((c_i - 1)\, u_i(a) - \epsilon_i(a)\Big)\right]$$

$$= \frac{1}{I^2}\mathrm{var}\left[\sum_{i\in\mathcal{I}}\Big((c_i - 1)\, u_i(a) - \epsilon_i(a)\Big)\right]$$

$$\underset{(b)}{=} \frac{1}{I^2}\sum_{i\in\mathcal{I}}\Big(\mathrm{var}\,[(c_i - 1)\, u_i(a)] + \mathrm{var}\,[\epsilon_i(a)]\Big)$$

$$\underset{(c)}{=} \frac{1}{I^2}\sum_{i\in\mathcal{I}}\Big(u_i(a)^2\cdot \mathrm{var}\,[c_i] + \mathrm{var}\,[\epsilon_i(a)]\Big)$$

$$\underset{(d)}{\leq} \frac{1}{I^2}\sum_{i\in\mathcal{I}}\Big(u_i(a)^2\cdot \sigma_c^2 + \sigma_\epsilon^2\Big)$$

$$= \frac{\sigma_c^2}{I}\left(\frac{1}{I}\sum_{i\in\mathcal{I}} u_i(a)^2\right) + \frac{1}{I^2}\left(\sum_{i\in\mathcal{I}}\sigma_\epsilon^2\right) \underset{(e)}{\leq} \frac{\sigma_c^2 M^2}{I} + \frac{\sigma_\epsilon^2}{I},$$

as desired. Here, (a) is by Eq. (11), while (b) is because $\{\epsilon_i(a)\}_{i\in\mathcal{I}} \cup \{c_i\}_{i\in\mathcal{I}}$ are all jointly independent random variables, and (c) is because $\{u_i(a)\}_{i\in\mathcal{I}}$ are constants. Finally, (d) is by assumptions (U1) and (U2), while (e) is by (U3). $\diamond$ Claim 2

**Claim 3** *Let* $q \in (0, 1)$*, and suppose* $I \geq \overline{I}(\delta, q)/A$*. Then for all* $a \in \mathcal{A}$*, we have* $\mathrm{Prob}\left[|V_{\mathcal{I}}(a) - U_{\mathcal{I}}(a)| > \frac{\delta}{2}\right] < q$.

*Proof* For any $a \in \mathcal{A}$, we have

$$\text{Prob}\left[|V_{\mathcal{I}}(a) - U_{\mathcal{I}}(a)| > \frac{\delta}{2}\right] \underset{(*)}{<} \frac{\text{var}[U_{\mathcal{I}}(a) - V_{\mathcal{I}}(a)]}{(\delta/2)^2} \underset{(\dagger)}{\leq} 4\,\frac{M^2 \cdot \sigma_c^2 + \sigma_\epsilon^2}{I\,\delta^2} \underset{(\diamond)}{\leq} q,$$

as desired. Here, $(*)$ is by Claim 1 and Chebyshev's inequality, $(\dagger)$ is by Claim 2, and $(\diamond)$ is because $I \geq \overline{I}(\delta, q)/A = 4\,\dfrac{M^2 \cdot \sigma_c^2 + \sigma_\epsilon^2}{q\,\delta^2}$. ◇ Claim 3

Set $q = p/A$ in Claim 3. Then $\overline{I}(\delta, p) = \overline{I}(\delta, A\,q) = \overline{I}(\delta, q)/A$. Thus, if $I \geq \overline{I}(\delta, p)$, then for any particular $a \in \mathcal{A}$, Claim 3 says that $\text{Prob}\left[|V_{\mathcal{I}}(a) - U_{\mathcal{I}}(a)| > \frac{\delta}{2}\right] \leq q$. Since $A = |\mathcal{A}|$, this implies that

$$\begin{aligned}\text{Prob}\left[\|U_{\mathcal{I}} - V_{\mathcal{I}}\|_\infty > \frac{\delta}{2}\right] &= \text{Prob}\left(|V_{\mathcal{I}}(a) - U_{\mathcal{I}}(a)| > \frac{\delta}{2} \text{ for some } a \in \mathcal{A}\right)\\ &\leq A \cdot q \;=\; p,\end{aligned}$$

as desired. □

*Remark* If the error vectors $\{\boldsymbol{\epsilon}(a)\}_{a\in\mathcal{A}}$ were all jointly independent, then the events $\{|V_{\mathcal{I}}(a) - U_{\mathcal{I}}(a)| > \frac{\delta}{2}\}$ would be jointly independent (for all $a \in \mathcal{A}$). In this case, we would actually have $\text{Prob}\left[|V_{\mathcal{I}}(a) - U_{\mathcal{I}}(a)| > \frac{\delta}{2} \text{ for some } a \in \mathcal{A}\right] \leq 1 - (1-q)^A$, which is a tighter estimate. From this, we could deduce that $\text{Prob}\left[U_{\mathcal{I}}(a_V^*) < U_{\mathcal{I}}^* - \delta\right] < p$ whenever $I \geq \widehat{I}(\delta, p)$, where $\widehat{I}(\delta, p) := 4\,(M^2\,\sigma_c^2 + \sigma_\epsilon^2)/\delta^2\,(1 - (1 - \frac{p}{2})^{1/A})$. In general, $\widehat{I}(\delta, p) < \overline{I}(\delta, p)$. However, if $p$ is close to zero, then $\widehat{I}(\delta, p)/\overline{I}(\delta, p) \approx 1$, so we do not obtain any significant improvement in practice from this much more restrictive independence hypothesis and the more complicated population bound it yields.

*Proof of Theorem 2* Let $b^* \in \text{argmax}_{\mathcal{A}}(U_{\mathcal{I}})$; thus, $U_{\mathcal{I}}(b^*) = U_{\mathcal{I}}^*$. Then

$$\begin{aligned}0 \leq U_{\mathcal{I}}^* - U_{\mathcal{I}}(a_V^*) &= U_{\mathcal{I}}(b^*) - U_{\mathcal{I}}(a_V^*)\\ &= U_{\mathcal{I}}(b^*) - V_{\mathcal{I}}(b^*) + V_{\mathcal{I}}(b^*) - V_{\mathcal{I}}(a_V^*) + V_{\mathcal{I}}(a_V^*) - U_{\mathcal{I}}(a_V^*)\\ &\leq U_{\mathcal{I}}(b^*) - V_{\mathcal{I}}(b^*) \;+\; V_{\mathcal{I}}(a_V^*) - U_{\mathcal{I}}(a_V^*),\end{aligned}$$

where the last step is because $V_{\mathcal{I}}(b^*) - V_{\mathcal{I}}(a_V^*) \leq 0$ because $a_V^* \in \text{argmax}_{\mathcal{A}}(V_{\mathcal{I}})$. Thus, if $U_{\mathcal{I}}^* - U_{\mathcal{I}}(a_V^*) > \delta$, then either $U_{\mathcal{I}}(b^*) - V_{\mathcal{I}}(b^*) > \frac{\delta}{2}$ or $V_{\mathcal{I}}(a_V^*) - U_{\mathcal{I}}(a_V^*) > \frac{\delta}{2}$. Thus,

$$\begin{aligned}\text{Prob}\left[U_{\mathcal{I}}^* - U_{\mathcal{I}}(a_V^*) > \delta\right] &\leq \text{Prob}\left(U_{\mathcal{I}}(b^*) - V_{\mathcal{I}}(b^*) > \tfrac{\delta}{2} \text{ or } V_{\mathcal{I}}(a_V^*) - U_{\mathcal{I}}(a_V^*) > \tfrac{\delta}{2}\right)\\ &\leq \text{Prob}\left(|V_{\mathcal{I}}(a) - U_{\mathcal{I}}(a)| > \tfrac{\delta}{2} \text{ for some } a \in \mathcal{A}\right) < p,\end{aligned}$$

where the last inequality is by Proposition A1. □

*Proof of Theorem 1* Let $\Delta$ be as in assumption (U4). Let $\delta < \Delta$; then for any $a \in \mathcal{A}$, we have

$$\Big(U_{\mathcal{I}}(a) > U^*_{\mathcal{I}} - \delta\Big) \iff \Big(a \in \operatorname{argmax}_{\mathcal{A}}(U_{\mathcal{I}})\Big).$$

But Theorem 2 implies $\lim_{I\to\infty} \text{Prob}\Big(U_{\mathcal{I}}(a) > U^*_{\mathcal{I}} - \delta \text{ for all } a \in \operatorname{argmax}_{\mathcal{A}}(\mathcal{V}_{\mathcal{I}})\Big) = 1.$ □

For any $\delta > 0$ and $p \in (0, 1)$, let $\widetilde{I}(\delta, p)$ be as defined below inequality (10). Theorems 5 and 6 are a consequence of the following result.

**Proposition A2** *Let $U_{\mathcal{I}}$ and $V_{\mathcal{I}}$ be as in Theorem 6, and let $\delta > 0$ and $p \in (0, 1)$. If $I \geq \widetilde{I}(\delta, p)$, then* $\text{Prob}\left[\|U_{\mathcal{I}} - V_{\mathcal{I}}\|_\infty > \frac{\delta}{2}\right] < p$.

*Proof* The proof strategy is very similar to Proposition A1. For all $a \in \mathcal{A}$, the quantity $U_{\mathcal{I}}(a) - V_{\mathcal{I}}(a)$ is a random variable. We will first compute its expected value and the variance of its distribution.

**Claim 1** *For all $a \in \mathcal{A}$,* $\mathbb{E}[U_{\mathcal{I}}(a) - V_{\mathcal{I}}(a)] = 0$.

*Proof* Same as Claim 1 of Proposition A1. ◇ Claim 1

Let $\overline{\sigma}_I$ be as defined below equation (9).

**Claim 2** *For all $a \in \mathcal{A}$,* $\text{var}[U_{\mathcal{I}}(a) - V_{\mathcal{I}}(a)] \leq (M+1)^2\,\overline{\sigma}_I$.

*Proof* Let $\mathbf{S}^c_{\mathcal{I}} := [s^c_{i,j}]_{i,j\in\mathcal{I}}$ be the $I \times I$ covariance matrix of the $I$-dimensional random vector $\mathbf{c}$, as described in assumption (U1′). Let $\mathbf{S}^a_{\mathcal{I}} := [s^a_{i,j}]_{i,j\in\mathcal{I}}$ be the $I \times I$ covariance matrix of the $I$-dimensional random error vector $\boldsymbol{\epsilon}(a)$, as described in assumption (U2′). Finally, let $\mathbf{S}^{c,a}_{\mathcal{I}} := [s^{c,a}_{i,j}]_{i,j\in\mathcal{I}}$ be the $I \times I$ covariance matrix of $\mathbf{c}$ and $\boldsymbol{\epsilon}(a)$, as described in assumption (U0′). Then

$$\begin{aligned}
\text{var}[U_{\mathcal{I}}(a) - V_{\mathcal{I}}(a)] &\underset{\text{(a)}}{=} \text{var}\left[\frac{1}{I}\sum_{i\in\mathcal{I}}\Big((c_i - 1)\,u_i(a) - \epsilon_i(a)\Big)\right] \\
&\underset{\text{(b)}}{=} \frac{1}{I^2}\left(\sum_{i,j\in\mathcal{I}} u_i(a)^2\, s^c_{i,j} + 2\sum_{i,j\in\mathcal{I}} u_i(a)\, s^{c,a}_{i,j} + \sum_{i,j\in\mathcal{I}} s^a_{i,j}\right) \\
&\underset{\text{(c)}}{\leq} \frac{1}{I^2}\left(\sum_{i,j\in\mathcal{I}} u_i(a)^2\, |s^c_{i,j}| + 2\sum_{i,j\in\mathcal{I}} |u_i(a)|\, |s^{c,a}_{i,j}| + \sum_{i,j\in\mathcal{I}} |s^a_{i,j}|\right) \\
&\underset{\text{(d)}}{\leq} \frac{1}{I^2}\left(\sum_{i,j\in\mathcal{I}} M^2\, |s^c_{i,j}| + 2\sum_{i,j\in\mathcal{I}} M\, |s^{c,a}_{i,j}| + \sum_{i,j\in\mathcal{I}} |s^a_{i,j}|\right) \\
&\underset{\text{(e)}}{=} \frac{1}{I^2}\left(M^2\,\|\mathbf{S}^c\|_1 + 2M\,\|\mathbf{S}^{c,a}\|_1 + \|\mathbf{S}^a\|_1\right) \underset{\text{(f)}}{\leq} (M^2 + 2M + 1)\,\overline{\sigma}_I = (M+1)^2\,\overline{\sigma}_I,
\end{aligned}$$

as desired. Here, (a) is by Eq. (11) from the proof of Proposition A1. (b) is because $\{u_i(a)\}_{i \in \mathcal{I}}$ are all constants, and for any random variables $X$ and $Y$ and constant $u$, we have $\text{var}(uX + Y) = u^2 \text{var}(X) + 2u \text{cov}(X, Y) + \text{var}(Y)$.[18] Next, (c) is by the triangle inequality, (d) is by hypothesis (U3′), (e) is by defining Eq. (8), and (f) is by the definition of $\overline{\sigma}_I$. ◇ Claim 2

**Claim 3** *Let* $q \in (0, 1)$, *and suppose* $I \geq \widetilde{I}(\delta, A\,q)$. *Then for all* $a \in \mathcal{A}$, *we have* $\text{Prob}\left[|V_{\mathcal{I}}(a) - U_{\mathcal{I}}(a)| > \frac{\delta}{2}\right] < q$.[19]

*Proof* For any $a \in \mathcal{A}$, we have

$$\text{Prob}\left[|V_{\mathcal{I}}(a) - U_{\mathcal{I}}(a)| > \frac{\delta}{2}\right] \underset{(*)}{<} \frac{\text{var}[U_{\mathcal{I}}(a) - V_{\mathcal{I}}(a)]}{(\delta/2)^2} \underset{(\dagger)}{\leq} \frac{4\,(M+1)^2\,\overline{\sigma}_I}{\delta^2} \underset{(\diamond)}{\leq} q,$$

as desired. Here, $(*)$ is by Claim 1 and Chebyshev's inequality, $(\dagger)$ is by Claim 2, and $(\diamond)$ is by inequality (10), because $I \geq \widetilde{I}(\delta, A\,q)$. ◇ Claim 3

Set $q = p/A$ in Claim 3. Then $\widetilde{I}(\delta, p) = \widetilde{I}(\delta, A\,q)$. Thus, if $I \geq \widetilde{I}(\delta, p)$, then for any $a \in \mathcal{A}$, Claim 3 says $\text{Prob}\left[|V_{\mathcal{I}}(a) - U_{\mathcal{I}}(a)| > \frac{\delta}{2}\right] \leq q$. Since $A = |\mathcal{A}|$, this implies that

$$\begin{aligned}\text{Prob}\left[\|U_{\mathcal{I}} - V_{\mathcal{I}}\|_\infty > \frac{\delta}{2}\right] &= \text{Prob}\left(|V_{\mathcal{I}}(a) - U_{\mathcal{I}}(a)| > \frac{\delta}{2} \text{ for some } a \in \mathcal{A}\right) \\ &\leq A \cdot q = p,\end{aligned}$$

as desired. □

*Proof of Theorems 5 and 6* Theorem 6 follows from Proposition A2 by an argument identical to the proof of Theorem 2. Then Theorem 5 follows from Theorem 6 by an argument identical to the proof of Theorem 1. □

## References

Breit, W., & Culbertson, W., Jr. (1970, June). Distributional equality and aggregate utility; comment. *American Economic Review, 60*(3), 435–441.

Breit, W., & Culbertson, W. P., Jr. (1972, June). Distributional equality and aggregate utility: Reply. *American Economic Review, 62*(3), 501–502.

Clarke, E. (Fall 1971). Multipart pricing of public goods. *Public Choice, 11*, 17–33.

Groves, T. (1973). Incentives in teams. *Econometrica, 41*, 617–631.

[18] If we strengthen (U1′) to (U1), and strengthen (U0′) to the assumption that $\mathbf{c}$ and $\boldsymbol{\epsilon}(a)$ are independent, while weakening (U3′) to (U3), then this formula simplifies to $(M^2\,\sigma_c^2 + \sum_{i,j=1}^{I} s_{i,j}^a)/I^2$. This leads to the variant of Theorems 5 and 6 described in footnote 15.

[19] If $(\mathbf{c}, \boldsymbol{\epsilon})$ was a multivariate normal random variable, then $V_{\mathcal{I}}(a) - U_{\mathcal{I}}(a)$ would also be normal; then Claim 3 would hold for the variant of inequality (10) that is mentioned in footnote 14.

Kahneman, D., Diener, E., & Schwarz, N. (Eds.). (1999). *Well-being: The foundations of hedonic psychology*. New York: Russell Sage Foundation.

Krengel, U. (1985). *Ergodic theorems*. Berlin: Walter de Gruyter & Co.

Lerner, A. P. (1944). *The Economics Of Control*. New York: Macmillan.

Lerner, A. P. (1970, June). Distributional equality and aggregate utility: Reply. *American Economic Review, 60*(3), 442–43.

Loewenstein, G., & Schkade, D. (1999). Wouldn't it be nice? predicting future feelings. In: Kahneman et al. (1999), Chap. 5, pp. 85–105.

McCain, R. (1972, June). Distributional equality and aggregate utility: Further comments. *American Economic Review, 62*(3), 497–500.

McManus, M., Walton, G. M., & Coffman, R. B. (1972, June). Distributional equality and aggregate utility: Further comment. *American Economic Review, 62*(3), 489–496.

Núñez, M., & Pivato, M. (2016). Truth-revealing voting rules for large populations. Preprint.

Pivato, M. (2015). Condorcet meets Bentham. *Journal of Mathematical Economics, 59*, 58–65.

Pivato, M. (2016a). Asymptotic utilitarianism in scoring rules. *Social Choice and Welfare* (to appear)

Pivato, M. (2016b). Dominant strategy utilitarianism in large populations. Preprint.

Sen, A. (1999). *Commodities and capabilities*. New Delhi: Oxford University Press.

# Measuring Campaign Spending Effects in Post-Citizens United Congressional Elections

**Brandon Barutt and Norman Schofield**

Using pooled OLS analysis, early literature found that incumbent and challenger campaign spending exhibited asymmetrical marginal productivities. That is, early scholars found that challenger campaign spending was more productive than incumbent campaign spending. More contemporary literature found this asymmetry was the result of endogeneity bias in the campaign spending covariates. Using 2014 U.S. House election data, we examine the nature of now-ubiqitous independent expenditures and whether these expenditures also exhibit an asymmetry in pooled OLS analysis. In addition to finding that independent expenditures can be differentiated from traditional campaign expenditures in their scope, dispersion, and magnitude, we find that the marginal impact of incumbent and challenger independent expenditures in pooled OLS analysis is symmetrical. We attribute this finding to the limited scope of independent expenditures. Moreover, using our estimated campaign spending effects, we find that political expenditures are rarely pivotal in determining election outcomes.

B. Barutt (✉)
Department of Political Science, Washington University in St. Louis, Campus Box 1063, One Brookings Drive, St. Louis, MO 63130, USA
e-mail: b.barutt@go.wustl.edu

N. Schofield
Dr. William Taussig Professor of Political Economy, Washington University in St. Louis, Campus Box 1063, One Brookings Drive, St. Louis, MO 63130, USA

Department of Political Science, Washington University in St. Louis, Campus Box 1063, One Brookings Drive, St. Louis, MO 63130, USA
e-mail: schofield.norman@gmail.com



M. Gallego, N. Schofield (eds.), *The Political Economy of Social Choices*, Studies in Political Economy, DOI 10.1007/978-3-319-40118-8_9

## 1 Introduction

Since the Supreme Court's decision in *Citizens United* and the D.C. Circuit Court of Appeals' subsequent decision in *SpeechNOW*, politicians, commentators, and citizens alike have derided the impact of independent expenditures in the political process[1] (Pelosi and Sarbanes 2014; Dionne Jr. 2012; Op-Ed 2014; Egan 2014). Critics cite a number of objections to the campaign finance regime after the Supreme Court's decision in *Citizens United.* Massive political spending by independent expenditure groups could create confusion among the electorate about the policy positions of candidates, allow wealthy and well-connected individuals to have a disproportionate amount of influence in the political process, and potentially change the outcome of elections.

Prior literature has examined the impact of campaign expenditures[2] on election outcomes. This research has reached contradictory conclusions. Early scholars argued challenger campaign expenditures were more productive on the margin than incumbent campaign expenditures, arguing this fact was the result of challengers beginning the election cycle with less name recognition (Jacobson 1978). More contemporary scholars have argued that this asymmetry is the result of endogeneity bias. Determining the nature of this bias and its impact on substantive conclusions has proven difficult. Given *Citizens United* was decided only 6 years ago, there is uncertainty about the nature of independent expenditures and their impact on elections outcomes.

Using data from the 2014 U.S. House elections, this article will seek to reduce this uncertainty by examining the nature and impact of independent expenditures. This article will proceed as follows. Section Two will detail recent developments in campaign finance law; Section Three will review extant literature examining the impact of campaign expenditures on election outcomes; Section Four will examine the scope, dispersion, and magnitude of independent expenditures and how they compare to traditional campaign expenditures; Section Five will present estimation results from a number of multivariate models; Section Six will examine the impact of independent expenditures on election outcomes; and Section Seven will discuss the significance of the empirical results and make recommendations for future research.

[1] As defined by the Federal Election Commission in 11 CFR 100.16(a), an independent expenditure is a "communication expressly advocating the election or defeat of a clearly identified candidate that is not made in cooperation, consultation, or concert with, or at the request or suggestion of, a candidate, a candidate's authorized committee, or their agents, or a political party committee or its agents" (Coordinated Communications and Independent Expenditures 2015).

[2] In contrast to independent expenditures in 11 CFR 100.16(a), campaign expenditures are made by and/or coordinated with the campaigns themselves; they are filed with the Federal Election Commission and published in 2-year reports.

## 2 Campaign Finance After *Citizens United*

**Citizens United (2010)** Section 203 of the Bipartisan Campaign Reform Act of 2002 ("McCain-Feingold") prohibits corporations and unions from spending on "electioneering communications" that explicitly advocate for the election or defeat of a candidate for federal office. Overruling Austin v. Michigan Chamber of Commerce, 494 U.S. 652 (1990) and McConnell v. Federal Election Commission (FEC), 540 U.S. 93 (2003), the Supreme Court's ruling in Citizens United v. FEC, 558 U.S. 310 (2010) holds that limits on corporate financing of independent political advertising or broadcast is in violation of the First Amendment (Citizens United v. Federal Election Commission 2015). Given that such prohibitions curtail the First Amendment guarantee of free speech, the Court examined McCain-Feingold using strict scrutiny.[3] Accordingly, in order for McCain-Feingold to be constitutional, the government had to demonstrate a compelling government interest, and moreover, the means to pursue that interest must have been narrowly tailored.

The Court held that, since these expenditures are technically independent of candidates, an anti-corruption justification does not constitute a compelling government interest. Specifically, the benefits of regulating these purportedly[4] independent expenditures do not outweigh the costs of restricting the First Amendment guarantee of free speech. Moreover, a less publicized aspect of the ruling upheld McCain-Feingold's disclosure and disclaimer requirements enumerated in Sections 201 and 311 (Citizens United v. Federal Election Commission 2015).

**SpeechNOW (2010)** The Supreme Court's holding in *Citizens United* created the precedent for the D.C. Circuit Court of Appeals' holding in SpeechNOW.org v. FEC (2010). The D.C. Circuit Court of Appeals ruled that limits on contributions to independent expenditure groups are unconstitutional since the government's anti-corruption interest does not outweigh the damage to the First Amendment's guarantee of free speech (SpeechNOW.Org v. FEC 2015). Moreover, as in *Citizens United*, the ruling retained the reporting and organizational requirements on these independent expenditure groups.

**Practical Impact of Judicial Decisions** *Citizens United* and *SpeechNOW* have catalyzed the creation of prolific independent expenditures groups. These groups can raise and spend unlimited monies to expressly advocate for the election or defeat of a candidate. These expenditures, since they are considered independent

[3]United States v. Carolene Products Co. 304 U.S. 144 (1938) established the framework of different levels of judicial scrutiny. Cases involving a fundamental right or suspect class invoke strict scrutiny. Under such scrutiny, in order to be constitutional, the government must demonstrate a compelling government interest, and moreover, the means to pursue that interest must be narrowly tailored. Most economic legislation is subject to a less stringent rational basis examination, which holds that the government must simply demonstrate that the regulation is rationally related to a legitimate state interest.

[4]While these expenditures are technically independent, these independent expenditure groups are frequently managed by close advisors of the candidate and/or party they support.

and not subject to contribution limits, may not be made in cooperation with the candidates and/or political parties they support. As previously noted, while these independent expenditure groups can spend unlimited monies in support or opposition to candidates, these expenditures must be reported in a timely manner to the FEC (Coordinated Communications and Independent Expenditures 2015). Prominent independent expenditure groups formed in the aftermath of these rulings include American Crossroads, a conservative independent expenditure group that spent $21,860,037 during the 2014 midterm elections, and Senate Majority PAC, a liberal independent expenditure group that spent $46,651,418 during the same period. In total, independent expenditure groups spent $345,117,042 during the 2014 election cycle (SuperPACs 2015).

**Summary** The Supreme Court's decision in *Citizens United* provided the precedent for the D.C. Circuit Court of Appeals' decision in *SpeechNOW*. These decisions have catalyzed the creation of prolific independent expenditures groups that spend large sums in congressional elections. Given these expenditures are unlimited and technically independent of the candidates and parties, their nature and impact on election outcomes could be different than traditional campaign expenditures. This article will probe these possibilities.

## 3 Literature Review

**Asymmetrical Marginal Productivity** Early literature found that incumbent and challenger campaign spending exhibited asymmetrical marginal productivities. That is, early scholars found that challenger campaign spending was more productive on the margin than incumbent campaign spending. Jacobson (1978) examined the effect of campaign spending in the 1972 and 1974 U.S House elections. Using Ordinary Least Squares (OLS), Jacobson (1978) estimated a weakly negative and statistically insignificant coefficient for incumbent campaign spending, finding such spending did not increase the vote shares of incumbents. Conversely, Jacobson (1978) estimated a positive and statistically significant coefficient for challenger campaign spending, finding a $10,000 increase in challenger campaign spending increases challenger vote shares by 1.12 percentage points.

Importantly, Jacobson (1978) noted a now-familiar problem: the OLS coefficients could suffer from endogeneity bias. In addition to spending levels affecting vote shares, expectations about vote shares affect spending levels. Therefore, there is bidirectional causality mediated by expectations about the closeness of an election. The OLS coefficients could be biased because this estimation procedure does not control for and deal with these candidate expectations about the vote (Erickson and Palfrey 2000). Incumbents spend more when faced with a close election; therefore, the OLS estimate of incumbent campaign spending could be biased downward. Similarly, challengers spend more when they believe they have an opportunity to win; therefore, the OLS estimate of challenger campaign spending could be biased upward[5] (Erickson and Palfrey 2000). The following literature will explore potential remedies to this endogeneity problem.

Jacobson (1978) attempts to remedy this problem using Two-Stage Least Squares (TSLS). This common two-stage estimation procedure purges the component of the endogenous variables (i.e., the campaign spending variables) correlated with the error term; doing so facilitates the estimation of the endogenous variables with less bias and consistency. To identify this two-stage model, Jacobson (1978) uses the following instrumental variables (IVs) for the campaign spending variables: challenger quality, incumbent experience, the challenger's expected vote, and whether the challenger ran in a primary. Using these IVs, Jacobson (1978) finds that the asymmetry persists; challenger campaign spending continued to have a greater marginal productivity than incumbent campaign spending. Jacobson (1978) explains this finding by arguing that the limited name recognition of challengers is responsible for the greater productivity of challenger campaign spending; early challenger campaign spending is highly productive as the electorate learns about the candidate.

There were a number of problems with Jacobson's (1978) model specification. First, the author utilizes a linear functional form; this functional form assumes the impact of spending is linear and additive. It is much more likely, however, that spending exhibits diminishing marginal returns—that is, the tenth dollar of campaign spending is less productive than first dollar of campaign spending. A more appropriate functional form would be quadratic or semi-log. Second, many of Jacobson's (1978) IVs, most notably challenger quality and incumbent experience, likely have a direct impact on vote shares. Therefore, these are inappropriate IVs. Proper IVs must be correlated with the endogenous variables, but uncorrelated with the outcome variable. For these reasons, the results of Jacobson (1978) are suspect.

Abramowitz (1988) findings largely comport with Jacobson (1978). While Abramowitz (1988) finds that both coefficients for challenger and incumbent campaign spending are statistically significant, he finds that the coefficient on the former is nearly three times as large; moreover, this difference is statistically significant. Abramowitz (1988) finds that this asymmetry persists despite purportedly dealing with endogeneity bias. Rather than use TSLS, the author tries to explicitly control for the expectations on the right-hand side of the model. To do so, Abramowitz

[5]While theory implies both the incumbent and challenger campaign spending covariates should be biased, the following literature demonstrates that while the coefficient on incumbent campaign spending is consistently biased downward, the estimates for the coefficient on challenger campaign spending do not appear overestimated. This contention is substantiated by a comparison of the OLS and TSLS models' estimates of the challenger campaign spending variable; the estimates for challenger campaign spending are typically not significantly attenuated in the TSLS model. This suggests there could be factors that prevent the theoretical expectation of upward bias from being an empirical reality. Erickson and Palfrey (2000) discuss a number of explanations for the challenger campaign spending covariate being insensitive to endogeneity bias; these include inexperienced challengers, challenger's expectations being less accurate than incumbent's expectations, and the fundamental asymmetry in the nature of the endogeneity bias for challengers and incumbents. For a thorough discussion of these explanations, see Erickson and Palfrey (2000, pp. 603–604).

(1988) uses pre-election forecasts published by *Congressional Quarterly* in October of the election year.[6] While this is a clever attempt at identifying the model, more contemporary research has suggested this approach is inadequate[7] (Erickson and Palfrey 2000).

**Symmetrical Marginal Productivity** Using data from the 1978 U.S. House elections, Green and Krasno (1988) refute the findings of Jacobson (1978) and Abramowitz (1988). The authors cite many of the problems associated with Jacobson's (1978) analysis, including, but not limited to, use of improper IVs, failure to control for the quality of the challenger, and failure to account for diminishing marginal returns in the campaign spending variables. Green and Krasno (1988) account for these issues.

For IVs, the authors use lagged incumbent campaign expenditures as an instrumental variable for incumbent campaign expenditures, arguing that doing so proxies an incumbent's general propensity to spend,[8] and early challenger receipts, as an instrumental variable for challenger campaign expenditures. The authors also use the natural logarithm of challenger campaign spending to account for diminishing marginal returns and include a measure of challenger quality on the right-hand side of the second-stage regression equation. Using this specification, Green and Krasno (1988) find that challenger political quality has a statistically significant direct effect on the vote. Moreover, the authors find that the impact of challenger and incumbent spending is essentially symmetrical—that is, the marginal productivities (i.e., magnitude of the coefficients) are statistically equal. This contradicts the OLS findings of Jacobson (1978) and Abramowitz (1988).

The methodology of Green and Krasno (1988) is not above reproach. The authors curiously neglect to estimate the natural logarithm of incumbent campaign expenditures. If challenger campaign expenditures are subject to diminishing marginal returns, surely incumbent campaign expenditures are subject to the same phenomenon. Second, the authors' instrumental variable for challenger campaign expenditures, early challenger receipts, are probably subject to the same expectations that bias the challenger campaign spending covariate. Despite these problems, Green and Krasno's (1988) paper is nevertheless notable because it finds that incumbent and challenger campaign spending are equally productive on the margin.

---

[6]This measure of expectations was likely used due to data availability. Election forecasts in October, which occur near the end of the election cycle, are likely partially comprised of year-to-date fundraising numbers. If this is the case, the impact of the campaign expenditure variables is likely mis-estimated.

[7]This article also explored this possibility using different expectations data. Rather than use *Congressional Quarterly's* October election forecast, this article used Stuart Rothenberg's election forecasts from April, August, and October. Unfortunately, confirming the findings of Erickson and Palfrey (2000), this approach did not solve the endogeneity problem.

[8]Since U.S. House elections are not staggered like U.S. Senate elections, the previous race almost certainly includes the same incumbent and/or challenger; a criticism could be that specific personal characteristics of the candidates are correlated with the error term.

Gerber (1998) examines the impact of campaign expenditures on Senate election outcomes; this is in contrast to the aforementioned literature that examined campaign spending in U.S. House elections. Noting that campaign spending levels respond to expectations about the competitiveness of the election, Gerber (1998) uses three innovative instrumental variables to identify a TSLS model. The author uses the wealth of the challenger, operationalized as a dummy variable whose value is determined according to *Congressional Quarterly's* description of the challenger's professional experience. Wealthier challengers have a greater ability to self-finance their campaign, and therefore, should have more campaign expenditures, *ceteris paribus*.

Gerber (1998) also uses state voting age population as an IV. Gerber (1998) writes that "if holding the office of senator allows incumbents to raise a fixed sum independent of their state population, then senators from small state can raise large sums per capita, while those from populous states will have more difficulty raising the same amount per capita" (Gerber, 405). Finally, Gerber (1998) uses lagged spending in Senate elections, operationalized as the total spending in the previous Senate election within a state. Gerber (1998) notes that, since Senate elections are staggered, the previous race likely does not include the same incumbent or challenger, thereby largely avoiding the criticism that there are specific personal characteristics of candidates correlated with the error term. Using these instrumental variables, Gerber (1998) estimates a TSLS model for all Senate elections between 1974 and 1992.

Using this identification strategy, Gerber (1998) finds that incumbent and challenger campaign expenditures have similar marginal effects; this contradicts the findings of previous literature, including Jacobson (1978) and Abramowitz (1988). Despite this finding, Gerber (1998) notes that increasing spending levels by a fixed amount still benefits the challenger because incumbents spend more, on average, and consequently are subject to greater diminishing marginal returns. However, given incumbents have greater spending levels on average, the total effect of campaign spending still provides incumbents an advantage. This finding raises an interesting question. Do independent expenditures exhibit the same relationship? This article will probe this question in the empirical analysis below.

While Gerber (1998) represents the most recent and best attempt to estimate campaign spending effects using TSLS, Erickson and Palfrey (2000) represent the authoritative literature on measuring campaign spending effects in congressional elections. The authors develop a game-theoretic model to explicate the previously-mentioned endogeneity issue. Their model suggests a non-linear relationship between the expected closeness of an election and spending by both the incumbent and challenger. Specifically, the authors argue that, in equilibrium, total campaign spending is continuously increasing in the expected closeness of the election. This fact substantially informs the endogeneity bias found in OLS models examining cross-sectional pooled election data. Erickson and Palfrey (2000) argue that given this fact, the simultaneity bias should be nearly non-existent in elections expected to be very close and increase in severity as the expected competitiveness of the election declines. The authors confirm this expectation using data from contested U.S. House elections between 1974–1980 and 1984–1990.

The findings of Erickson and Palfrey (2000) have substantial implications for this article. First, these findings imply that the best approach to estimate campaign spending effects in contested elections (i.e., the elections in which campaign spending could be pivotal) is to subset the data rather than include all elections in a pooled regression. Moreover, the findings suggest TSLS is not an appropriate remedy for the endogeneity problem. Erickson and Palfrey (2000, 606) write that "[w]e showed here, both theoretically and empirically, that the degree of bias [in the estimation of the spending variables] will vary with incumbent vulnerability in predictable ways, *and this is not accounted for in those methods* [TSLS or explicit modeling of expectations]" (emphasis added).

**Summary** This section has examined literature relevant to the research questions of this article. Early literature found that, even correcting for the aforementioned endogeneity issue using a variety of means, challenger campaign spending was more productive than incumbent campaign spending (Abramowitz 1988; Jacobson 1978). More recent literature, using different IVs, has found that the marginal productivity of incumbent campaign spending is equal to marginal productivity of challenger campaign spending (Gerber 1998; Green and Krasno 1988). Lastly, the most contemporary literature (Erickson and Palfrey 2000) has detailed the nature of the endogeneity issue more extensively than previous works, finding that a subsetting approach allows relatively unbiased estimation of the spending covariates in close elections and that previous identification strategies (e.g., TSLS and explicit modeling of expectations) do not properly account for the nature of the endogeneity problem.

In light of this literature, the forgoing analysis will not use TSLS. In addition to Erickson and Palfrey (2000) suggesting this approach is inadequate, there is a dearth of quality IVs for the campaign spending variables in the U.S. House. Moreover, given the newness of independent expenditures, there is currently no literature documenting IVs for the independent expenditure variables. Interestingly, however, the forgoing analysis provides suggestive evidence that IVs are not necessary for unbiased estimation of the independent expenditure covariates. Further, the forgoing analysis will not use subsetting. Since there have only been three election cycles since the decisions in *Citizens United* and *SpeechNOW*, there is insufficient data to do so. Rather, relying on knowledge about the nature of the endogeneity issue and independent expenditure covariates, this article will use OLS to provide suggestive evidence with respect to the aforementioned research questions.

This paper will proceed as follows. The following section will use simple descriptive statistics and cross-tabulations to determine the nature of independent expenditures. Thereafter, this article will present estimation results from a number of pooled OLS regressions. Informing our analysis with the insights of the literature, namely Erickson and Palfrey (2000), this article will present strong suggestive evidence that (1) independent expenditures do not suffer from severe endogeneity bias and (2) political spending, including independent expenditures, is not pivotal in congressional elections.

## 4 Nature of Independent Expenditures

**Scope** Independent expenditures and traditional campaign expenditures can be differentiated by their scope—that is, the extent to which these types of expenditures are present in races of varying competitiveness. Unsurprisingly, campaign expenditures are greatest in contested elections. In the most competitive elections detailed in the second row of Table 1 (i.e., those decided by fewer than five percentage points), candidates spend $4,929,403.78 on average. Comparatively, in the least competitive elections (i.e., those decided by more than 25 percentage points), candidates spend $1,051,526.67 on average. Citizens largely donate according to their expectations—that is, many citizens donate to their favored candidates if they believe the outcome of the election is at stake. Therefore, campaign contributions are more accessible and greater in contested races. Table 1 demonstrates this relationship. While campaign contributors appear quite particular about the races to which they contribute, independent expenditure groups are significantly more discriminatory.

Similar to campaign expenditures, independent expenditures are greatest in contested elections (See Table 2). Notably, however, the rate at which independent expenditures diminish as an election becomes less competitive is significantly greater than the rate observed for campaign expenditures. As previously noted, independent expenditures groups are quite selective about the races in which they intervene.[9] In fact, the rate at which these expenditures diminish is even concealed by the data in Table 2. Rather than utilize ex post election results, as in the

**Table 1** Campaign expenditures[a] by ex post margin of victory

| Margin | N | Incumbent expenditures | Challenger expenditures |
|---|---|---|---|
| 0–5.0 | 18 | $2,783,301.00 | $2,146,101.78 |
| 5.1–10.0 | 17 | $2,066,428.18 | $1,215,994.53 |
| 10.1–15.0 | 24 | $1,718,616.50 | $682,148.21 |
| 15.1–20.0 | 28 | $1,561,652.86 | $597,443.79 |
| 20.1–25.0 | 40 | $1,270,250.93 | $344,064.90 |
| 25.0 + | 229 | $956,466.97 | $95,057.70 |
| Average | | $1,236,075.13 | $359,361.04 |

[a]Campaign expenditure data for the 2014 U.S. House elections collected from the FEC campaign expenditure database. Data are released in 2-year reports

[9]Independent expenditure groups cannot contribute directly to campaigns; these groups are only allowed to spend monies independently of the candidates. One could argue that donors are making a choice between contributing to independent expenditure groups or campaigns; this is not supported anecdotally and is inconsistent with intuition, however. Donors likely use independent expenditure groups to influence election campaigns in excess of limits on contributions to candidates. The correlation between incumbent campaign spending and incumbent independent expenditures is $r = 0.45$. Similarly, the correlation between challenger campaign spending and challenger independent expenditures is $r = 0.65$. While these positive correlations aggregate over the individual-level story,

**Table 2** Independent expenditures[a] by ex post margin of victory

| Margin | N | Incumbent expenditures | Challenger expenditures |
|---|---|---|---|
| 0–5.0 | 18 | $2,342,660.56 | $2,863,790.61 |
| 5.1–10.0 | 17 | $1,404,895.39 | $1,419,933.85 |
| 10.1–15.0 | 24 | $671,696.33 | $683,056.77 |
| 15.1–20.0 | 28 | $190,347.07 | $220,242.49 |
| 20.1–25.0 | 40 | $47,888.02 | $13,672.06 |
| 25.0 + | 229 | $16,689.54 | $998.44 |
| Average | | $261,907.21 | $278,153.89 |

[a]Independent expenditure data for the 2014 U.S. House elections collected from the FEC independent expenditure database. Data are released in 2-year reports

**Table 3** Independent expenditures[a] in the Rothenberg Eight[b]

| State[c] | District | Incumbent | Challenger | Disparity[d] | Margin |
|---|---|---|---|---|---|
| AZ | 2 | $5,194,327.84 | $6,469,529.21 | ($1,275,201.37) | 0.07 |
| CA | 52 | $4,197,345.02 | $4,007,075.02 | $190,270.00 | 3.17 |
| WV | 3 | $4,410,964.51 | $5,260,994.35 | ($850,029.84) | 10.70 |
| IL | 10 | $3,252,410.64 | $6,876,645.70 | ($3,624,235.06) | 2.56 |
| FL | 26 | $2,894,687.52 | $6,264,547.33 | ($3,369,859.81) | 2.93 |
| NH | 1 | $3,159,150.63 | $5,457,959.45 | ($2,298,808.82) | 3.61 |
| AZ | 1 | $5,058,086.85 | $6,119,780.89 | ($1,061,694.04) | 5.22 |
| CO | 6 | $4,773,611.84 | $4,055,844.63 | $717,767.21 | 9.39 |
| Average | | $4,117,573.11 | $5,564,047.07 | ($1,446,473.96) | |

[a]Independent expenditure data for the 2014 U.S. House elections collected from the FEC independent expenditure database. Data is released in 2-year reports
[b]For the archive of Rothenberg's ratings, see http://www.rothenberggonzales.com/ratings/house/2014-house-ratings-october-3-2014
[c]The state acronyms are Arizona (AZ), California (CA), West Virginia (WV), Illinois (IL), Florida (FL), New Hampshire (NH), Arizona (AZ), and Colorado (CO), respectively. The particular districts identified by Rothenberg vary by election cycle and are not essential to the argument
[d]A disparity in parentheses suggests the incumbent has been outspent by the challenger

aforementioned tables, Table 3 provides independent expenditures data in elections rated as pure toss-ups or tilting in April, August, and October in Stuart Rothenberg's widely publicized election ratings.[10]

Compared to elections decided by fewer than five percentage points, independent expenditure groups were even more involved in elections considered most

they provide suggestive evidence that campaign contributions and contributions to independent expenditure groups are complements for major donors.

[10]During each election cycle, Stuart Rothenberg publishes election forecasts for congressional elections; these are well-known in the media and can be considered an adequate measure of expectations determining the level of spending observed in elections. For the archive of Rothenberg's ratings, see http://www.rothenberggonzales.com/ratings/house/2014-house-ratings-october-3-2014

competitive in Rothenberg's election rankings (hereafter known as the "Rothenberg Eight"). This finding underlines the fact that independent expenditure groups are more selective than campaign contributors. In fact, 40.29 % of all independent expenditures in the 2014 U.S. House elections occurred in the eight contests detailed below. The comparable figure for campaign expenditures is a meager 8.22 %.

**Dispersion** Independent expenditures and campaign expenditures can also be differentiated by their dispersion—that is, the proportion of these expenditures given to incumbents and challengers, respectively. Are they given in equal proportion? Or conversely, do incumbents enjoy an advantage with respect to independent expenditures? As demonstrated in Table 1, incumbents enjoy a spending advantage with respect to campaign expenditures. On average, incumbents outspend challengers by $876,714.09; this is the result of greater fundraising capacity. Of course, as Table 1 also demonstrates, this incumbent advantage diminishes as an election becomes more contested. In the least competitive elections, the incumbent enjoys a $861,409.27 spending advantage; this advantage shrinks to $637,199.22 in races decided by fewer than five percentage points. Again, rather than utilize the ex post election results, Table 4 provides campaign expenditures in the Rothenberg Eight. Restricting our attention to these highly-contested elections, the incumbency spending advantage diminishes further to $454,159.88. These findings indicate that although incumbents enjoy a spending advantage in elections at every level of competition, this advantage diminishes as the election became more competitive.

Perhaps surprisingly, in contrast to campaign expenditures, the incumbent does not enjoy an advantage with respect to independent expenditures (See Table 2). Similar to campaign expenditures, the incumbent advantage diminishes as the race becomes more competitive. Unlike campaign expenditures, however, the advantage erodes more quickly, and notably, in competitive elections, there is no incumbent

**Table 4** Campaign expenditures[a] in the Rothenberg eight[b]

| State | District | Incumbent | Challenger | Disparity[c] | Margin |
|---|---|---|---|---|---|
| AZ | 2 | $3,193,552.00 | $4,336,932.00 | ($1,143,380.00) | 0.07 |
| CA | 52 | $4,438,671.00 | $3,308,658.00 | $1,130,013.00 | 3.17 |
| WV | 3 | $2,619,432.00 | $1,625,225.00 | $994,207.00 | 10.70 |
| IL | 10 | $4,702,195.00 | $3,582,005.00 | $1,120,190.00 | 2.56 |
| FL | 26 | $3,755,709.00 | $2,343,433.00 | $1,412,276.00 | 2.93 |
| NH | 1 | $1,680,321.00 | $1,219,939.00 | $460,382.00 | 3.61 |
| AZ | 1 | $92,242.00 | $2,904.00 | $89,338.00 | 5.22 |
| CO | 6 | $4,666,935.00 | $5,096,682.00 | ($429,747.00) | 9.39 |
| Average | | $3,143,632.13 | $2,689,472.25 | $454,159.88 | |

[a]Campaign expenditure data for the 2014 U.S. House elections collected from the FEC campaign expenditure database. Data are released in 2-year reports

[b]For the archive of Rothenberg's ratings, see http://www.rothenberggonzales.com/ratings/house/2014-house-ratings-october-3-2014

[c]A disparity in parentheses suggests the incumbent has been outspent by the challenger

advantage at all. In elections decided by fewer than five percentage points, challengers actually significantly outspent incumbents. Moreover, in the Rothenberg Eight, challengers enjoyed a spending advantage in three-quarters of the races and maintained a $1,446,473.97 advantage on average (See Table 3). Moreover, challengers won 80 % of elections in which they held a spending advantage.

**Magnitude** Lastly, independent expenditures and campaign expenditures can be distinguished by their magnitude—that is, the overall level of spending for each type of expenditure. Even if this article uncovered aspects of independent expenditures that are normatively undesirable, it would make little difference unless these expenditures comprise a significant proportion of total spending.

In the aggregate, campaign expenditures accounted for a greater share of total spending in the 2014 U.S. House elections. Specifically, campaign expenditures totaled $567,975,279 while independent expenditures totaled $192,261,750 of the $760,237,029 spent on U.S. House elections in 2014. This is largely the result of independent expenditure groups intervening in fewer contests. Table 5 provides average independent expenditure levels by level of competition. As an election becomes more competitive, independent expenditures comprise a greater proportion of the total spending; in fact, in races decided by fewer than five percentage points, independent expenditures exceed campaign expenditures by a wide margin. This trend is highlighted by the fact that, in the Rothenberg Eight, independent expenditures comprised 62.40 % of total spending. Accordingly, while campaign expenditures are greater in the aggregate, independent expenditures are the dominant form of spending in contested races.

**Summary** This section has differentiated independent expenditures along three dimensions: scope, dispersion, and magnitude. Unlike campaign expenditures, which exhibit meaningful levels of spending across the preponderance of elections, particularly for incumbents, independent expenditures are limited to the most contested races; specifically, less than 25 elections received meaningful independent spending. Moreover, unlike campaign expenditures, which demonstrate a strong incumbency-bias, even in the most competitive races, independent expenditures

**Table 5** Magnitude and proportion of independent expenditures[a]

| Margin | N | Incumbent independent | Challenger independent | Proportion (%) |
|---|---|---|---|---|
| Rothenberg | 8 | $5,833,104.38 | $9,681,620.18. | 62.40 |
| 0–5.0 | 18 | $4,929,402.78 | $5,206,451.17 | 51.37 |
| 5.1–10.0 | 17 | $3,282,422.71 | $2,824,829.24 | 46.25 |
| 10.1–15.0 | 24 | $2,400,764.71 | $677,376.55 | 22.01 |
| 15.1–20.0 | 28 | $2,159,096.65 | $205,294.78 | 8.68 |
| 20.1–25.0 | 40 | $1,614,315.83 | $30,780.04 | 1.87 |
| 25.0 + | 229 | $1,051,524.67 | $8,843.99 | 0.83 |

[a]Independent expenditure data for the 2014 U.S. House elections collected from the FEC independent expenditure database. Data is released in 2-year reports

are more egalitarian and actually favor the challenger in the Rothenberg Eight. Finally, despite only being an important factor in fewer than five percent of races, independent expenditures often exceed campaign expenditures in these highly contested elections.

Now that we have established the basic nature of independent expenditures and how they compare to traditional campaign expenditures, the following section will present estimation results of a number of multivariate models that will further inform our understanding of independent expenditures. There are a number of questions that cannot be answered using descriptive statistics and cross-tabulations. Do independent expenditures behave similar to campaign expenditures in multivariate models? That is, does endogeneity bias cause the impact of incumbent independent expenditures to be underestimated? Are political expenditures changing the outcome of elections? In other words, are political expenditures an important determinant of election outcomes? Or alternatively, are election outcomes driven more by fundamentals such as partisanship, national trends, and a non-monetary incumbent advantage? This following section will seek to answer these questions and enhance our understanding of Post-Citizens United congressional elections.

## 5 Multivariate Models

**Data and Methods** Data[11] were collected for U.S. House elections occurring in 2014 meeting the following conditions: the election featured an incumbent (i.e., open seats are excluded) and the election was contested (i.e., there was a challenger). There were 356 elections meeting these conditions. The basic empirical model has the following form:

$$\begin{aligned}\text{IncumbentShare}_i = \alpha &+ \beta_1 \text{IncumbentCampaign}_i \\ &+ \beta_2 \text{ChallengerCampaign}_i \\ &+ \beta_3 \text{IncumbentIndependent}_i \\ &+ \beta_4 \text{ChallengerIndependent}_i \\ &+ \beta_5 \text{Z}_i + \epsilon_i\end{aligned}$$

where `IncumbentShare` is the incumbent's percentage of the two-party vote, `IncumbentCampaign` are campaign expenditures by the incumbent in the 2-year election cycle, `ChallengerCampaign` are campaign expenditures

[11] Spending data were collected from the Federal Election Commission. The FEC has an independent expenditure database providing independent expenditures spent on behalf of all candidates for federal election. Moreover, the FEC has a database of campaign disbursements in each 2 year election cycle for all federal candidates.

by the challenger in the 2-year election cycle, `IncumbentIndependent` are general election independent expenditures on behalf of the incumbent, `ChallengerIndependent` are general election independent expenditures on behalf of the challenger, and `Z` is a vector of control variables discussed below.

U.S. House[12] districts are roughly equal in population. The mean and median populations of a congressional district are 723,725 and 723,250, respectively. The minimum and maximum populations of a congressional district, however, are 523,170 and 925,749, respectively. In light of this substantial variation at the tails of the distribution, all spending variables are population-adjusted. Accordingly, the spending variables can be interpreted as spending per-person in each congressional district by the candidates.

`Partisan` is a measure of partisanship in each congressional district, operationalized as the Cook Partisan Voting Index (PVI), which measures how strongly Republican or Democratic a district is relative to the nation as a whole. The index used in this article was constructed using voting results from the 2012 and 2008 presidential campaigns.[13] `Lagged Share` is a measure of the normal vote in each congressional district; it can also be interpreted as a measure of the non-monetary incumbency advantage. This variable is operationalized as the incumbent's percentage of the two-party vote received in the previous election (i.e., in 2012).

`Legislator` is a measure of the quality of a challenger, operationalized as a dummy variable coded as 1 if the challenger has previously held an elected legislative office (e.g., state legislator or former U.S. Representative) and a 0 if the challenger has not held an elected legislative office. `Republican Incumbent` is a dummy variable, coded as a 1 if the incumbent is a member of the Republican Party and a 0 if otherwise. As anticipated by election forecasters, the 2014 midterm U.S. House elections favored the Republican Party, which gained 13 seats in the U.S. House of Representatives and nine seats in the U.S. Senate.

Finally, `Unemployment` is the unemployment rate in each congressional district measured at the beginning of the election cycle. The data were extracted from the American Community Survey conducted by the United States Census.[14]

Early spending models (Jacobson 1978) relied on a linear and additive functional form. Such a functional form assumes political spending exhibits constant productivity (i.e., the second dollar per person spent yields the same return as the first dollar per person). This is almost certainly a poor assumption. Particularly in high spending elections like those in the Rothenberg Eight, districts become saturated

[12]Data for U.S. House districts were collected from the 2013 American Community Survey sponsored by the U.S. Census.

[13]For additional information regarding the PVI, see http://cookpolitical.com/story/5604. Of course, a more direct measurement of district-level partisanship is desirable. Such a measurement would be more direct and would allow lagged presidential vote shares to be used as an instrumental variable. Unfortunately, such a measurement was unavailable for this project. Future research could use Multi-level Modeling with Post-Stratification to develop such a measurement.

[14]Data can be retrieved from http://www.census.gov/mycd/

with advertising and political activity; citizens' votes become hardened late in the election cycle as little new information is provided to the electorate. In short, a functional form that allows the spending variables to exhibit diminishing marginal returns is necessary; this necessity has been supported by prior literature (Gerber 1998; Green and Krasno 1988).

The primary options are a quadratic functional form and a semi-log functional form. The latter is desirable because the natural logarithm ensures positive returns at high-levels of spending; the quadratic functional form could yield negative returns to spending at high spending levels. Conversely, the former is desirable because it can estimate variables with values of zero (e.g., elections that do not feature independent expenditures). In order to estimate these observations using the semi-log functional form, one must add a small value $\epsilon$ to each observation. Doing so could yield substantive changes. In light of these competing factors, it will be important to estimate both functional forms to determine whether the conclusions drawn are dependent on model specification. Finally, heteroskedasticity is a common problem in cross-sectional data. In this article, the inclusion of districts of widely varied levels of competition makes the model susceptible to non-constant variance of residuals with one or more of the campaign spending covariates. Therefore, all models are corrected for heteroskedasticity using White's heteroskedasticity-consistent standard errors (Summary statistics are provided in Table 6 below).

**Results**[15] We estimated the empirical model delineated above using OLS. As previously noted, given the inadequacy of TSLS (Erickson and Palfrey 2000) and the dearth of strong instruments, we have chosen to use OLS and discuss our results

**Table 6** Summary statistics[a]

| Variable | N | Mean | Median | Std. Dev. | Min | Max |
|---|---|---|---|---|---|---|
| Incumbent share | 356 | 66.81 | 65.56 | 10.35 | 40.40 | 98.13 |
| Incumbent independent | 356 | 0.37 | 0.00 | 1.38 | 0.00 | 9.76 |
| Challenger independent | 356 | 0.40 | 0.00 | 1.57 | 0.00 | 9.60 |
| Incumbent campaign | 356 | 1.72 | 1.23 | 1.49 | 0.13 | 11.82 |
| Challenger campaign | 356 | 0.50 | 0.06 | 1.12 | 0.00 | 8.14 |
| Partisan | 356 | 12.52 | 11.00 | 9.22 | −9.00 | 43.00 |
| Lagged share | 356 | 66.02 | 63.68 | 11.16 | 50.20 | 100.00 |
| Legislator | 356 | 0.08 | 0.00 | 0.28 | 0.00 | 1.00 |
| Republican incumbent | 356 | 0.53 | 1.00 | 0.50 | 0.00 | 1.00 |
| Unemployment | 356 | 8.58 | 8.30 | 2.53 | 2.60 | 19.10 |
| Population (thousands) | 356 | 723.73 | 723.25 | 34.72 | 523.17 | 925.75 |

[a]All forgoing tables are created using the `stargazer` package in LaTeX

[15]Statistical significance is determined using one-tailed probability values when there is a theoretical expectation about the sign of the coefficient.

using the nature of the endogeneity bias as support. The estimation results for the linear, quadratic, and semi-log[16] functional forms are provided in the subsequent tables. For ease of the reader, we have divided the estimation results into two parts. Table 7 provides the estimation results for non-spending determinants of the vote. Table 8 provides the estimation results for the spending determinants of the vote; these are our primary explanatory variables of interest. R-Squared values exceed 0.7 in all models, suggesting the empirical model is adequately explaining variance in the outcome variable regardless of functional form.

*Non-Spending Determinants of the Vote* `Partisan` is statistically significant in all models at any conventional level of significance. Moreover, the estimate is quite stable, between 0.52 and 0.59. Since only the spending variables are logged, the coefficients for the non-spending determinants are comparable across all models. This suggests that a one point increase in Cook's PVI is associated with an approximately half percentage point increase in the incumbent's vote share. Given the average PVI for an incumbent is 12.52, a partisanship advantage contributes approximately seven percentage points to an incumbent's vote share.

**Table 7** Non-spending determinants of the vote

| | *Dependent variable: incumbent vote share* | | |
|---|---|---|---|
| | Linear | Semi-log | Quadratic |
| Partisan | 0.59*** | 0.53*** | 0.57*** |
| | (0.08) | (0.07) | (0.05) |
| LaggedShare | 0.26*** | 0.21*** | 0.25*** |
| | (0.04) | (0.07) | (0.04) |
| Legislator | −3.43*** | −2.33*** | −3.23*** |
| | (0.97) | (0.87) | (1.23) |
| RepublicanIncumbent | 3.62*** | 3.28*** | 3.52*** |
| | (0.65) | (0.59) | (0.63) |
| Unemployment | −0.02 | −0.09 | −0.03 |
| | (0.11) | (0.10) | (0.12) |
| Constant | 41.22*** | 41.18*** | 42.53*** |
| | (4.23) | (3.83) | (2.88) |
| Observations | 356 | 356 | 356 |
| $R^2$ | 0.73 | 0.78 | 0.74 |
| Adjusted $R^2$ | 0.72 | 0.77 | 0.73 |
| F statistic (df = 13; 342) | 101.78*** | 132.61*** | 74.99*** |

*Note:* $^{*}p < 0.1$; $^{**}p < 0.05$; $^{***}p < 0.01$

[16]For clarification, it is important to note that the semi-log model only takes the natural logarithm of the campaign spending variables. The non-spending determinants of the vote remain in level form. Since the outcome variable is the same across models, coefficients for these covariates are comparable across models.

**Table 8** Spending determinants of the vote

| | *Dependent variable: incumbent vote share* | | |
|---|---|---|---|
| | Linear | Semi-Log[a] | Quadratic |
| IncumbentIndependent | 1.41*** | 1.42*** | 3.14*** |
| | (0.55) | (0.32) | (1.17) |
| $\text{IncumbentIndependent}^2$ | | | −0.22* |
| | | | (0.12) |
| ChallengerIndependent | −1.26*** | −1.35*** | −2.97*** |
| | (0.46) | (0.58) | (1.13) |
| $\text{ChallengerIndependent}^2$ | | | 0.21* |
| | | | (0.12) |
| IncumbentCampaign | −0.13 | 0.38 | 0.33 |
| | (0.28) | (0.49) | (0.57) |
| $\text{IncumbentCampaign}^2$ | | | −0.03 |
| | | | (0.06) |
| ChallengerCampaign | −0.98*** | −1.74*** | −3.80*** |
| | (0.63) | (0.24) | (0.91) |
| $\text{ChallengerCampaign}^2$ | | | 0.45*** |
| | | | (0.13) |
| Observations | 356 | 356 | 356 |
| $R^2$ | 0.73 | 0.78 | 0.74 |
| Adjusted $R^2$ | 0.72 | 0.77 | 0.73 |
| F statistic (df = 13; 342) | 101.78*** | 132.61*** | 74.99*** |

[a]This semi-log model uses $\epsilon = 0.01$ to facilitate estimation of observations of the campaign spending variables whose value is zero. The results are robust to varying this $\epsilon$

`LaggedShare` is statistically significant in all models at any conventional level of significance. Moreover, the estimate is quite stable, between 0.21 and 0.26. This suggests a one percentage point increase the incumbent's lagged vote share is associated with an approximately quarter percentage point increase in the incumbent's vote share. Given the average incumbent received 66 % of the vote in 2012, a non-monetary incumbency advantage contributes approximately sixteen percentage points to an incumbent's vote share. The contribution of partisanship and the non-monetary incumbency advantage, coupled with the intercept, suggests the average incumbent begins with approximately 63 % of the vote.

`Legislator` is statistically significant in all models. The estimate varies from −2.33 in the semi-log model to −3.43 in the linear model. Quality challengers, potentially due to greater political experience and higher name recognition, reduce the vote share of incumbents, *ceteris paribus*. Moreover, `Republican Incumbent` is also statistically significant in all models at any conventional level of significance. The estimate varies from 3.28 in the semi-log model to 3.62 in the linear model. Unsurprisingly, incumbents in the Republican Party were associated with increased vote shares; this was a so-called "wave" election for the Republican Party.

Finally, while the sign is negative, as anticipated, `Unemployment` is statistically insignificant in all three models. Voters do not appear to hold incumbents accountable for economic conditions in their districts. Another possibility is that voters only hold incumbents of the president's party accountable. We tested this possibility by interacting `Unemployment` and `RepublicanIncumbent`; the interaction was statistically insignificant and therefore this hypothesis is unsupported by the data.

*Spending Determinants of the Vote*[17] The estimation results for the spending covariates are provided in Table 8. These estimation results will be used to examine whether independent expenditures behave like traditional campaign expenditures in a multivariate model.

`IncumbentCampaign` is statistically insignificant in all models. Increases in incumbent campaign expenditures are not associated with increased incumbent vote shares. As repeatedly argued in prior literature, this is likely the result of endogeneity bias. In addition to campaign expenditures affecting vote shares, expectations about vote shares impact the level of campaign contributions, and consequently, campaign expenditures; there is bidirectional rather than unidirectional causality. Since incumbents spend the most in contested elections, this endogeneity bias yields an underestimated coefficient. The limited impact of incumbent campaign expenditures is demonstrated by the shape of the value function for the incumbent in Fig. 1. Relative to the value function for the challenger, the value function for the incumbent has nearly no slope and its value straddles zero across the domain.

`ChallengerCampaign` is statistically significant in all models. Increases in challenger campaign expenditures are associated with decreased incumbent vote shares. Moreover, as indicated by the statistically significant squared term in the quadratic model and constituent term in the semi-log model, these campaign expenditures are subject to diminishing marginal returns. Notably, in the quadratic model, the value function for the challenger is maximized at $4.22 per person (See Fig. 1).

The findings with respect to the campaign spending variables are consistent with prior literature. Early literature attributed this asymmetrical productivity to the minimal visibility of challengers; early spending for challengers was purported to be more productive because it significantly boosted their name recognition with the electorate. Is a similar relationship observed with independent expenditures spent on behalf of the candidates? Surprisingly, the findings with respect to independent expenditures are quite different.

Perhaps unexpectedly, `IncumbentIndependent` is statistically significant in all models. Increases in independent expenditures on behalf of incumbents are associated with increased incumbent vote shares. Moreover, as indicated by the

[17]As previously mentioned, the functional forms described above have notable advantages and drawbacks; accordingly, it was important to estimate the model with a variety of functional forms, and determine whether the conclusions drawn are model dependent. Fortunately, the estimation results suggest the conclusions are not dependent on functional form.

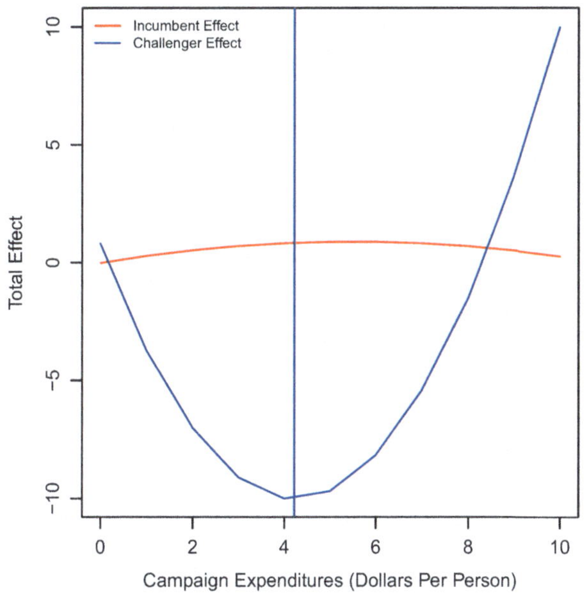


**Fig. 1** Impact of campaign expenditures

statistically significant squared term in the quadratic model and constituent term in the semi-log model, these independent expenditures are subject to diminishing marginal returns. Notably, in the quadratic model, the value function for the incumbent is maximized at $7.14 per person (See Fig. 2).

Similarly, `ChallengerIndependent` is statistically significant in all models. Increases in independent expenditures on behalf of challengers are associated with decreased incumbent vote shares. Moreover, as indicated by the statistically significant squared term in the quadratic model and constituent term in the semi-log model, these independent expenditures are subject to diminishing marginal returns. Notably, in the quadratic model, the value function for the challenger is maximized at $7.07 per person (See Fig. 2); nine challengers had independent expenditures exceeding this level of spending.[18]

Interestingly, the marginal impacts of these independent expenditure variables are similar in all three models; this can be demonstrated using Wald Tests. In the linear model, the coefficient on `IncumbentIndependent` is 1.41 while the coefficient on `ChallengerIndependent` is −1.25. When subject to a Wald

[18]As previously noted, a problem with the quadratic model is the possibility of having a negative marginal impact, or even a negative total impact, at high levels of spending. The figure provided above suggests the maximum effect on vote share occurs at approximately seven dollars of independent expenditures per person. There are a smattering of observations exceeding this threshold, unfortunately. Moreover, the total impact becomes negative at approximately 14 dollars of independent expenditures per person; there are no observations with this level of spending.

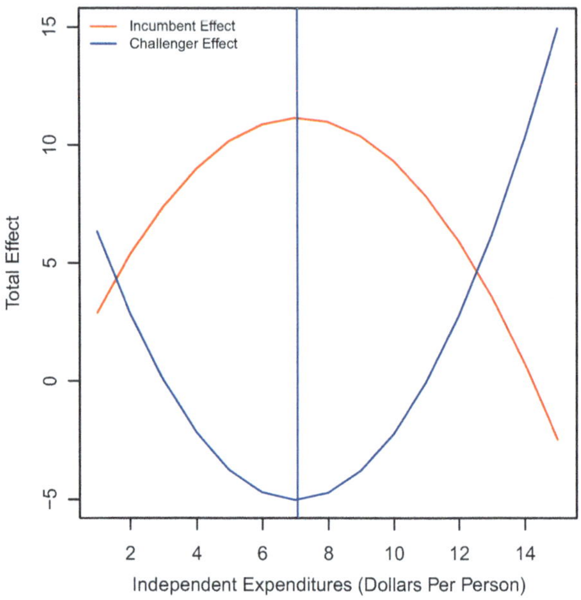


**Fig. 2** Impact of independent expenditures

Test, the absolute value of these coefficients are equal ($t = -0.41$; $p = 0.6852$). Similarly, the coefficient on `IncumbentIndependent` in the semi-log model is 1.41 while the coefficient on `ChallengerIndependent` is −1.35. When subject to a Wald Test, the absolute value of these coefficients are equal ($t = -0.31$; $p = 0.7599$). Finally, the coefficient on `IncumbentIndependent` in the quadratic model is −3.14 and the coefficient on its squared term is −0.22; the coefficient on `ChallengerIndependent` is −2.97 and the coefficient on its squared term is 0.21. When subject to a Wald Test, the absolute value of the main coefficients ($t = 0.16$; $p = 0.8711$) and the squared terms ($t = -0.07$; $p = 0.9424$) are statistically equivalent.

These results imply that the independent expenditure variables are equally productive on the margin. This is in contrast to the campaign expenditure variables that exhibit asymmetrical marginal effects in a pooled OLS analysis. It seems that the endogeneity problem fails to bias the estimates of the independent expenditure variables; if they were biased, we would observe asymmetrical productivities. Why are these independent expenditure coefficients unaffected? The answer is found in the nature of independent expenditures. As noted in Section Four, independent expenditures are limited to the most competitive contests—that is, they are limited in scope. *The result of this fact is that the OLS estimates provided above are being almost entirely identified off these highly contested elections; this is statistically synonymous to using the subsetting procedure propagated by Erickson and Palfrey (2000) to estimate campaign spending effects in the most competitive elections.*

**Summary** This subsection has reached a number of notable conclusions. All of these conclusions are robust to different model specifications. Using pooled OLS analysis because of the deficiencies in TSLS, our estimation results with respect to the campaign spending variables are consistent with prior literature; we observe an asymmetrical relationship. Specifically, challenger campaign spending is more productive than incumbent campaign spending. In contrast, we observe a symmetrical relationship with respect to the independent expenditure variables. Across different model specifications, incumbent independent expenditures and challenger independent expenditures are equally productive. Why does it seem that the estimates of the independent expenditure variables are not undermined by the endogeneity issue? We argue the scope of independent expenditures accounts for this fact. Since these expenditures are limited to the most competitive contests, the pooled OLS coefficients are identified off the most competitive elections, thereby facilitating unbiased estimation consistent with the findings of Erickson and Palfrey (2000).

## 6 Pivotality of Political Expenditures

Are political expenditures changing the outcome of congressional elections? Examining this inquiry with the previously-presented OLS models is sub-optimal because the estimates, particularly of the campaign expenditure variables, are contaminated by endogeneity bias. Despite this fact, with an acknowledgement of the difficulties, a careful attempt can be made. To do so, an unrestricted and restricted version of the semi-log model will be used to construct a back-of-the-envelope calculation. The unrestricted model will use the coefficients (as estimated) to determine the contribution of spending to each candidate's vote share. The restricted model, consistent with the previously detailed results of the Wald Tests, will consider the contribution of spending by equalizing the impact of each type of spending.

**Unrestricted Model** The contribution of spending to vote shares in the unrestricted model is calculated by multiplying the estimated marginal impact of each type of spending by the level of each type of spending for each candidate. For example, in the AZ-2 (that is, the Second U.S. House District in Arizona), independent expenditure groups spent $7.23 per person on behalf of incumbents (See Table 9 above). The semi-log model estimated that the marginal impact of one dollar of independent expenditures on behalf of incumbents equals 1.415 (See Table 8 above). Therefore, the contribution of incumbent independent expenditures to vote share in the AZ-2 is equal to 1.415 multiplied by the natural logarithm of 7.23; this equals 2.80. This calculation is made for each type of spending in each 2014 U.S. House election.

A changed outcome is defined as an election in which the contribution of spending to vote share for the winning candidate (provided in column 7 of Table 9) exceeds the margin of victory for the winning candidate. The impact of spending in highly contested elections (i.e., the Rothenberg Eight) is provided in Table 9

**Table 9** Contribution of spending variables in unrestricted model

| State | District | II Cont.[a] | CI Cont. | IC Cont.[b] | CC Cont. | Total Cont. | Margin[c] |
|---|---|---|---|---|---|---|---|
| AZ | 2 | 2.80 | −2.96 | 0.56 | −3.13 | −2.72 | −0.07 |
| CA | 52 | 2.51 | −2.32 | 0.69 | −2.67 | −1.79 | 3.17 |
| WV | 3 | 2.80 | −2.90 | 0.55 | −1.71 | −1.26 | −10.70 |
| IL | 10 | 2.14 | −3.04 | 0.71 | −2.80 | −2.99 | −2.56 |
| FL | 26 | 1.93 | −2.87 | 0.61 | −2.01 | −2.34 | −2.93 |
| NH | 1 | 2.21 | −2.84 | 0.35 | −1.07 | −1.34 | −3.61 |
| AZ | 1 | 2.75 | −2.87 | −0.75 | 7.42 | 6.55 | 5.22 |
| CO | 6 | 2.59 | −2.25 | 0.68 | −3.30 | −2.27 | 9.39 |
| Avg[d] | | 2.43 | −2.74 | 0.59 | −2.38 | −2.10 | −1.04 |

[a]II Cont. refers to the contribution to vote shares of independent expenditures spent on behalf of the incumbent. Similarly, CI Cont. refers to the contribution to vote shares of independent expenditures spent on behalf of the challenger

[b]IC Cont. refers to the contribution to vote shares of incumbent campaign expenditures. Similarly, CC Cont. refers to the contribution to vote shares of independent expenditures spent on behalf of the challenger

[c]A negative margin suggests the incumbent lost

[d]The averages exclude data from the AZ-1 since this district is plagued by the statistical artifact mentioned in footnotes 19 and 20

above. For example, in the AZ-2, the challenger's spending resulted in a 2.72 percentage point contribution to spending. The incumbent lost by 0.07 percentage points. Therefore, this is considered a changed outcome. In the 2014 U.S. House election cycle, the unrestricted models finds five election outcomes were changed by political expenditures; these were the AZ-2, IL-10, AZ-1,[19] FL-26, and NY-25.[20] Out of these elections, three challengers and two incumbents won elections that they would have otherwise lost without political expenditures[21] (i.e., all political expenditure variables equal zero). On average, political expenditures accounted for a 3.28 percentage point change in vote shares across all elections.

The impact of political expenditures in the Rothenberg Eight is smaller, calculated at 2.10 percentage points in the Rothenberg Eight. While the average impact is small, a small impact can be decisive given that these are the most competitive elections in the cycle. It is important to note that this average impact is largely the result of differences in campaign expenditures rather than independent expenditures. As Table 9 illustrates, this average difference is largely driven by the average contribution of challenger campaign spending, calculated at −2.38.

[19]The outcome of the AZ-1 being changed by political expenditures is the result of a statistical artifact. In addition to the natural logarithm of zero being undefined, the natural log of very small values is strongly negative; given the dearth of campaign spending in the AZ-1, this latter problems skews the findings; in particular, minimal challenger campaign spending yielded a 7.42 percentage point *increase* in incumbent vote shares; this is obviously an inaccurate depiction of reality.

[20]The NY-25 suffers from the same statistical artifact detailed in the previous footnote.

[21]This could be considered a counter-factual society in which campaign expenditures are outlawed.

These conclusions must be qualified, however. Since this is the restricted model, we are assuming an asymmetrical impact of incumbent and challenger campaign spending. Given our findings with respect to independent expenditures, and the findings of Erickson and Palfrey (2000) and Gerber (1998), there is little reason to think this is an accurate depiction of reality. Therefore, we can resort to a less theoretical, but more realistic and illustrative restricted model.

**Restricted Model** In the restricted model, the contribution of spending to vote share is calculated by multiplying a uniform marginal impact of spending by the level of each type of spending for each candidate. The uniform marginal impact of spending is determined by averaging the marginal effect of `IncumbentIndependent`, `ChallengerIndependent`, and `ChallengerCampaign`. Pairwise Wald Tests demonstrate these marginal impacts are statistically equivalent. We have not included `IncumbentCampaign` in the average because, consistent with the findings of Gerber (1998), Erickson and Palfrey (2000), and Section Five, it is significantly underestimated due to endogeneity bias. The average of the relevant coefficients is 1.50. Therefore, the contribution of incumbent independent expenditures to vote share in the AZ-2 is equal to 1.50 multiplied by the natural logarithm of 7.23; this equals 2.97. This calculation is made for each type of spending in each 2014 U.S. House election.

Using the same definition of changed outcome, the restricted model finds two election outcomes were changed by political expenditures; these are the AZ-2 and the NY-25.[22] In the AZ-2, the challenger had \$9.01 per person in campaign expenditures and \$6.04 per person in independent expenditures for a total of \$15.05 per person in total political expenditures. Meanwhile, the incumbent had \$7.24 per person in campaign expenditures and \$4.45 in independent expenditures for a total of \$11.69 per person in total political expenditures. The challenger's spending advantage of \$3.36 per person yielded a 0.79 percentage point gain, exceeding the margin of victory, 0.07 percentage points.

Table 10 provides the total impact of political expenditures for the restricted model in the Rothenberg Eight. The impact of political expenditures is more limited in the restricted model than the unrestricted model; this is demonstrated by the average total contribution of spending equaling −0.17 percentage points on average and not exceeding 0.79 percentage points in absolute value (excluding the AZ-1, which is plagued by the statistical artifact in the semi-log model). Given the estimated contributions to vote shares, political expenditures are only pivotal in the *closest* elections.

**Summary** This section has examined the importance of political expenditures in congressional elections and whether they are pivotal in determining outcomes. Under the unrestricted model that assumes an asymmetry between incumbent and challenger campaign spending, we find five election outcomes were altered, although as previously noted, two of these outcome changes are the result of a

[22] As previously noted, the estimate for the NY-25 is a statistical artifact.

**Table 10** Contribution of spending variables in restricted model

| State | District | II Cont. | CI Cont. | IC Cont. | CC Cont. | Total Cont. |
|---|---|---|---|---|---|---|
| AZ | 2 | 2.97 | −3.30 | 2.24 | −2.70 | −0.79 |
| CA | 52 | 2.66 | −2.59 | 2.74 | −2.30 | 0.51 |
| WV | 3 | 2.97 | −3.23 | 2.19 | −1.48 | 0.45 |
| IL | 10 | 2.27 | −3.39 | 2.82 | −2.42 | −0.71 |
| FL | 26 | 2.05 | −3.20 | 2.44 | −1.73 | −0.45 |
| NH | 1 | 2.35 | −3.17 | 1.40 | −0.92 | −0.34 |
| AZ | 1 | 2.92 | −3.20 | −2.98 | 6.40 | 3.14 |
| CO | 6 | 2.75 | −2.50 | 2.71 | −2.85 | 0.11 |
| Avg | | 2.57 | −3.05 | 2.36 | −1.24 | −0.17 |

statistical artifact. The asymmetry assumption is likely unrealistic given the findings of Gerber (1998), Erickson and Palfrey (2000), and Section Five. Therefore, we resorted to a restricted model wherein the impact of political expenditures is equalized at 1.50 in the semi-log model. In this model, we find only two election outcomes were altered, one of which is the result of the statistical artifact. We identify only the AZ-2 as an election whose outcome was altered by political expenditures. Due to the back-of-the-envelope, interpolated nature of our estimates, we have purposely de-emphasized the numerical estimates of the contributions of spending to vote shares. *Our finding is limited to the fact that political expenditures are only important in the most contested elections; and even in those, they are rarely pivotal.*

## 7 Conclusion

This article has examined the nature of independent expenditures and their impact on election outcomes. The scope of independent expenditures is different than campaign expenditures. While meaningful levels of campaign expenditures are present in most elections, even uncompetitive elections, independent expenditures are limited to the most competitive elections. In 2014, less than 25 elections (out of the 356 included in the sample) featured significant independent expenditures.

Independent expenditures and campaign expenditures can also be differentiated by their dispersion. While incumbents enjoy an advantage with respect to campaign expenditures, even in the most competitive elections, challengers enjoy an advantage with respect to independent expenditures. In the most competitive elections of the 2014 election cycle, identified by Stuart Rothenberg, challengers enjoyed a \$1,446,473.97 advantage on average with respect to independent expenditures.

Lastly, the magnitude of independent expenditures is different than campaign expenditures. While there are more campaign expenditures in the aggregate, independent expenditures are the dominant form of spending in contested races. Specifically, in the Rothenberg Eight, independent expenditures comprised 62.40 % of all political expenditures.

Independent expenditures also behave differently in a multivariate setting. In a counter-intuitive, yet anticipated finding, campaign expenditures have different marginal productivities in pooled OLS models; challenger campaign expenditures are significantly more productive than incumbent campaign expenditures. This finding is likely the result of endogeneity bias (Erickson and Palfrey 2000). Perhaps surprisingly, independent expenditures exhibit symmetrical marginal productivities—that is, when subject to a Wald Test, the absolute value of the coefficients for incumbent independent expenditures and challenger independent expenditures are statistically equivalent.

Why do independent expenditures behave differently than campaign expenditures in multivariate models? The endogeneity problem fails to bias the estimates of the independent expenditure covariates. If they were substantially biased, we would likely observe asymmetrical productivities—that is, the coefficient for challenger independent expenditures would be greater than the coefficient for incumbent independent expenditures. We likely do not observe bias because independent expenditures are overwhelmingly limited to the most competitive contests. Therefore, our estimates of the independent expenditure coefficients in the pooled OLS models are overwhelmingly identified off those most competitive races, thereby attenuating or eliminating the bias in accordance with the discussion in Erickson and Palfrey (2000).

Despite the unusual nature of independent expenditures, their impact on election outcomes is limited. Back-of-the-envelope calculations suggest political expenditures, particularly independent expenditures, are rarely pivotal in determining election outcomes. This is likely the result of a number of factors. Since incumbent and challenger political expenditures are equally productive, given the findings of Erickson and Palfrey (2000) and Section Five, gaining an advantage via political expenditures requires a significant advantage in total spending. Empirically, few candidates are able to garner a significant advantage. The mean and median differences in total political expenditures between incumbents and challengers are $1.19 and $0.99 per person, respectively. These values illustrate that few candidates are able to garner a significant total spending advantage. Similarly, the mean and median differences in campaign expenditures are $1.21 and $0.99, respectively. Given the similarity between these values and the differences with respect to total political expenditures, it is clear that the minor disparity in total expenditures is driven by differences in campaign expenditures, not independent expenditures. This is confirmed by the mean and median differences in independent expenditures, which are $0.02 and $0.00, respectively.

Independent expenditures, while different in their characteristics, do not seem to exercise significant influence over election outcomes. If anything, they are less influential than traditional campaign expenditures. The failure of independent expenditures to alter election outcomes does not necessarily mean the outcry over *Citizens United* is misplaced or incorrect. Changing the outcome of general elections is only one outlet for *Citizens United* to exercise undue influence on the political process. Perhaps independent expenditures are more pivotal in primary elections; if this is true, independent expenditures could be determining the choices of the general electorate. Alternatively, it would be interesting to investigate whether the policy positions of elected officials are influenced by the few big-money donors that finance many independent expenditure groups. Finally, examining whether independent expenditures have a detrimental impact on democratic efficacy or turnout could be fruitful.

Additional investigation of independent expenditures is warranted. Future research should overcome limitations with this article. Ideally, more than one election cycle should be examined to determine whether our results are generalizable. Moreover, after more elections are conducted under the new campaign finance regime, scholars could use the subsetting approach propagated by Erickson and Palfrey (2000) to estimate the spending effects of independent expenditures in elections with varying levels of competition. Doing so could confirm our argument about the unusual scope of independent expenditures facilitating the unbiased estimation of spending effects in the pooled OLS analysis used above.

## Appendix

See Table 11

**Table 11** Correlation matrix

| | Inc. Share[a] | II[b] | CI[c] | IC[d] | CC[e] | Party[f] | Lagged[g] | Leg[h] | Rep. Inc.[i] | Unemploy[j] | Pop[k] |
|---|---|---|---|---|---|---|---|---|---|---|---|
| Inc. Share | 1 | −0.412 | −0.409 | −0.394 | −0.477 | 0.790 | 0.732 | −0.367 | 0.069 | 0.140 | 0.134 |
| II | −0.412 | 1 | 0.916 | 0.453 | 0.696 | −0.338 | −0.313 | 0.480 | −0.186 | −0.019 | −0.176 |
| CI | −0.409 | 0.916 | 1 | 0.407 | 0.649 | −0.308 | −0.287 | 0.435 | −0.178 | −0.013 | −0.157 |
| IC | −0.394 | 0.453 | 0.407 | 1 | 0.604 | −0.397 | −0.316 | 0.319 | 0.047 | −0.118 | −0.110 |
| CC | −0.477 | 0.696 | 0.649 | 0.604 | 1 | −0.405 | −0.391 | 0.363 | −0.055 | −0.034 | −0.081 |
| Party | 0.790 | −0.338 | −0.308 | −0.397 | −0.405 | 1 | 0.777 | −0.268 | −0.141 | 0.258 | 0.081 |
| Lagged | 0.732 | −0.313 | −0.287 | −0.316 | −0.391 | 0.777 | 1 | −0.281 | −0.162 | 0.205 | 0.078 |
| Leg | −0.367 | 0.480 | 0.435 | 0.319 | 0.363 | −0.268 | −0.281 | 1 | −0.102 | −0.023 | −0.123 |
| Rep Inc. | 0.069 | −0.186 | −0.178 | 0.047 | −0.055 | −0.141 | −0.162 | −0.102 | 1 | −0.319 | 0.040 |
| Unemplcy | 0.140 | −0.019 | −0.013 | −0.118 | −0.034 | 0.258 | 0.205 | −0.023 | −0.319 | 1 | −0.068 |
| Pop | 0.134 | −0.176 | −0.157 | −0.110 | −0.081 | 0.081 | 0.078 | −0.123 | 0.040 | −0.068 | 1 |

[a]Incumbent vote share
[b]Incumbent independent expenditures
[c]Challenger independent expenditures
[d]Incumbent campaign expenditures
[e]Challenger campaign expenditures
[f]Partisan voter index
[g]Lagged incumbent vote share
[h]Legislator dummy variable
[i]Republican incumbent dummy variable
[j]District unemployment rate
[k]District population

## References

Abramowitz, A. (1988, June). Explaining senate election outcomes. *The American Political Science Review, 82*(2), 385–403.

Bipartisan Campaign Reform Act of 2002 ("McCain-Feingold"), Section 203 (2002). *Federal Election Commission*. Accessed 23 July 2015. http://www.fec.gov/press/bkgnd/bcra_overview.shtml

Citizens United v. Federal Election Commission. Oyez: Chicago-Kent College of Law at Illinois Tech. Accessed 23 July 2015. https://www.oyez.org/cases/2008/08-205

Coordinated Communications and Independent Expenditures. *Federal Election Commission*. Accessed 23 July 2015. http://www.fec.gov/pages/brochures/indexp.shtml

Dionne, E. J., Jr. (2012, February 5). The Citizens United catastrophe. *Washington Post*.

Egan, T. (2014, October 23). The disgust election. *New York Times*.

Erickson, R., & Palfrey, T. (2000, September). Equilibria in campaign spending games: Theory and data. *The American Political Science Review, 94*(3), 595–609.

Gerber, A. (1998, June). Estimating the effect of campaign spending on senate election outcomes using instrumental variables. *The American Political Science Review, 92*(2), 401–411.

Green, D., & Krasno, J. (1988, November). Salvation of the spendthrift incumbent: Re-estimating the effects of campaign spending in house elections. *American Journal of Political Science, 31*(4), 894–907.

Jacobson, G. (1978, June). The effects of campaign spending in congressional elections. *The American Political Science Review, 72*(2), 469–491.

Op-Ed. (2014, October 13). A trickle down effect of Citizens United. *New York Times*.

Pelosi, N., & Sarbanes, J. (2014, February 4). Reversing the grievous error of Citizens United. *Washington Post*.

SpeechNOW.Org v. FEC. *Federal Election Commission*. Accessed 23 July 2015. http://www.fec.gov/law/litigation/speechnow.shtml

Super PACs. OpenSecrets.org. *Center for Responsive Politics*. Accessed 24 July 2015. https://www.opensecrets.org/pacs/superpacs.php

# Spatial Model of U.S. Presidential Election in 2012

**Jeong Hyun Kim and Norman Schofield**

**JEL Classification** H11, C1

## 1 Introduction

In this research paper, we explore a following set of research questions. First, what are the relative role of ideological position and valence in determining vote choices in 2012 U.S. presidential election? Second, is there a Nash equilibrium for the position of candidates in 2012 election? Third, is the effect of ideological position robust to the inclusion of demographic characteristics of voters?

This study is not the first attempt applying spatial model to the study of U.S. presidential elections. For instance, using the 2008 American National Election Study (ANES), Schofield and Gallego (2011) show that there exists a local Nash equilibrium for candidates' policy positions (see also Schofield et al. 2011a, b). However, studying 2012 election itself is academically interesting for many reasons. In particular, it provides much implication to the study on the role of activists in U.S. elections. In January 2010, there was a historical the U.S. Supreme Court's *Citizens United v. Federal Elections Commission* ruling. This Supreme Court's decision resulted in the removal of the limits on campaign contribution in the U.S. elections. President Barack Obama criticized the *Citizens United* decision for "opening the floodgates for special interests to spend without limit in our elections." In addition to an unprecedented increase in political spending, the *Citizens United* decision also led the creation of "Super PACs," that can accept unlimited amounts of political

J.H. Kim • N. Schofield (✉)
Washington University in St. Louis, Campus Box 1063, One Brookings Drive, St. Louis, MO 63130-4899, USA
e-mail: schofield.norman@gmail.com



M. Gallego, N. Schofield (eds.), *The Political Economy of Social Choices*, Studies in Political Economy, DOI 10.1007/978-3-319-40118-8_10

funds from contributors (Hansen et al. 2015). 2012 election is the first presidential election that was held after this significant event. Thus, analyzing 2012 election and comparing the results with those of previous elections will allow us to test the exogenous effect of campaign contribution and activist influence on electoral outcomes. In this regard, this research examines how did an increase in campaign contribution that is prompted by the Supreme Court's decision affect vote choice and policy position of candidates.

## 2 The Electoral Effect of Campaign Contribution

In American politics, there is a large literature examining the impact of candidate spending on electoral outcomes (see Abramowitz 1991; Gerber 1998; Jacobson 1990; Krasno and Green 1988). Yet, only a few studies have investigated the electoral effect of the campaign contribution law. Among them, La Raja and Schaffner (2014) find that there is only minimal effect of the campaign spending bans on electoral outcomes such as partisan control of government and incumbency re-election rates. Their study suggests that the *Citizen United* decision may not have significant electoral consequences. On the other hand, Klumpp et al. (2012) demonstrate that the *Citizens United* decision has increased the electoral prospects of Republican candidates in state elections. Although these recent studies address whether a change in campaign contribution law have electoral consequences, their findings are mainly from state elections. The question of whether and how campaign contribution affects election outcomes at the federal level remains.

## 3 Applying Spatial Model of Elections to 2012 U.S. Presidential Election

### *3.1 Spatial Model of Elections*

In order to examine the role of ideological position in determining vote decisions in 2012 election, we start the analysis with a pure spatial model, M(λ, β). The data of the spatial model is a distribution, $\{x_i \in X\}_{i \in N}$, of voter ideal points for the members of the electorate, N, of size n. And, each of the candidates in the set P = {1, ..., j, ..., p} chooses a policy, $z_j \in X$, to declare prior to the specific election to be modeled. In this model, the utility of voter *i* with position $x_i$ gets from voting for candidates *j* with position $z_j$ is:

$$\mu_{ij}\left(x_i, z_j, \lambda_j\right) = \lambda_j - \beta \parallel x_i - z_j \parallel^2 + \epsilon_j \tag{1}$$

Where $\lambda_j$, denotes to the intrinsic or exogenous valence of candidate j, the term β is the spatial parameter, giving the importance of policy difference defined in terms of a metric induced from the Euclidean norm, and the vector $\epsilon = (\epsilon_1, \cdots, \epsilon_j, \cdots \epsilon_p)$ is the stochastic error, whose multivariate cumulative distribution is the Type 1 extreme value distribution.

## 3.2 *Data: The American Panel Survey (TAPS)*

In this research, we examine the role of policy positions and valence in determining vote choices and estimate Local Nash equilibrium of candidate positions by utilizing The American Panel Survey (TAPS) data. TAPS is a monthly online survey of a nationally representative sample from a panel of about 2000 adults in the U.S., started November 2011. The data include responses about vote choice at 2012 presidential election as well as the panels' policy positions and demographic characteristics. Unlike ANES data, the panels of TAPS are asked to identify their residential area, including state and congressional district they live in. This information allows us to investigate whether the effect of ideological position vary by state or congressional district the respondents belong to, and how the varying effect can aggregate to the national level.

Miller and Schofield (2003, 2007) have argued that American politics has a stable two-dimensional nature that consists of economic and social issues. Following this, we begin by conducting a factor analysis based on a set of questions asking about respondents' positions on economic and social policies in TAPS data. The set of questions used and the factor loadings of each item are presented in appendix.

Figure 1 illustrates the position of voters with the x-axis corresponding to the economic dimension and the y-axis corresponding to the social dimension. A movement from left to right on the $x$-axis indicates favoring less government regulation on economic activities while a movement from south to north on the $y$-axis means more liberal attitudes on social issues.

The variance on both axes is very similar to each other; the variance on x-axis is 0.722, while the variance on y-axis is 0.751. The covariance between the two dimensions is −0.128. The voter variance-covariance matrix can be presented as following:

$$\nabla = \begin{bmatrix} 0.722 & -0.128 \\ -0.128 & 0.751 \end{bmatrix} \tag{2}$$

with trace $(\nabla) = 1.47$. The negative sign of covariance between the two dimensions seems plausible since an increase in $x$-axis refers to a movement towards economic conservatism, while an increase in $y$-axis indicates a move to social liberalism. However, the magnitude of covariance between the two dimensions is not considerably high.

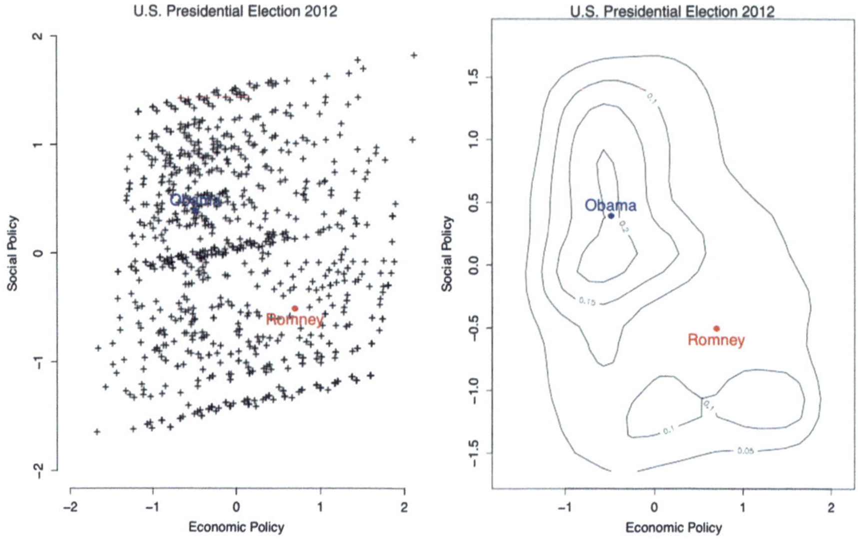


**Fig. 1** Voter distribution and candidate positions in 2012 U.S. presidential election

The positions of the major presidential candidates, Romney and Obama, in 2012 are estimated by taking the means of policy positions of respondents who voted for each candidate. These positions are denoted ($z^*_{\text{Obama}}$, $z^*_{\text{Romney}}$), and the candidate position matrix is given by:

$$z^* = \begin{bmatrix} Candidate & \text{Obama} & Romney \\ \text{x : Economic} & -0.48 & 0.71 \\ y : Social & 0.39 & -0.50 \end{bmatrix} \tag{3}$$

From the matrix, we see that Obama's estimated position is more liberal than Romney's position in terms of both social and economic policy area, as expected.

Next, we estimate a multinomial logit model (MNL) to examine the effect of ideological position and valence of candidates on vote choice. Table 1 summarizes the results. From the first column in Table 1, we see that the spatial coefficient is 1.117 and statistically significant. Here, the reference category is Vote for Obama. The relative valence of Romney is –0.044. Compared to the 2008 presidential election (Schofield and Gallego 2011), the magnitude of spatial coefficient is greater in 2012 election. However, unlike 2008 election, the valence term in 2012 election is not statistically significant. The spatial coefficient remains statistically significant when we condition demographic variables, such as ethnicity, education, age, and gender in the model. From the second column in Table 1, we see that black respondents are less likely to vote for Romney, and respondents with college education are more likely to vote for Romney than Obama.

**Table 1** Spatial model of 2012 U.S. presidential election

| Variable | (1) Pure spatial | (2) Spatial + traits | (3) Spatial + traits + socios |
|---|---|---|---|
| Romney valence | −0.044 (0.117) | −0.422* (0.192) | −2.899 (5.757) |
| **Spatial $\beta$** | 1.117*** (0.075) | 0.575*** (0.116) | 0.721*** (0.156) |
| Obama trait | | −3.237*** (0.354) | −3.640*** (0.481) |
| Romney trait | | 2.844*** (0.321) | 3.070*** (0.413) |
| Ethnicity (black) | | | −2.503 (5.809) |
| Ethnicity (Hispanic) | | | 0.878 (5.669) |
| Ethnicity (white) | | | 3.287 (5.641) |
| Ethnicity (other) | | | 2.623 (5.699) |
| Education (less than high school) | | | 4.939*** (1.451) |
| Education (high school) | | | 0.689 (0.702) |
| Education (college) | | | 0.555 (0.572) |
| Age | | | −0.013 (0.014) |
| Gender (female) | | | 0.606 (0.466) |
| N | 834 | 819 | 692 |
| Log likelihood | −244.65 | −106.59 | −73.606 |
| McFadden $R^2$ | 0.57 | 0.81 | 0.85 |

*Note*: Baseline category is *Vote for Obama*, Standard error in parentheses
*p<0.05; **p<0.01; ***p<0.001

Also, from the results, we can calculate the probability that a voter chooses each candidate, when all candidates are located at the electoral mean of policy positions. First, the probability that a voter chooses Obama is

$$\rho_{Obama} = \frac{\exp(0)}{\exp(0) + \exp(-0.044)} = 0.511 \tag{4}$$

Using the same calculation, the probability that a voter chooses Romney is

$$\rho_{Romney} = \frac{\exp(-0.044)}{1 + \exp(-0.044)} = 0.489 \tag{5}$$

The results of calculation are close to the actual result from the sample. While 56 % of the respondents answered they voted for Barack Obama, 44 % of them answered that they voted for Mitt Romney at 2012 presidential election.

In order to check whether the joint origin, $z_0$ is a Local Nash Equilibrium (LNE), we examine the Hessians of the vote share utility functions. The Hessian, or the characteristic matrix of Romney's vote share function at $z_0$ is given by:

$$\begin{aligned} c_{Romney} &= 2\beta\left(1 - 2\rho_{Romney}\right)\nabla - \mathbf{I} \\ &= 2 \times 1.117 \times (1 - 2 \times 0.489)\begin{bmatrix} 0.752 & -0.128 \\ -0.128 & 0.751 \end{bmatrix} - \mathbf{I} \\ &= \begin{bmatrix} -0.963 & -0.006 \\ -0.006 & -0.964 \end{bmatrix} \end{aligned} \tag{6}$$

The eigenvalues of the characteristic matrix are −0.957 with the eigenvector (−0.747, 0.665), and −0.970 with the eigenvector (0.665, 0.747). According to Schofield (2007), negative eigenvalues of the characteristic matrix is a necessary condition for the electoral mean to be SLNE.

The convergence coefficient can be calculated by using the following formula:

$$\begin{aligned} c &= 2\beta\left(1 - 2\rho_{Romney}\right) Tr\left[\nabla\right] \\ &= 2 \times 1.117 \times (1 - 2 \times 0.489) \times 1.473 = 0.073 \end{aligned} \tag{7}$$

Schofield et al. (2011a, b) show that $c < 1$ is a sufficient condition for convergence to z0 in the pure spatial model. Using simulations, we find a LNE with the following candidate positions:

$$z_1 = \begin{bmatrix} Candidate & Obama & Romney \\ Economic: & 0.032 & 0.032 \\ Social: & 0.004 & 0.004 \end{bmatrix} \tag{8}$$

As we see, the candidate positions in this LNE are very close to the joint electoral origin.

We extend the spatial model by including perceptions of candidate traits. Using the respondents' evaluations of candidates, we constructed a trait index by factor analysis (see the appendix for factor loadings). From Table 1, we find that the spatial coefficient is still significant in models with candidate trait variables.

## 4 Discussion

From the analysis of voter perceptions in 2012 U.S. presidential election, we can draw the following preliminary findings. First, as in the case of previous presidential elections, ideological distance between candidate and voters plays a significant role in determining vote choice in 2012 election. The magnitude of spatial coefficient turns out to be even greater than the previous election. Second, the valence of candidate in 2012 election is not a statically significant predictor of vote choice.

These findings provide implications about the influence of increased campaign contribution, resulting from Supreme Court decision in 2011. That is, the exogenous increase in campaign contribution has emphasized the role of ideological distance in voting behavior, and decreased the effect of valence.

## Appendix 1: Questions for the 2012 TAPS

1. Indicate your level of agreement with this statement: Federal personal income taxes for individuals with incomes higher than $250,000 should be raised.
2. Which actions are you in favor of and which are you against: less government regulation of business?
3. Indicate your level of agreement: Incomes should be more equal because everyone's contribution to society is equally important.
4. Do you consider your view of the federal government recognizing same-sex marriages liberal, moderate, or conservative?
5. Do you consider your view of federal funding for abortion liberal, moderate, or conservative?
6. Do you consider your view of the banning possession of handguns liberal, moderate, or conservative?
7. Do you consider your view of programs designed to help minorities get better jobs and education liberal, moderate, or conservative?

## Appendix 2: Factor Loadings for Economic and Social Policy

See Tables 2 and 3.

**Table 2** Factor loadings for economic and social policy

| Question | Economic policy | Social policy |
|---|---|---|
| Personal income tax | 0.78 | 0.27 |
| Government regulation of business | 0.65 | 0.30 |
| Income inequality | 0.57 | 0.15 |
| Gay marriage | 0.20 | 0.77 |
| Abortion | 0.19 | 0.79 |
| Gun control | 0.26 | 0.57 |
| Helping minorities | 0.27 | 0.62 |

**Table 3** Factor loadings for candidate traits

| | | Obama trait | Romney trait |
|---|---|---|---|
| Obama | Optimistic | 0.724 | −0.192 |
| | Fair | 0.893 | −0.294 |
| | Strong | 0.833 | −0.320 |
| | Honest | 0.873 | −0.291 |
| | Trustworthy | 0.882 | −0.295 |
| | Experienced | 0.835 | −0.315 |
| | Knowledgeable | 0.823 | −0.277 |
| | Inspiring | 0.796 | −0.301 |
| | Decisive | 0.763 | −0.208 |
| | Moral | 0.842 | −0.262 |
| Romney | Optimistic | −0.177 | 0.708 |
| | Fair | −0.320 | 0.802 |
| | Strong | −0.358 | 0.790 |
| | Honest | −0.280 | 0.830 |
| | Trustworthy | −0.304 | 0.864 |
| | Experienced | −0.286 | 0.747 |
| | Knowledgeable | −0.242 | 0.774 |
| | Inspiring | −0.337 | 0.751 |
| | Decisive | −0.304 | 0.675 |
| | Moral | −0.196 | 0.774 |

# References


Abramowitz, A. I. (1991). Incumbency, campaign spending, and the decline of competition in US House elections. *Journal of Politics, 53*(1), 34–56.

Gerber, A. (1998). Estimating the effect of campaign spending on senate election outcomes using instrumental variables. *American Political Science Review, 92*(2), 401–411.

Hansen, W. L., Rocca, M. S., & Ortiz, B. L. (2015). The effects of Citizens United on corporate spending in the 2012 presidential election. *Journal of Politics, 77*(2), 535–545.

Jacobson, G. C. (1990). The effects of campaign spending in House elections: New evidence for old arguments. *American Journal of Political Science, 34*, 334–362.

Klumpp, T., Mialon, H. M., & Williams, M. A., et al. (2012). *Money talks: The impact of Citizens United on state elections*. Emory University. Journal of Law and Economics, forthcoming.

Krasno, J. S., & Green, D. P. (1988). Preempting quality challengers in House elections. *Journal of Politics, 50*(4), 920–936.

La Raja, R. J., & Schaffner, B. F. (2014). The effects of campaign finance spending bans on electoral outcomes: Evidence from the states about the potential impact of Citizens United v. FEC. *Electoral Studies, 33*, 102–114.

Miller, G., & Schofield, N. (2003). Activists and partisan realignment in the United States. *American Political Science Review, 97*(2), 245–260.

Schofield, N. (2007). The mean voter theorem: Necessary and sufficient conditions for convergent equilibrium. *Review of Economic Studies, 74*(3), 965–980.

Schofield, N., & Gallego, M. (2011). Leadership or Chaos. Berlin: Springer.

Schofield, N., Claassen, C., Gallego, M., & Ozdemir, U. (2011a). Empirical and formal models of the US presidential elections in 2000 and 2004. In N. Schofield & G. Caballero (Eds.), *The political economy of institutions, democracy and voting* (pp. 217–258). Berlin: Springer.

Schofield, N., Claasen, C., Ozdemir, U., & Zakharov, A. (2011b). Estimating the effects of activists in two-party and multi-party systems: A comparison of the United States in 2008 and Israel in 1996. *Social Choice and Welfare, 36*(3–4), 483–518.

Schofield, N., & Miller, G. (2007). Elections and activist coalitions in the United States. *American Journal of Political Science, 51*(3), 518–531.

# Modeling Elections and Referenda in Ireland

**Norman Schofield and William Simoneau**

## 1 Introduction

The period between 2007 and 2009 was a tumultuous time for the Republic of Ireland. During this time the Irish economy was fully exposed to the global financial crisis. In turn, Irish households saw an overall drop in net worth of e453 billion between 2007 and 2010 (Cussen and Phelan 2010). In January of 2008 the Irish unemployment rate started a steady climb from 4 % up to 15.1 % in 2012 (CSO Data). The financial crisis marked the end of the Celtic Tiger, a period of economic growth and modernization that had begun in the early 1990s. During this sharp negative downturn the Irish electorate went to the polls on three separate occasions. In 2007, there was the general election for the lower house, the Dáil Éireann, or Dáil, in 2008, a referendum on the ratification of the Lisbon Treaty and in 2009 a second referendum on the Lisbon Treaty.

Signed by EU member states in December 2007, the Lisbon Treaty, proposed changes to European Union institutions and attempted to streamline the internal workings of the EU after enlargement in 2004 and 2007 (Quinlan 2009). The Lisbon Treaty was the result of the redrafting of the failed attempt at a Constitutional Treaty. The Constitutional Treaty, agreed upon in 2004, was the first attempt to streamline the decision making process in the wake of the 2004 EU enlargement that added ten new member states. The Constitutional Treaty failed in the ratification process when Dutch and French voters voted down the Treaty in a national referendum. In the wake of the failed ratification process, the Lisbon Treaty was developed and passed. Although there were differences between the two treaties their goals of streamlining the institutions remained the same. Although these differences were disputed,

N. Schofield (✉) • W. Simoneau
Washington University in Saint Louis, Campus Box 1063, One Brookings Drive, St. Louis, MO 63130-4899, USA
e-mail: schofield.norman@gmail.com



M. Gallego, N. Schofield (eds.), *The Political Economy of Social Choices*, Studies in Political Economy, DOI 10.1007/978-3-319-40118-8_11

former French President Valery Giscard d'Estaing said, "The Treaty of Lisbon is the same as the rejected constitution. Only the format has been changed to avoid referendums" (Bonde 2010). Further, Alexander Stubb, former Finnish minister of foreign affairs, stated that 99 % of the Constitution had been kept in the Lisbon Treaty (Bonde 2010). One key difference was the ratification process. The Lisbon Treaty was passed by all member states, except Ireland, through Parliamentary votes instead of national referenda. The Irish Constitution necessitated a referendum in order to pass the treaty.

According to Article 46 of the Irish Constitution any provision of the Constitution may be amended, whether by way of variation, addition, or repeal. It also provides that every proposal for amendment must be initiated in the Dáil as a Bill. Having been passed or deemed to have been passed by both houses of the Oireachtas, the Bill is submitted by referendum to the decision of the people. If the government desires to enact a treaty, such as Lisbon, there needs to be a public referendum. In 2008, the first Lisbon Treaty referendum was held, and resulted in its rejection by the electorate. The following year the Lisbon Treaty was again, put forward to the electorate. This time it passed. There were three opportunities for the electorate to express their preferences, and each time the result differed from the previous. The initial vote approved of the Fianna Fáil government, the second, was a vote against the supported position of Fianna Fáil and the third was a direct overturn of the previous referendum decision. This paper plans to develop a greater understanding of what happened between those elections through spatial models of the electorate and these elections.

The analysis of these elections will center on the idea of valence for the parties in the general election and the campaign sides in the referenda. Valence originally proposed by Stokes (1963) is considered the non-policy attractiveness of the candidate, or the quality of the candidate (Schofield and Gallego 2011; Gallego and Schofield 2013). These elections will provide two different analysis of valence. One type of analysis is the role of party valence and how that determines the ideological position a party takes. The results from the 2007 general election illustrate how the lowest valence party, in this case the Progressive Democrats, establish a position that does not converge toward the electoral mean position. The second analysis allows for a direct comparison of the valence levels of the campaign sides in the 2008 and 2009 referenda. Valence of the Yes campaign decreased between these two referenda, and the results from the 2009 referendum show the importance of the financial crisis in the Yes campaign winning the election.

## 2 2007 General Election

The 2007 election resulted in a coalition government similar to the incumbent coalition. Fianna Fáil was still the leading party of the government, that consisted of Fianna Fáil, the Progressive Democrats, and the Green Party. The change from the previous coalition government was the addition of the Green party to the cabinet.

The election results netted Fianna Fáil and the Progressive Democrats a loss of eight seats in the Dáil. The Progressive Democrats contributed to most of that loss, losing six out of the eight seats. These seats were replaced with six from the Green party. Much of the coalition program, 76 %, was taken from the Fianna Fáil manifesto, and the Green Party was considered to have weak terms of commitment (O'Malley 2007). The Green party, a Euro-skeptic party, was particularly challenged to make policy concessions as the government accepted a redrafted European Union treaty, the Lisbon Treaty[1] (O'Malley 2007). This treaty was then turned to the electorate in 2008 via referendum.

In order to model policy positions of the parties during the election a factor analysis with a varimax rotation was used with 2007 data from the Irish National Election Survey (INES, see the reference list for details on this data set)[2] using the following questions.

- *It would be better if more people with strong religious beliefs held public office.* The responses were on a 5 point scale of agreement.
- *Same sex marriages should be prohibited by law.* The responses were on a 5 point scale of agreement.
- *People who fully agree that there should be a total ban on abortion in Ireland would give a score of 0. People who fully agree that abortion should be freely available in Ireland to any woman who wants to have one would give a score of 10. Other people would place themselves in between these two views.* Where would you place yourself on this scale?
- *Most of business and industry should be owned by the state ... Most of business and industry should be privately owned.* The responses were a 10 point scale on level of agreement.
- *Where would you place yourself on these scales? The first scale is as follows Business and industry should be strictly regulated by the state. Business and industry should be entirely free from regulation by the state.* The responses were a 10 point scale on level of agreement.

The results from the factor analysis of the INES data, seen in Chart 1, show that factor one includes our economic variables and factor two includes the social variables. Using the factor loadings and individual voter's response to the above survey questions, we are able to estimate each voter's position along the $x_i$, social position, and $y_i$, economic position, axes.

Using these factors, the mean position $(x_i,y_i)$ of each party's voters are listed in Table 1. Figure 1 presents the smoothed[3] electoral distribution of the electorate's positions and the positions of the parties.

Using these points, we can now measure the distance of each voter, noted at $(x_i,y_i)$ from the positions of each party $(z_j)$. For the purpose of this paper we will use a pure spatial model, $M(\lambda, \beta)$ and full model with a pure spatial model and

[1]The Lisbon Treaty was considered a redrafted version of the European Constitutional Treaty.

[2]INES Data made available through Irish Social Science Data Archive.

[3]Using a Kernel Density Estimation.

| Issue | Economic Factors | Social Factors |
|---|---|---|
| Religious Politicians | | .342 |
| Gay Marriage | | .661 |
| Abortion | .101 | .163 |
| Business Ownership | .986 | .148 |
| Business Regulation | .248 | |
| SS loadings | 1.059 | .612 |
| Proportion Var | .212 | .122 |
| Cumulative Var | .212 | .334 |

**Chart 1** Factor analysis results for 2007 general election

**Table 1** Economic and social positions of political parties

| | | FF | FG | GR | LB | PD | SF | IN |
|---|---|---|---|---|---|---|---|---|
| $z^*=$ | x: Social | .018 | .02 | −.06 | −.16 | .24 | −.28 | .18 |
| | y: Economic | .007 | .02 | .23 | .04 | −.07 | −.17 | −.11 |

*Abbreviations*: *FF* Fianna Fáil, *FG* Fine Gael, *GR* Green, *LB* Labour, *PD* Progressive Democrats, *SF* Sinn Fein, *IN* Independent

**Fig. 1** 2007 Irish electorate

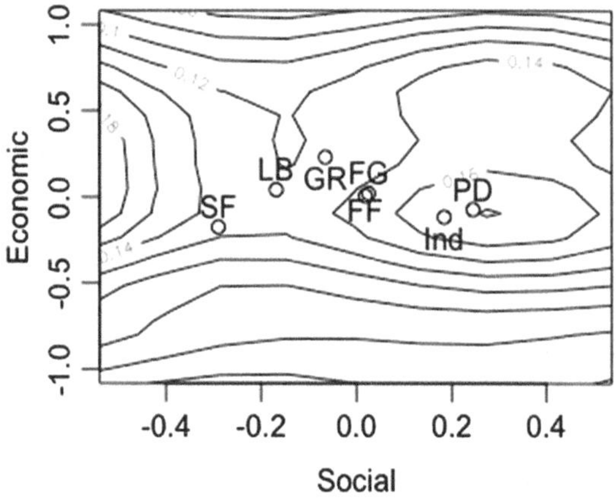


sociodemographic traits M(λ, θ, β). This model assumes that the implicit utility of voter *i* for party *j* has the below form, and that the spatial coefficient, β, must be positive (Schofield and Gallego 2011; Schofield and Sened 2006; Gallego and Schofield 2013).

$$u_{ij}\left(x_i, z_j\right) = \lambda_{ij} - \beta \parallel x_i - z_j \parallel^2 + \theta_j T_{\eta i} \quad (1)$$

Here $\theta_j{}^T \eta_i$ models the effect of the sociodemographic characteristics $\eta_i$ of voter *i* in making a political choice. Further, $\theta_j$ is a vector that specifies how the various sociodemographic variables affect choice for party (Schofield and Sened 2006). Table 2 shows the valence estimates and spatial coefficients using a multinomial logit model, socioeconomic coefficients can be seen in the Appendix. The results

**Table 2** 2007 General election logit model results

| | Dependent variable |
|---|---|
| | Vote (spatial model) |
| $\lambda fg$ (intercept) | −0.421*** (0.098) |
| $\lambda gr$ (intercept) | −2.505*** (0.223) |
| $\lambda in$ (intercept) | −1.839*** (0.167) |
| $\lambda lb$ (intercept) | −1.385*** (0.138) |
| $\lambda pd$ (intercept) | −2.735*** (0.251) |
| $\lambda sf$ (intercept) | −2.223*** (0.199) |
| $\beta$ | 0.794*** (0.203) |
| Observations | 615 |
| $R^2$ | 0.008 |
| Log likelihood | −917.838 |
| LR Test | 15.675** (df = 7) |

*Abbreviations*: $\lambda fg$ valence of Fine Gael, $\lambda gr$ valence of Green Party, $\lambda in$ valence of Independents, $\lambda lb$ valence of Labour, $\lambda pd$ valence of Progressive Democrats, $\lambda sf$ valence of Sinn Fein

*Note*: *p < 0.1; **p < 0.05; ***p < 0.01

from the model show that the valence ($\lambda$) terms for all parties and the spatial coefficient ($\beta$), are statistically significant in the pure spatial model. Once we condition for demographic variables such as education, sex, and mass attendance the spatial coefficient and all but the valence for Fine Gael ($\lambda_{fg}$) remain significant. Moreover, we see that voters with higher levels of education are more likely to vote for the Greens or Labour and level of mass attendance positively affects the likelihood of voting for Labour and Sinn Fein.[4]

Because of the stochastic assumption of the model, voter behavior is modeled as a probability vector; the probability that voter *i* votes for party *j* is the probability that the voter's utility from voting for party j is greater than all other parties[5] (Schofield 2007). Since a logit model was used we assume the cumulative distribution function, $\Psi$, is *Type I extreme value distribution*[6] that takes on the closed form:

$$\Psi(h) = \exp(-(\exp -h)) \tag{2}$$

[4]See results of full model in the Appendix.

[5]Or, $\rho ij(z) = \Pr[uij(xi, zj) > (uil(xi, zl), \forall l \neq j$.

[6]Variance of $\Psi = \frac{1}{6}\pi^2$.

Following from Train (2003) the probability that voter *i* will vote for party *j* (Schofield 2007):

$$\rho_{\mathrm{ij}} = \frac{\exp\left(u^*_{ij}\left(\mathrm{x_i}, \mathrm{z_j}\right)\right)}{\sum_{k=1}^{p}\left(\exp\left(u^*_{ik}\left(\mathrm{x_i}, \mathrm{z_k}\right)\right)\right.} \tag{3}$$

From the pure spatial model the utility of voter *i* is dependent on the valence of party *j*, $\lambda_j$ and the policy distance between themselves and party j, $\beta \parallel x_i - z_j \parallel^2$. We assume that a voter is positioned at the electoral mean, $z_0$, with 0 representing the origin, so that $u^*_{ij}\left(x_i, z_j\right) = \lambda_j$. Using the valence coefficients, $\lambda_j$, we can estimate the probability that a voter, positioned at the electoral mean, $z_0$ selects the lowest valence party, which in this case is the Progressive Democrats (noted by $\lambda_{pd}$). The actual vote share for the Progressive Democrats was .027, and our estimation of $\rho$pd is .028. The results of the model inputed into the below formula will yield $\rho$pd.

$$\rho_{\mathrm{pd}} = \frac{e^{\lambda_{pd}}}{\sum e^{\lambda_{\forall j}}}$$

$$= \frac{e^{\lambda_{pd}}}{e^{\lambda_{ff}} + e^{\lambda_{fg}} + e^{\lambda_{gn}} + e^{\lambda_{in}} + e^{\lambda_{lb}} + e^{\lambda_{pd}} + e^{\lambda_{sf}}}$$

$$= \frac{\exp\left(-2.73\right)}{\sum \exp(0), \exp\left(0 - .42\right), \exp\left(-2.5\right), \exp\left(-1.83\right), \exp\left(0 - 1.38\right), \exp\left(-2.73\right), \exp\left(-2.22\right)}$$

$$\rho_{\mathrm{pd}} = \frac{.065}{2.32} = .028$$

The covariance matrix, $\nabla_0$ is a 2 by 2 matrix giving the variance of the electoral distribution on each axis. The variance for the *x*-axis is .505, and the variance for the *y*-axis is .984. The covariance between these variables is .073. The positive value for the covariance is expected since higher values correspond to a more conservative stance both economically and socially. For the INES sample this is:

$$\nabla = \begin{bmatrix} .505 & .073 \\ .073 & .984 \end{bmatrix}$$

With $\rho_{\mathrm{pd}}$ and $\nabla$ we can calculate convergence to the electoral mean. Below is the calculation of the "convergence coefficient", which characterizes the local equilibrium in two-dimensional space. The Valence Theorem states that the sufficient condition for convergence to the electoral mean $z_0$ is $c < 1$. The necessary condition for $z_0$ to be a strict local Nash Equilibrium (SLNE) is that $c < 2$, and the necessary and sufficient second order condition for SLNE at $z_0$ is that the characteristic matrix, or Hessian, $C_{pd}$, has negative eigen values (Schofield and Gallego 2011; Schofield 2007; Gallego and Schofield 2013).

$$c = 2\beta\left(1 - 2\rho_{\mathrm{pd}}\right) \operatorname{trace} \nabla$$
$$c = 2 * (.794) * (1 - 2 * (.028)) * (.504 + .984) = 2.23$$

Since $\beta$ and $\rho_{pd}$ are stochastic variables we can apply a confidence interval of 95 % to the variables in the convergence coefficient. The upper bound at this interval is 3.47 and the lower bound is .89. The lower bound of the convergence coefficient (.89), and we cannot conclude that the value is significantly greater than 1 at the 95 % confidence interval, but it is significantly greater than 1 at the 90 % interval.

$$
\begin{aligned}
c &= 2\beta\left(1 - 2\rho_{pd}\right) \text{trace } \nabla \\
c &= 2 * (1.27) * (1 - 2 * (.040)) * (.504 + .984) = 1.13 \\
c &= 2\beta\left(1 - 2\rho_{pd}\right) \text{trace } \nabla \\
c &= 2 * (.313) * (1 - 2 * (.018)) * (.504 + .984) = .89
\end{aligned}
$$

The Hessian, or characteristic matrix for the Progressive Democrats is:

$$
\begin{aligned}
C_{pd} &= 2\beta\left(1 - 2\left(\rho_{pd}\right)\right) \nabla\text{–I} \\
C_{pd} &= 2(.794)\Big(1 - 2(.028)\begin{bmatrix} .505 & .073 \\ .073 & .984 \end{bmatrix} - I \\
C_{pd} &= \begin{bmatrix} -.244 & .109 \\ .109 & .475 \end{bmatrix}
\end{aligned}
$$

Our convergence coefficient is above the threshold of $c < 2$, but since the confidence interval includes vales less than 1 and greater than 2 we cannot conclude that we do not have a SLNE at the 95 % interval, but if we relax the interval to the 90 % interval we can conclude that we do not have an SLNE. And we also find that for the Hessian, $C_{pd}$, there is one negative eigen value (.49, –.26) for $C_{pd}$, which gives us a saddlepoint. We can conclude, with a convergence coefficient of 2.23, and a lower bound of 1.16 at the 90 % confidence interval, that the Progressive Democrats do not converge toward the mean since a value of $c < 1$ would not meat the sufficient threshold for convergence. This lack of convergence is expected for a low valence party that should occupy space on the periphery (Schofield and Sened 2006).

## 3 2008 Lisbon Treaty

On June 12, 2008 the Irish electorate went to the polls for the second time in as many years. The referendum was an amendment to the constitution, a necessary step in the adoption of the Lisbon Treaty. In the wake of the rejection of the Constitutional Treaty by the French and Dutch electorate, the Lisbon Treaty was proposed as an alternate form of continued integration for the European Union. The Lisbon Treaty proposed various changes, among them: the increase of the number of MEP's in the European Parliament, that the President of the European Council would be elected by a qualified majority, the use of Qualified Majority Voting would become the norm in the Council of Ministers, the strategic interests of the EU would be determined by the European Council, among other institutional changes (Laursen 2012). What did these proposed institutional changes mean for the Irish electorate? One of the issues taken up by the No campaign was the lack of clarity provided to the electorate regarding the consequences of this decision (Quinlan 2009; O'Brennan 2009). The Yes campaign provided a campaign that failed to inform the electorate, and lack of success by the Yes campaign was attributed to lack of information from the Yes campaign (Quinlan 2009). Perhaps the most damning moment of the Yes campaign came when Fianna Fáil Taoiseach Brian Cowen admitted that he had not fully read the treaty. This lack of clarity contributed to much of the rhetoric from the No campaign. Moreover, the Irish electorate were concerned that the treaty would put an end to Irish military neutrality, it would undermine Ireland's abortion prohibition, and feared possible tax harmonization to higher continental European rates (Dinan 2009). We use the stochastic model for the referendum to show the effect of valence on the probability of a Yes vote. Valence, as defined earlier, is the non-policy attributes of a candidate, or in this case campaign. Even though in this example of direct democracy the decision seems straight forward, "Should Ireland ratify this treaty?", the actual decision is actually much more complex. The 2008 campaign for the Lisbon Treaty referendum essentially was a one in which both the Yes and No campaigns, asked the voter to trust their interpretation of the ramifications of ratifying the treaty. Due to the complexities of the treaty the question asked, the electorate relied on the valence of those presenting the arguments since both sides were campaigning for what is best for the country (Stokes 1963). Complexities of referenda make it difficult for the voters to completely understand what they are voting on (Blais et al. 1996), so the valence of the campaigns plays role in light of policy muddiness.

## 4 Model of 2008 Lisbon Treaty

The 2008 Lisbon Treaty was modeled using data from Irish Social Science Data Archive. The questionnaire for the 2008 referendum included questions on both social and economic questions. The responses to the below questions are out of a 5 point scale from (5) Very Important to (1) Not at all important. The statements are listed below:

- *Ireland retaining full control over its corporation tax*
- *Ireland retaining full control over its abortion law*
- *Ensuring that workers' rights are protected*
- *Ireland retaining control over its public services such as health and education*
- *Ensuring that there is not too much EU regulation*

| Issue | Social Factors | Economic Factors |
|---|---|---|
| Business Regulation | .339 | .657 |
| Corporate Tax | .416 | .378 |
| Abortion | .341 | .435 |
| Public Services | .452 | .325 |
| Workers Rights | .995 | |
| SS loadings | 1.598 | .877 |
| Proportion Var | .266 | .146 |
| Cumulative Var | .266 | .412 |

**Chart 2** Factor analysis results for 2008 Lisbon Treaty referendum

Using responses to these statements a factor analysis using a varimax rotation was used to create two variables. The results from the factor analysis are seen in Chart 2.

The results from the First Lisbon Treaty survey do not clearly separate into different factors as the previous factor analysis for the 2007 General Election, and as you will see for the 2009 Lisbon Treaty. Using the highest factor-score of each factor, and the other corresponding factors we would conclude that factor 2 consists of economic factors and factor 1 consisting of social elements.[7] Using the factors from this analysis we are able to create a map of the Irish Electorate, seen in Fig. 2.[8]

The results from Fig. 2 provide a picture of two separate groups of voters with their means coalescing near the origin of both the social and economic axes. The plot of the voters generated voters who were near the mean voter, but there were no voters that actually fell on the mean. So when we think of the average voter it is actually a culmination of all voters on these opposing sides.

The average Yes voter was older male who was more educated and attended mass less often. He did not differ much from his younger, less educated, more religious female counterpart, but the differences were present. These voters occupy policy places on both sides of the mean voter rather than two distinct camps of Yes and No voters. To model this referendum we use a pure positional binomial logit model. According to this positional model, voter i, with preferred position ($x_i$,$y_i$) is estimated with probability to Vote yes. The model can be found below, and the results can be found in Table 3.

$$\Pr(\text{Vote} = \text{Yes}) = \lambda + \text{b}x_i + \text{c}x_i \tag{4}$$

[7]The lack of variation in the responses may be due to the differing nature of the questions. This lack of clarity leads to a more difficult time separating voters on the axis. Other options included using differing questions, but these questions would allow for zero comparability between the elections, as they cover quite different topics.

[8]As you can see on the social axis, a large group of voters coalesce near 1, but do not reach this limit. So the figure may appear to cut off these voters, but in actuality it is voters grouped near this value.

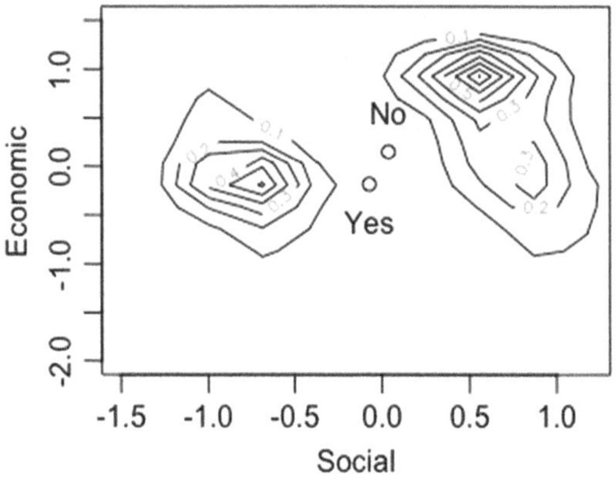


**Fig 2** 2008 Referendum voters

**Table 3** 2008 Lisbon Treaty referendum logit model results

| | Dependent variable |
|---|---|
| | Yes vote |
| Economic | 0.383***<br>(0.072) |
| Social | 0.318***<br>(0.060) |
| $\lambda_{yes}$ | 0.318***<br>(0.058) |
| Observations | 1258 |
| Log likelihood | −830.372 |
| Akaike Inf. Crit. | 1666.743 |

*Note*: *p<0.1; **p<0.05; ***p<0.01

Using the above coefficients for λ, b, c (.318, .318, .383). Setting the equiprobability line to $\frac{1}{2}$ we derive the cleavage line using the below equation and above coefficients:

$$\rho_{\text{yes}} = \frac{\exp(\lambda + \text{b}x_i + \text{c}y_i)}{1 + \exp(\lambda + \text{b}x_i + \text{c}y_i)} \tag{5}$$

$$= \frac{\exp(\lambda + \text{b}x_i + \text{c}y_i)}{1 + \exp(\lambda + \text{b}x_i + \text{c}y_i)} = \frac{1}{2}$$

$$2\exp(\lambda + \text{b}x_i + \text{c}y_i) = 1 + \exp(\lambda + \text{b}x_i + \text{c}y_i)$$

$$1 = \exp(\lambda + \text{b}x_i + \text{c}y_i)$$

$$0 = (\lambda + \text{b}x_i + \text{c}y_i)$$

$$-3.83y = .318x + .318$$

$$y = . - .83 - .83x$$

**Fig 3** 2008 Referendum voters

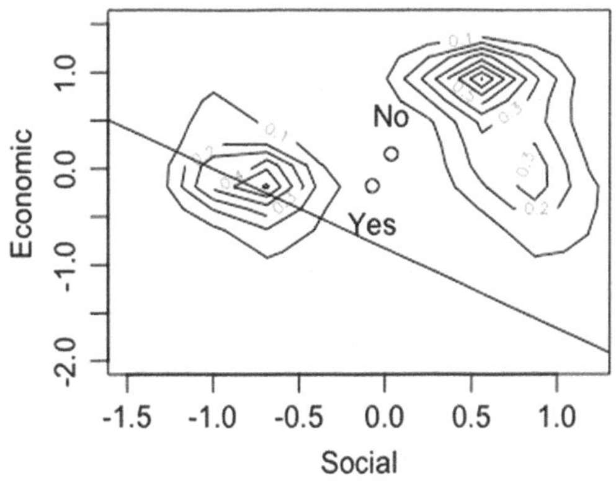


The cleavage line misses the mean, and crosses at (0, −.83). The cleavage line is a helpful way of showing the separation between the Yes and No campaigns; the estimated cleavage line shows the voters that are indifferent between the Yes and No campaigns (Gallego and Schofield 2013). Voters on either side of the cleavage line will be more likely to support one campaign than the other. If the cleavage line crossed the mean, which is determined by policy position, it would indicate that a voter would base their decision on policy position, and that neither side had a valence advantage since the likelihood of their vote would be dependent on the policy mappings. The coefficient $\lambda_{yes}$ (.318) is a measure of the relative valence of the Yes campaign compared to the No campaign.[9] The positional model does not explicitly include the campaign positions, and cannot determine political equilibrium (Fig. 3).

The solid line is the cleavage line including valence. The position of the cleavage line shows a slight valence advantage of the Yes campaign since those positioned above the line are more likely to vote for the Yes campaign. Once valence is taken into account along with policy preferences a greater proportion of the electorate would fall into the group that we would expect to vote Yes. The advantage contradicts previous literature on the issues and lack of trust of the Yes campaign (Sinnott and Elkink 2010). Although the Yes campaign had a valence advantage, they still lost the election. This would lead us to believe that in 2008 the rejection of the treaty centered more on the issue preferences of the electorate, and not a negative view of the Yes campaign, as was originally thought. Despite this, the rejection of the treaty only lasted for a year. In 2009, the Irish electorate, in the midst of the global financial crisis, were asked to vote on the treaty again.

There was precedent in Ireland for a redo on a European referendum; the Irish initially rejected the Nice Treaty in 2001, but approved it in a second referendum in 2002. In order to make a second attempt at ratification of the Lisbon Treaty, the EU Council presented a package to the Irish voters, similar to what they did to

[9]This is true because we have statistical significance for $\lambda_{yes}$ at the .01 level as seen in Table 3.

Danish voters after their rejection of the Maastricht Treaty in 1993 (Tonra 2009). The Danish option included legally binding decision that ameliorates the concerns of the Irish voters on the role of the EU in Ireland's ability to legislate on an EU Council Decision to maintain a single Commissioner from each member state; three legal guarantees on taxation, social issues and neutrality (which were registered with the United Nations and which are to be appended as protocols to the next EU accession treaty); and a Declaration on workers' rights (Tonra 2009; Barrett 2008). It was after establishing these assurances the Irish government felt they could resubmit the Treaty to the people. A legal argument behind the resubmission of the treaty was that the need for unanimity among member states puts pressure on a state that may be holding up something that all other member states were in agreement, as was the case in Lisbon (de Brca 2009). This pressure, precedent, the clarifications of the Danish solution, and the financial crisis justified the Irish governments resubmission of the Treaty. Moreover, the ongoing financial crisis provided greater prospects for more favorable result for the Yes campaign (Dinan 2011, see Millward Brown and Landsdowne 2009a, b for more details).

## 5 2009 Lisbon Treaty

Using the ISSDA data from 2009 Referendum we performed a factor analysis with a varimax rotation to create two district variables in order to plot the electorate. The responses to the questions use a scale of one to ten. The respondent selected the number closest to their view. The following questions were included in the analysis with the two extreme ends of the scale offered to the respondent.

- *There should be a total ban on abortion or Abortion should be freely available in Ireland to any woman who wants to have one.*
- *There should be more emphasis on the protection of workers' rights or There should be less emphasis on the protection of workers' rights.*
- *Ireland should have power to fully control its corporate taxation rates or The European union should have the power to set a common corporate tax rate throughout the Union.*
- *The European Union should be less involved in the regulation of small companies and enterprises or The European Union should be more involved in the regulation of small companies and enterprises.*
- *People coming to live here from other countries has made Ireland a much worse place to live or People coming to live here from other countries has made Ireland a much better place to live.*

The results from the factor analysis, shown in Chart 3, illustrate that factor 1 includes our economic variables and factor 2 includes the social questions. As compared to the 2008 responses, the results from this analysis created two more

| Issue | Economic Factors | Social Factors |
|---|---|---|
| Corporate Tax Rate | .981 | .180 |
| Business Regulation | .235 | .376 |
| Workers Rights | .207 | .312 |
| Immigration | .113 | .333 |
| Abortion | | .422 |
| SS loadings | 1.073 | .560 |
| Proportion Var | .215 | .112 |
| Cumulative Var | .215 | .327 |

**Chart 3** Factor analysis results for 2009 Lisbon Treaty referendum

**Fig 4** Lisbon referendum 2009

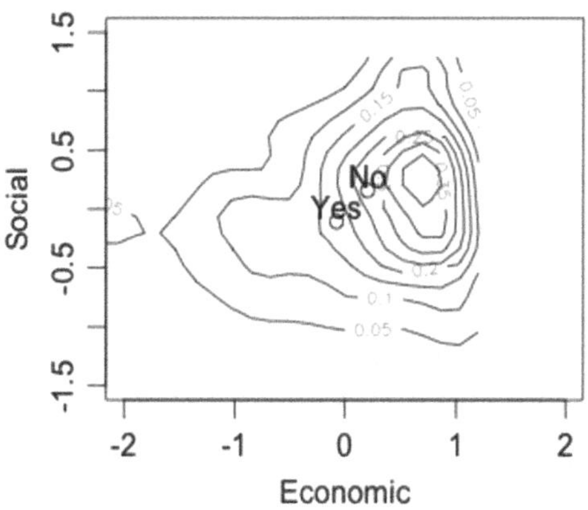


distinct variables.[10] Using these factors we are able to use a density plot for the electorate, consisting of a social and economic axis, seen in Fig. 4.

Using the social and economic variables we run another pure positional binomial logit model based on the probability of a yes vote. The model used to estimate the coefficients are as follows:

$$\text{Pr}\,(\text{Vote} = \text{Yes}) = \lambda + \text{b}x_i + \text{c}x_i \tag{6}$$

Using the coefficients from the binomial logit model, as seen above in Table 3, for λ, b, c (−.68, .67, .26) we can set the equiprobability line to $\frac{1}{2}$ and derive the

[10]The increased possible response range from 2008 to 2009, (5–10) allowed for greater variability. Moreover, the results from the factor analysis yielded a clearer distinction on the left-right scale between the economic and social factors. Both these differences contributed to clearer results.

**Table 4** 2009 Lisbon Treaty referendum logit model results

| | Estimate | Std. error | z value | Pr (> \|z\|) |
|---|---|---|---|---|
| $\lambda_{yes}$ | −0.6769 | 0.0890 | −7.61 | 0.0000 |
| Social | 0.6714 | 0.1518 | 4.42 | 0.0000 |
| Economic | 0.2556 | 0.0992 | 2.58 | 0.0100 |

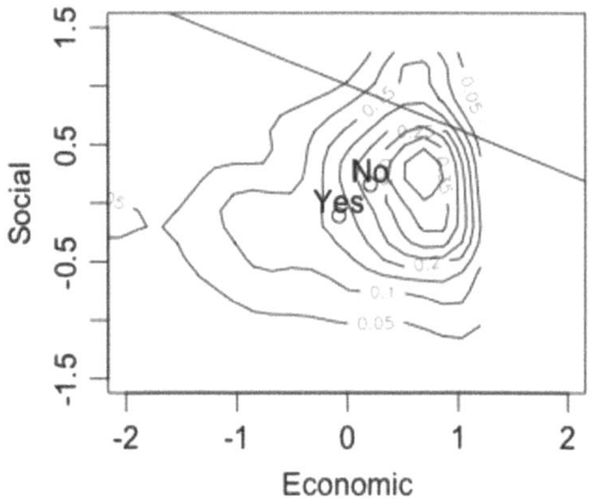


**Fig 5** Lisbon referendum 2009 with cleavage lines

cleavage line using the above equation and coefficients (Table 4):

$$\rho\text{yes} = \frac{\exp(\lambda + bx_i + cy_i)}{1 + \exp(\lambda + bx_i + cy_i)} = \frac{1}{2}$$

$$= 2\exp(\lambda + bx_i + cy_i) = 1 + \exp(\lambda + bx_i + cy_i)$$

$$1 = \exp(\lambda + bx_i + cy_i)$$

$$0 = (\lambda + bx_i + cy_i)$$

$$-.67y = .25x - .67$$

$$y = 1 - .38x$$

The cleavage line, shown in Fig. 5, (*solid*) misses the mean and crosses the axis at (0, 1), that shows a valence advantage of the No campaign. The 2009 referendum was, like 2008, an opportunity to provide a protest vote to the pro-treaty government. By time of the 2009 referendum the economy was in full implosion and the government had endured a year of headlines regarding corruption by the previous Taoiseach, Bertie Ahern. In light of this the electorate still voted Yes. Who was the voter who switched from No to Yes? They were most likely a middle aged married woman with a child or children with a secondary education level who

**Table 5** 2009 Logit model results with economic control

| | Estimate | Std. error | z value | Pr(> \|z\|) |
|---|---|---|---|---|
| $\lambda_{yes}$ | −5.6813 | 0.5972 | −9.51 | 0.0000 |
| Social variable | 0.3311 | 0.2404 | 1.38 | 0.1684 |
| Economic variable | 0.6060 | 0.1580 | 3.84 | 0.0001 |
| Economic effect | 1.8574 | 0.2124 | 8.75 | 0.0000 |

attended mass more than occasionally. Even though Yes voters were more likely employed than no voters employment overall had been decreasing. Further, a yes vote was seen as a way to ensure a European Central Bank funded rescue plan to rescue Irish banks and the economy. Many believe that despite all of the negative feelings toward the government the people feared the economic future of Ireland without integrating further with Europe. In order to try and capture this in the model we include a variable that measures the electorate feelings toward the economic effect of the treaty (Table 5).

The results from the model show that the Economic Effect variable has a strong effect and is highly significant. Further, the relative valence of the Yes campaign significantly decreases (−5.68). This decrease would lead us to assume that the higher negative valence, when economic effect was not controlled for, was simply incorporating the electorate's fear of negative economic consequences and not actual good sentiment toward the Yes campaign. This supports the notion that the switch in votes was centered on the financial crisis.

## 6 Conclusion

The valence of the Yes campaign significantly decreased between the 2008 and 2009 referenda, and it significantly decreased even more once the electorate's view on the economic effect of the treaty was controlled for. In 2009 the Yes campaign increased their campaign, which ameliorated the confusion of the electorate that marred their first referendum (Sinnott and Elkink 2010). It appears that this increased campaign did not improve the valence of the Yes campaign. The success of the Yes campaign in 2009 actually lied with the electorate's fear of weathering the financial crisis without help from Europe. The government correctly thought that the ongoing financial crisis provided a more favorable atmosphere for ratification despite a very low valence.

# Appendix

| | Dependent variable | |
|---|---|---|
| | Vote | |
| | (Spatial model) | (Full model) |
| $\lambda fg$ (intercept) | −0.421*** | −0.584 |
| | (0.098) | (0.428) |
| $\lambda gr$ (intercept) | −2.505*** | −5.604*** |
| | (0.223) | (1.201) |
| $\lambda in$ (intercept) | −1.839*** | −2.140*** |
| | (0.167) | (0.714) |
| $\lambda lb$ (intercept) | −1.385*** | −2.795*** |
| | (0.138) | (0.640) |
| $\lambda pd$ (intercept) | −2.735*** | −4.408*** |
| | (0.251) | (1.300) |
| $\lambda sf$ (intercept) | −2.223*** | −2.770*** |
| | (0.199) | (0.879) |
| $\beta$ | 0.794*** | 0.662*** |
| | (0.203) | (0.213) |
| fg:educ | | 0.135* |
| | | (0.071) |
| gr:educ | | 0.570*** |
| | | (0.184) |
| pd:educ | | 0.622*** |
| gr:mass | | 0.211* |
| | | (0.113) |
| in:mass | | 0.163* |
| | | (0.085) |
| lb:mass | | 0.260*** |
| | | (0.067) |
| sf:mass | | 0.337*** |
| | | (0.087) |
| Observations | 615 | 615 |
| $R^2$ | 0.008 | 0.039 |
| Log likelihood | −917.838 | −889.289 |
| LR Test | 15.675** (df = 7) | 72.773*** (df = 25) |

*Note*: *p < 0.1; **p < 0.05; ***p < 0.01